BORATE GLASSES, CRYSTALS, & MELTS:

7th International Conference

Proceedings of the Seventh International Conference on
Borate Glasses, Crystals and Melts: Borate7,
held at Dalhousie University, Halifax, Nova Scotia, Canada,
on 21-25 August 2011.

Edited by:
A. C. Hannon, J. W. Zwanziger,
S. Kroeker, L. Cormier & R. Youngman

Borate7
HALIFAX, CANADA · August 21 – 25, 2011

Society of Glass Technology
Sheffield, 2014

This conference and proceedings are dedicated to
Professor Tsutomu Minami to honour his achievements in
Glass Science, and in particular Borate Glasses.

Professor Tsutomu Minami with Professor Adrian Wright
of Reading University (right)

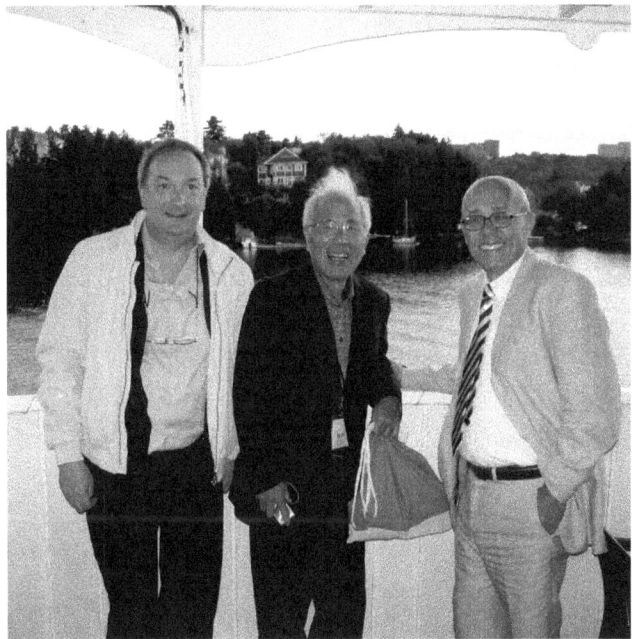

Professor Tsutomu Minami with Professor Francesco Rocca (left)
and Professor Giuseppe Dalba (right) of Trento University

Borate7
Proceedings of the Seventh International Conference on Borate Glasses, Crystals and Melts: New techniques and practical applications held at Dalhousie University, Halifax, Nova Scotia, Canada on 21–25 August 2011
A collected volume of papers from the Seventh International Conference, the papers were originally published in *Physics and Chemistry of Glasses: European Journal of Glass Science and Technology Part B*

ISBN 978-0-900682-73-5

The objects of the Society of Glass Technology are to encourage and advance the study of the history, art, science, design, manufacture, after treatment, distribution and end use of glass of any and every kind. These aims are furthered by meetings, publications, the maintenance of a library and the promotion of association with other interested persons and organisations.

Society of Glass Technology
9 Churchill Way
Chapeltown
Sheffield S35 2PY, UK
Tel +44(0)114 263 4455
Email info@sgt.org
Web http://www.sgt.org
The Society of Glass Technology is a registered charity no. 237438.

Foreword

The *Seventh International Conference on Borate Glasses, Crystals and Melts: Borate7* was held during the period 21–25 August, 2011 at Dalhousie University in Halifax, Nova Scotia, Canada. This followed previous conferences held at Alfred, USA (1977); Abingdon, United Kingdom (1996); Sofia, Bulgaria (1999); Cedar Rapids, USA (2002); Trento, Italy (2005); and Himeji, Japan (2008), and it was the third time that the Conference was held in North America. The Seventh Conference was dedicated to Professor Tsutomu Minami to honour his achievements in the study of Glassy Materials, and in particular Ionic Glasses and their application as Battery Materials.

In total, 63 scientists from 15 countries gathered in Halifax to discuss recent advances in the study of borate glasses, crystals, and melts, and their applications. The participants came from: Armenia (1), Austria (1), Brazil (2), Canada (10), Czech Republic (1), France (5), Germany (4), Greece (2), Hungary (1), Italy (2), Japan (13), Republic of Korea (1), United Kingdom (4), United States of America (15), and Uruguay (1). The first invited talk was by Norimasa Umesaki who summarised the contents of the previous Conference. The lecture in honour of Professor Tsutomu Minami, entitled "Introduction of T. Minami", was presented by Akitoshi Hayashi, who described the major role played for many years by Professor Minami in the study of Ionic Conducting Glasses. This was followed by a plenary talk entitled "New Borate Glasses for Ionics", given by Professor Minami. The other invited talks were by: David Schubert (US Borax, USA), Steve Jung (Mo-Sci, USA), Hubert Huppertz (University of Innsbruck, Austria), S. K. Lee (Seoul National University, South Korea), Hellmut Eckert (University of Münster, Germany), Bruce Aitken (Corning, Inc., USA), Stratos Kamitsos (National Hellenic Research Foundation, Greece), Guillaume Ferlat (University Pierre & Marie Curie, France), John Mauro (Corning, Inc., USA), and Richard Brow (Missouri University of Science and Technology, USA).

The conference structure involved the following sessions: Honoring Prof. Minami's Achievements (Chairman A. C. Wright), Ionics (Chairman A. Takada), Mixed Glass Formers (Chairman S. Kroeker), The Supply Side (Chairman S. Kroeker), Pressure (Chairman N. Umesaki), Structure 1 (Chairman N. Umesaki), Diffraction (Chairman D. Möncke), Structure 2 (Chairman R. Youngman), Crystallization (Chairman L. Cormier), Medical Applications (Chairman Y. Messaddèq), Theory and Simulations (Chairman J. Zwanziger), Optical Response (Chairman F. Rocca), Structure 3 (Chairman A. C. Hannon), and Structure 3 continued (Chairman L. Koudelka). In total, there were 38 oral presentations. In addition, a poster session was held, with a total of 17 posters.

I am most grateful to the members of the Scientific Organizing Committee: Josef Zwanziger (Dalhousie University, Canada), Laurent Cormier (Université Pierre & Marie Curie, France), Alex Hannon (Rutherford Appleton Laboratory, UK), Scott Kroeker (University of Manitoba, Canada), Younes Messaddèq (Université Laval, Canada), Sabyasachi Sen (University of California-Davis, USA), Masahiro Tatsumisago (Osaka Prefecture University, Japan), and Randy Youngman (Corning, Inc., USA).

Thanks are due to the chairman of the International Organising Committee, Adrian Wright (University of Reading, UK), and the members of the committee: Alexis Clare (Alfred University, USA), Guiseppe Dalba (University of Trento, Italy), Yanko Dimitriev (University of Chemical Technology and Metallurgy, Bulgaria), Doris Ehrt (University of Jena, Germany), Steve Feller (Coe College, USA), Alex Hannon (Rutherford Appleton Laboratory, UK), Stratos Kamitsos (National Hellenic Research Foundation, Greece), Norimasa Umesaki (SPRING-8, Japan), Natalia Vedishcheva (Institute of Silicate Chemistry, Russia), and Josef Zwanziger (Dalhousie University, Canada).

Generous financial contributions in support of the conference from Corning Inc. and Dalhousie University were very important in ensuring the success of the conference, and are gratefully acknowledged. The

The Borate7 Conference Chairman, Josef Zwanziger, during the boat tour of the Halifax harbour area

support of the Society of Glass Technology was essential for the publication of the Conference Proceedings. I thank the team of Editors for dealing with the reviewing of the papers for the Conference Proceedings: Alex Hannon, Josef Zwanziger, Laurent Cormier, Scott Kroeker, and Randy Youngman. In particular I should like to thank Alex Hannon for his role in the later stages of the editorial process.

I am also very grateful to my students, colleagues, and administrative staff, who did so much to make the conference a success. These include Courtney Calahoo, Sebastien Chenu, Justine Galbraith, Yiannis Kipouros, Ulrike Werner-Zwanziger, and Shelley Dorey.

Dalhousie University and the city of Halifax deserve special mention for providing a very pleasant location for the conference. The city of Halifax has a unique Scottish heritage, and this, together with the high walkability and the maritime atmosphere, made for a delightful environment in which the conference delegates could focus on the science of borates and relax afterwards.

The location of the next Conference was considered by the International Organising Committee, and it was decided to hold the next conference in the Czech Republic. Professor Ladislav Koudelka will chair the *Eighth International Conference on Borate Glasses, Crystals and Melts: Borate8*, which will take place in Pardubice on 29 June to 4 July, 2014, in conjunction with the *International Conference on Phosphate Glasses*. The Eighth Conference will be held in honour of Professor Stanislav Filatov, recognising his achievements in Crystallography, and in particular the structure of Borate Crystals. All of the participants present in Halifax are cordially invited to attend.

Josef Zwanziger (Dalhousie University), Borate7 Conference Chairman

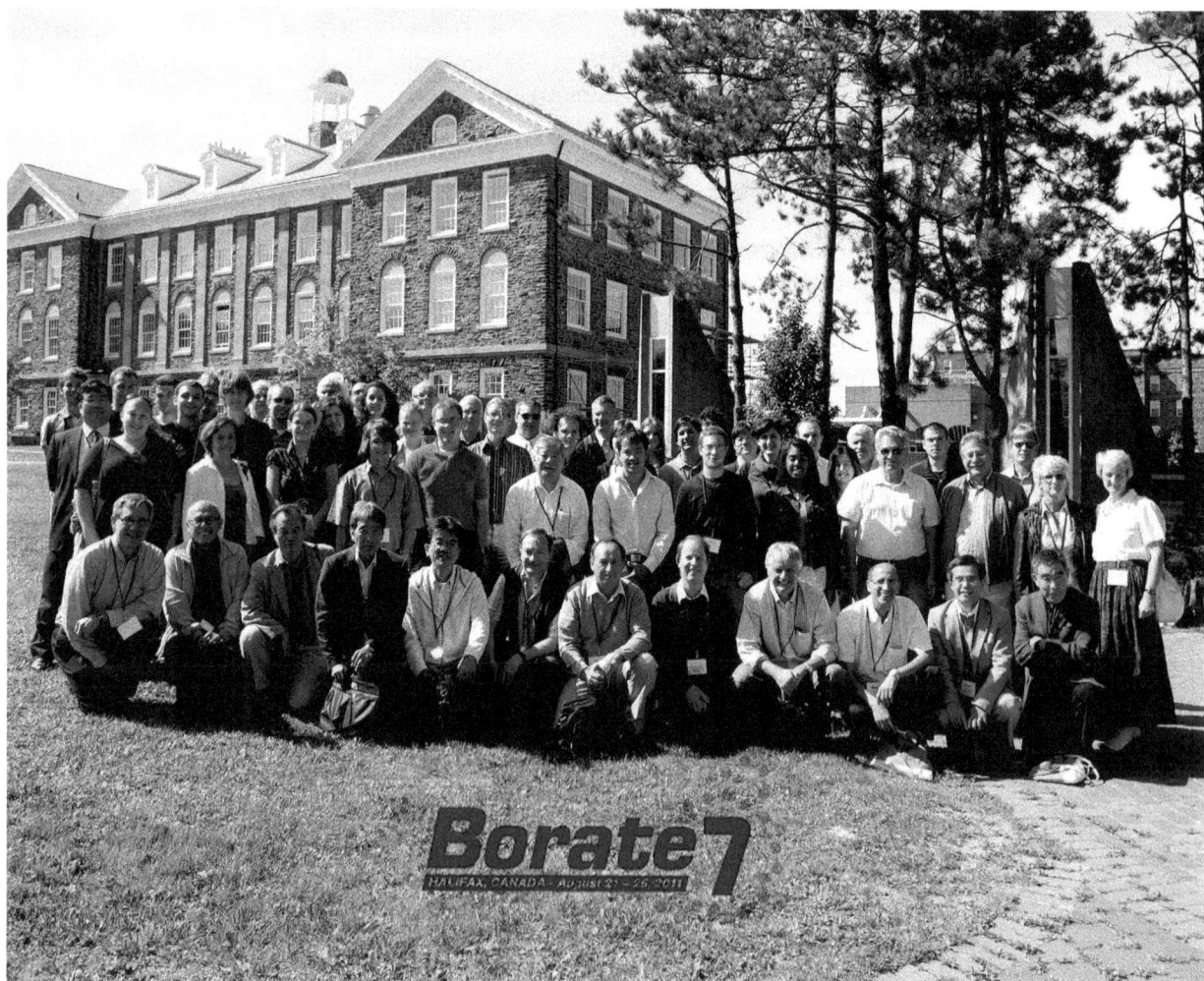

Contents

Proceedings of the Seventh International Conference on Borate Glasses, Crystals and Melts: Borate7

v   Foreword
vii   Contents
x   Programme
xv   Conference Participants
xvii Author Index
xxi   Subject Index

The journal references for the papers are given using these journal abbreviations:
PC=*Physics and Chemistry of Glasses: European Journal of Glass Science and Technology Part B;*

Page	Title and author(s)	Journal reference
1	New borate glasses for ionics T. Minami	PC 2012, **53**(2), 17
11	The research history of Professor Tsutomu Minami A. Hayashi	PC 2012, **53**(2), 52
20	Computational study of four-fold coordinate boron in borates: assignment of edge-shared structures J. W. Zwanziger	PC 2012, **53**(1), 7
24	SpectraFit: a new program to simulate and fit distributed ^{10}B powder patterns application to symmetric trigonal borons V. Khristenko, K. Tholen, N. Barnes, E. Troendle, D. Crist, M. Affatigato & S. Feller	PC 2012, **53**(3), 121
31	Mechanochemical synthesis of $BaO–B_2O_3$ glass and glass-ceramic phosphor powders containing europium ions A. Shinomiya, A. Hayashi, K. Tadanaga & M. Tatsumisago	PC 2012, **53**(3), 128
35	Double rotation ^{11}B NMR applied to polycrystalline barium borates O. L. G. Alderman, D. Iuga, A. P. Howes, D. Holland & R. Dupree	PC 2012, **53**(3), 132
44	A neutron diffraction study of $2M_2O.5B_2O_3$ (M=Li, Na, K, Rb, Cs Ag) and $2MO.5B_2O_3$ (M=Ca Ba) glasses A. C. Wright, R. N. Sinclair, C. E. Stone, J. L. Shaw, S. A. Feller, T. J. Kiczenski, R. B. Williams, H. A. Berger, H. E. Fischer & N. M. Vedishcheva	PC 2012, **53**(5), 191
58	Structure and properties of barium and calcium borosilicate glasses S. Feller, T. Mullenbach, M. Franke, S. Bista, A. O'Donovan-Zavada, K. Hopkins, D. Starkenberg, J. McCoy, D. Leipply, J. Stansberry, E. Troendle, M. Affatigato, D. Holland, M. E. Smith, S. Kroeker, V. K. Michaelis & J. E. C. Wren	PC 2012, **53**(5), 210
67	Structure and properties of lead borophosphate glasses doped with molybdenum oxide L. Koudelka, I. Rösslerová, Z. Černošek, P. Mošner, L. Montagne, B. Revel & G. Tricot	PC 2012, **53**(6), 245
76	Clustering in borate-rich alkali borophosphate glasses: a ^{11}B and ^{31}P MAS NMR study V. K. Michaelis, P. Kachhadia & S. Kroeker	PC 2013, **54**(1), 20

83   Spectroscopic study of manganese-containing borate and borosilicate glasses: PC 2013, **54**(1), 42
cluster formation and phase separation
D. Möncke, D. Ehrt & E. I. Kamitsos

93   Formation of an outer borosilicate glass layer on Late Bronze Age Mycenaean   PC 2013, **54**(1), 52
blue vitreous relief fragments
D. Möncke, D. Palles, N. Zacharias, M. Kaparou, E. I. Kamitsos &
L. Wondraczek

101   Radiation pattern control by the sidewall angle of Bi_2O_3–B_2O_3 based glass   PC 2013, **54**(1), 60
phosphor doped with Yb^{3+} and Nd^{3+}
S. Fuchi, S. Kobayashi, K. Oshima & Y. Takeda

105   Structure and properties of lithium borate glass electrolytes synthesised by   PC 2013, **54**(3), 109
a mechanochemical technique
A. Hayashi, D. Furusawa, Y. Takahashi, K. Minami & M. Tatsumisago

111   Luminescence efficiency of B_2O_3–Sb_2O_3–Bi_2O_3 glass phosphor doped with   PC 2013, **54**(5), 195
Sm^{3+}
K. Oshima, S. Fuchi, S. Kobayashi & Y. Takeda

115   Glass formation and crystalline phases in the ternary systems CaO–Bi_2O_3–   PC 2013, **54**(5), 199
B_2O_3 and SrO–Bi_2O_3–B_2O_3
A. H. Barseghyan, R. M. Hovhannisyan, B. V. Petrosyan, H. A. Aleksanyan
& V. P. Toroyan

122   Zinc and manganese borate glasses phase separation, crystallisation,   PC 2013, **54**(2), 65
photoluminescence and structure
D. Ehrt

133   Crystallisation of a lead borate glass and its influence on the thermolumi-   PC 2013, **54**(6), 241
nescence response
M. Rodríguez Chialanza, E. Castiglioni, J. Castiglioni & L. Fornaro

139   Increase of light extraction from Bi_2O_3-B_2O_3 based glass phosphor doped   PC 2013, **54**(6), 247
with Yb^{3+} and Nd^{3+}
S. Kobayashi, S. Fuchi, K. Oshima & Y. Takeda

146   Velocity of sound in and elastic properties of alkali metal borate glasses   PC 2014, **55**(1), 1
M. Kodama & S. Kojima

SEVENTH INTERNATIONAL CONFERENCE ON BORATE GLASSES, CRYSTALS AND MELTS: BORATE7

Dalhousie University, Halifax, Nova Scotia, Canada,
21-25 August 2011

PROGRAMME

Sunday August 21

17:00–20:00:   Welcome Reception and Registration, lobby of the Marion McCain Building.

Monday, August 22

Session 1: Honoring Prof. Minami's Achievements   *Chair: A. Wright*
9:00–9:10   J. W. Zwanziger: Welcoming remarks
9:10–9:30   N. Umesaki: Recap of *Borate6*
9:30–9:50   A. Hayashi: Introduction of T. Minami
9:50–10:30   T. Minami: [T1] New borate glasses for ionics

Session 2: Ionics   *Chair: A. Takada*
11:00–11:30   F. Rocca: [T2] EXAFS points of view on the local structure of AgI- containing silver borate glasses
11:30–12:00   A. Hayashi: [T3] Structure and properties of lithium borate glass electrolytes synthesized by a mechanochemical technique

Session 3: Mixed Glass Formers   *Chair: S. Kroeker*
13:30–14:10   B. Aitken: [T4] Structure and Properties of unmodified network thioborate glasses
14:10–14:40   A. Pradel: [T5] New insights into the mixed glass former effect in lithium boro-phosphate glasses: a solid state NMR point of view
14:40–15:10   L. Koudelka: [T6] Structural studies of boron and tellurium coordination in borophosphate glasses by ^{11}B MAS NMR and Raman spectroscopy

Session 4: The Supply Side   *Chair: S. Kroeker*
15:10–15:50   D. Schubert: [T7] The glass industry from a borate suppliers perspective
              E.Steffee:

Session 5: Posters   *Chair: J. Zwanziger*

Tuesday, August 23

Session 6: Pressure   *Chair: N. Umesaki*
9:00–9:40   H. Huppertz: [T8] High-pressure solid state chemistry in the field of borates
9:40–10:10   S. Reibstein: [T9] Heterogeneity and structural relaxation of compressed borate-silicate glasses
10:10–10:50   S. K. Lee: [T10] Pressure-induced structural transitions in borate glasses using solid-state NMR and synchrotron inelastic x-ray scattering

Session 7: Structure 1
11:15–11:45   S. Feller: [T11] Structure and properties of alkaline earth borosilicates
11:45–12:15   R. Youngman: [T12] Network former speciation and connectivity in alkali zn borophosphate glasses

Session 8: Diffraction   *Chair: D. Möncke*
13:30–14:00   A. Wright: [T13] A neutron diffraction study of $2M_2O.5B_2O_3$ (M = Li, Na, K, Rb, Cs and Ag) and $2MO.5B_2O_3$ (M = Ca and Ba) glasses

| 14:00–14:30 | M. Fabian: | [T14] Network structure of sodium borosilicate glasses in dependence of B_2O_3 content: neutron and x-ray diffraction and reverse Monte Carlo modelling |
| 14:30–15:00 | F. Michel: | [T15] An *in situ* neutron scattering study of boron coordination of sodium and soda-lime borosilicate glasses/melts |

Session 9: Structure 2   *Chair: R. Youngman*

15:30–16:00	S. Kroeker:	[T16] Clustering and phase separation in borate-rich alkali borophosphate glasses: An ^{11}B and ^{31}P MAS NMR study
16:00–16:30	G. Lelong:	[T17] A temperature induced boron coordination change in alkali borates: an inelastic x-ray scattering study
16:30–17:00	G. Sharma:	[T18] Radiation Damage and Structure of Borate Glasses
17:00–17:30	V. Khristenko:	[T19] SpectraFit: A new program to simulate distributed [10]B powder patterns

Wednesday, August 24

Session 10: Crystallization   *Chair: L. Cormier*

9:00–9:30	D. Ehrt:	[T20] Phase separation, crystallization, photoluminescence and structure of zinc and manganese borate glasses
9:30–10:00	K. Goetschius:	[T21] Glass formation and crystallization of Na_2O–CaO–B_2O_3 melts
10:00–10:30	D. Möncke:	[T22] Spectroscopic study of phase separated manganese containing borate and borosilicate glasses

Session 11: Medical Applications   *Chair: Y. Messaddèq*

| 10:50–11:30 | S.Jung: | [T23] Applications for borate glasses in medicine |
| 11:30–12:10 | R. Brow: | [T24] Dissolution behavior of borate glasses |

Session 12: Theory and Simulations   *Chair: J. Zwanziger*

13:30–14:10	G. Ferlat:	[T25] Hidden polymorphism in B_2O_3: a consistent explanation of the crystallisation and glass anomalies
14:10–14:40	A. Takada:	[T26] Structural modeling of sodium borate invert glass by computer simulation
14:40–15:20	J. Mauro:	[T27] Temperature-dependent constraint model of borate glasses

18:30: Banquet/Boat Tour

Thursday, August 25

Session 13: Optical Response   *Chair: F. Rocca*

9:00–9:40	S. Kamitsos:	[T28] Thermally poled glasses with nonlinear optical properties
9:40–10:10	I. Chermiti:	[T29] Femtosecond laser waveguide writing on sodium borotungstate glasses
10:10–10:40	M. Rodriguez:	[T30] Nucleation and growth mechanisms in a lead borate glass and their influence on thermoluminescence response

Session 14: Structure 3   *Chair: A. Hannon*

| 11:00–11:40 | H. Eckert: | [T31] New magnetic resonance approaches to structural studies of luminescent borate glasses |
| 11:40–12:10 | O. Alderman: | [T32] B-11 double rotation spin diffusion NMR as a probe of local and intermediate structure in borate crystals and glasses |

Session 14: Structure 3 continued   *Chair: L. Koudelka*

13:30–14:00	T. Okajima:	[T33] Local structures of iron ions in borosilicate glass: theoretical calculations and high resolution measurement of Fe K-edge XANES spectra
14:00–14:30	J. Matsuoka:	[T34] Infrared absorption spectra of borosilicate glass melts
14:30–15:00	N. Laorodphan:	[T35] Amorphous structure in thallium borate binary glasses
15:00–15:30	J. W. Zwanziger	[T36] Structural aspects of the photoelastic response in lead borate glasses

POSTERS

[P1] Velocity of Sound in and elastic properties of alkali borate glasses, Masao Kodama & Seiji Kojima
Graduate School of Pure and Applied Sciences, University of Tsukuba, Tsukuba, Ibaraki 305-8573, Japan.

[P2] Boron isotope effect on the thermal conduction of B_2O_3 glass, Jun Matsuoka, Taro Kimura, Toru Sugawara and Satoshi Yoshida
Univ. Shiga Pref.

[P3] High-temperature neutron scattering study of sodium borosilicate glasses in the immiscibility region, Laurent Cormier,[1] Florent Michel[1,2] & Brigitte Beuneu[3]
[1]Institut de Minéralogie et de Physique des milieux condensés, Université Pierre et Marie Curie, CNRS UMR7590, 4 place jussieu, 75005 Paris, France
[2]Saint-Gobain-Recherche, 39, quai Lucien-Lefranc BP 135, 93303 Aubervilliers Cedex, France
[3]Laboratoire Léon Brillouin, 91191 Gif-sur-Yvette Cedex, France

[P4] Radiation pattern control by sidewall angle of Bi_2O_3–B_2O_3 based glass phosphor doped with Yb^{3+} and Nd^{3+} Shingo Fuchi, Shunichi Kobayashi, Koji Oshima, and Yoshikazu Takeda
Department of Crystalline Materials Science, Graduate School of Engineering, Nagoya University

[P5] Investigation of formation and precipitation of Ag Nanoparticles on phosphate- tungsten glass system by solid state NMR, Silvia Santagneli,[1] Marcelo Nalin,[2] Mauricio Caiut,[3] Sidney J. L. Ribeiro,[1] Lino Mesoguti[4] & Younès Messaddeq[5]
[1]Instituto de Qumica, UNESP, Araraquara-SP, Brazil
[2]Universidade Federal de São Carlos, São Carlos-SP, Brazil
[3]Faculdade de Filosofia e Letras, USP, Ribeiro Preto-SP, Brazil
[4]Instituto de Fsica, USP So Carlos-SP, Brazil
[5]Centre d'optique, Photonique et laser, Université Laval, Quebec, Canada

[P6] Correlation between doping and thermoluminescence of the PbB_4O_7 glass- ceramic, M. Rodriguez,[1] A. Cárdenas,[1] I. Galain,[1] E. Castiglioni,[2] J. Castiglioni[3] & L. Fornaro[4]
[1]Grupo de Semiconductores Compuestos, Facultad de Qumica, Montevideo, Uruguay
[2]Laboratorio de datación por luminiscencia, Facultad de Ciencias, Montevideo, Uruguay
[3]Departamento de Experimentación y Teora de la Estructura de la Materia y sus Aplicaciones, Facultad de Qumica, Montevideo, Uruguay
[4]Grupo de Semiconductores Compuestos, Centro Universitario de la Región Este (CURE), Rocha, Uruguay

[P7] Glass formation and crystalline phases in the ternary CaO–Bi_2O_3–B_2O_3 and SrO–Bi_2O_3–B_2O_3 systems, A. H. Barseghyan,[1] R. M. Hovhannisyan,[1] B. V. Petrosyan,[1] H. A. Aleksanyan[1] & V. P. Toroyan[2]
[1]Scientific Production Enterprise of Material Science, 17 Charents st., 0025 Yerevan, Armenia
[2]Institute of General and Inorganic Chemistry of NAS RA, Argutyan st., district 2,10,0051 Yerevan, Armenia

[P8] Mechanochemical synthesis of BaO–B_2O_3 glass and glass-ceramic phosphor powders containing europium ions, Atsuko Shinomiya, Kiyoharu Tadanaga & Masahiro Tatsumisago
Department of Applied Chemistry, Graduate School of Engineering, Osaka Prefecture University

[P9] Physical properties and structure of boron oxide glass as a function of cooling rate, Deborah Watson,[1,2] Tiffany Myers,[1,2] Kevin Tholen,[1] Victor Khristenko,[1] Mario Affatigato,[1] Steve Feller[1] & Steve Singleton[1]
[1]Physics Department, Coe College, Cedar Rapids, Iowa, USA 52402
[2]Chemistry Department, Coe College, Cedar Rapids Iowa, USA 52402

[P10] Optical and structural investigations on sodium borotungstate and fluoroborotungstate glasses and glass-ceramics, Yannick Ledemi, Imen Chermiti, Mohammed El Amraoui & Younès Messaddeq
Centre d'Optique, Photonique et Laser, Université Laval, 2375 rue la Terrasse, local 2131, Québec, G1V 0A6, Canada

[P11] Kinetic analysis of dissolution behavior of bio-active borate and borosilicate glasses in aqueous solutions, Jaime George & Richard Brow
Missouri University of Science and Technology, Rolla, Missouri, USA

[P12] Luminescence efficiency of B_2O_3–Sb_2O_3–Bi_2O_3 glass phosphor doped with Sm^{3+}, Koji Oshima, Shingo Fuchi, Shunichi Kobayashi & Yoshikazu Takeda
Department of Crystalline Materials Science, Graduate School of Engineering, Nagoya University, Nagoya, Furo-cho, Chikusa-ku, 464-8603, Japan

[P13] Increase of light extraction efficiency of Bi_2O_3–B_2O_3 based glass phosphor doped with Yb^{3+} and Nd^{3+},

Shunichi Kobayashi, Shingo Fuchi, Koji Oshima & Yoshikazu Takeda
Department of Crystalline Materials Science, Graduate School of Engineering, Nagoya University, Furo-cho, Chikusa-ku, Nagoya 464-8603, Japan

[P14] Low Temperature Calorimetric Study of Potassium Germanate Glasses, Seiichi Mamiya, Tamotsu Koyano, Yasuhisa Yamamura & Seiji Kojima
Graduate School of Pure and Applied Sciences, University of Tsukuba

[P15] Formation of an outer borosilicate glass layer on Late Bronze Age Mycenaean vitreous relief fragments, Doris Möncke,[1,2] Dimitrios Palles,[1] Nikos Zacharias,[3] Maria Kaparou,[3,4] Efstratios I. Kamitsos[1] & Lothar Wondraczek[2]
[1]Theoretical and Physical Chemistry Institute, National Hellenic Research Foundation, 48 Vasileos Constantinou Avenue, 11635 Athens, Greece
[2]Institute of Glass and Ceramics, Dept. Materials Science, Friedrich-Alexander-University Erlangen-Nuremberg, Martensstr. 5, 91058 Erlangen, Germany
[3]Laboratory of Archaeometry, Dept. of History, Archaeology and Cultural Resources Management, University of Peloponnese, Old Camp, 24 100 Kalamata, Greece
[4]Institute of Materials Science, N.C.S.R. Demokritos, 15 310 Agia Paraskevi, Attiki, Greece

[P16] Structural studies of yttrium aluminoborate laser glasses using solid state NMR e electron spin echo envelope modulation spectroscopy, A. S. S. de Camargo,[1] J. F. Lima,[1] C. Magon,[1] C. N. Santos,[1] A. C. Hernandez,[1] C.R. Ferrari,[1] H. Deters[2] & H. Eckert[2]
[1]Physics Institute of Sao Carlos, University of Sao Paulo (IFSC/USP), Brazil
[2]Institut für Physikalische Chemie, WWU Münster, Germany

[P17] Influence of PbF_2 in thermal, optical and structural properties in lead borate glasses, R. G. Fernandes,[1] A. S. S. de Camargo,[1] H. Eckert[1,2] & A. C. Hernandes[1]
[1]Physics Institute of Sao Carlos, University of Sao Paulo (IFSC/USP), Brazil
[2]Institut für Physikalische Chemie, WWU Münster, Germany

Phys. Chem. Glasses: Eur. J. Glass Sci. Technol. B, April 2012, 53 (2), 17–26

New borate glasses for ionics

Tsutomu Minami

Osaka Prefecture University, Professor emeritus, Technical Research Institute of Osaka Prefecture, 2-7-1 Ohnodai, Osaka-Sayama, Osaka 589-0023, Japan

Manuscript received 28 October 2011
Revised version received 9 March 2012
Accepted 23 March 2012

Many new glasses composed of AgI, Ag_2O and oxides, M_xO_y, where M_xO_y is B_2O_3, SiO_2, GeO_2, P_2O_5, V_2O_5, As_2O_5, CrO_3, SeO_3 or MoO_3, were found. Among these oxides, the B_2O_3-based glasses, that is, the $AgI–Ag_2O–B_2O_3$ glasses have several unique properties such as (a) very wide glass forming regions, (b) high glass transition temperatures, (c) network structure with BO_3 and BO_4 groups, and (d) high conductivities of 10^{-2} to 10^{-7} $S cm^{-1}$ at room temperature. The high temperature form of AgI, α-AgI, was stabilized in the borate glasses in two ways; one was the rapid melt quenching, and the other was the annealing of glasses at a suitable temperature below the α–β transformation temperature (147°C) for AgI. The latter technique is now usefully applied for the development of glass ceramics as the solid electrolytes for all solid state lithium secondary batteries. Tin oxide containing borate and borophosphate glasses were prepared by the mechanical milling technique as well as melt quenching. These new glasses were proved to work as large capacity negative electrodes for Li^+ ion secondary batteries. The capacities and also the glass transition temperatures were strongly related to the amount of four coordinated boron.

1. Introduction

There are very few examples where glasses are used for the active parts in electrical devices, because glasses are usually electric insulators. Such a circumstance has changed in these two or three decades thanks to the development of new glasses with very high ionic conductivities. These glasses are called "fast ion conducting glasses" or "superionic conducting glasses".[1–3]

Ions generally cannot move quickly in solids because of their large size and because the interaction between cations and anions, of which the solid is composed, is strong. However, if some structural conditions are fulfilled, ions can move quickly in solids. The field where the movement of ions in solids is studied and exploited is called "solid state ionics" or simply "ionics," this name is the counterpart of "electronics" where the movement of electrons in solids is studied and exploited. In the field of solid state ionics, three types of materials are well known; (a) α-AgI type crystals, which have a random distribution of cations in the anion lattice, (b) β-Al_2O_3 type crystals, which have a layered structure which is favourable for cation movement, and (c) stabilized ZrO_2 type crystals, which have Koch–Wagner lattice defects for anion transport.[2] The materials belonging to these three categories are, of course, all crystalline solids. In contrast, vitrification or glass formation has been shown to be very useful for achieving higher conductivities than those found in crystalline solids.[4–37]

There are two explanations as to why glasses show

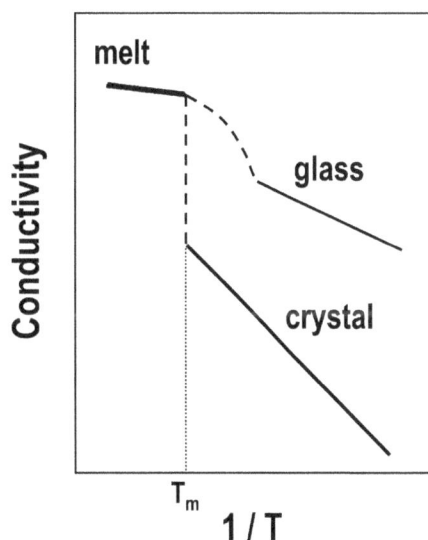

Figure 1. Schematic illustration of conductivity versus inverse absolute temperature to explain the higher conductivities in glasses than in crystals

higher conductivities than crystals do. The first one comes from the disordered, random and thus open structure of glasses. Ions like Ag^+, Cu^+, Li^+ and Na^+ can move much more easily in glasses with such an open structure than in crystals.

Another explanation is based on the process of formation of glasses from melts. Figure 1 schematically shows the relation between the ionic conductivity and the inverse absolute temperature.[33] The melt of a compound usually exhibits high conductivities and low activation energies for conduction at high temperatures. However, the conductivities drop very dramatically at the transformation temperature

Email thminami@livedoor.com
Original version presented at VII Int. Conf. on Borate Glasses, Crystals and Melts, Dalhousie University, Halifax, Nova Scotia, Canada, 21–25 August 2011

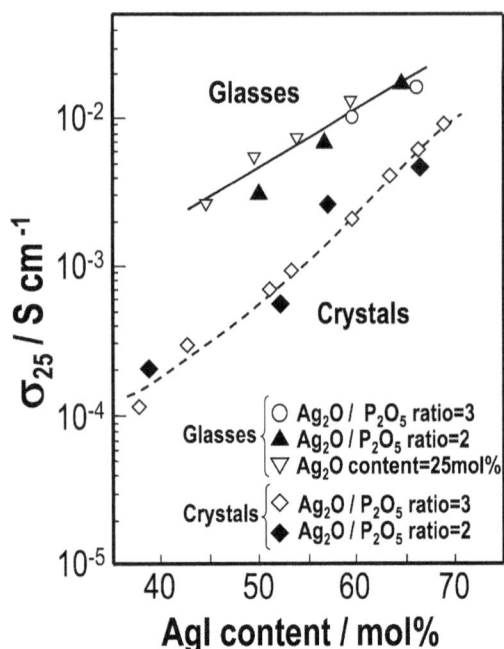

Figure 2. Comparison of conductivities between glasses and crystals for the system AgI–Ag$_2$O–P$_2$O$_5$

from the melt to the crystal, that is, at the melting temperature T_m, and the crystal exhibits low conductivities and high activation energies for conduction. Glasses are usually made by quenching a melt, and thus the disordered and open structure is preserved. The physical properties like conductivity are also quenched in giving values between those of melts and crystals as shown schematically in the figure. As a result, we can expect the conductivity of glasses to be higher than that of crystals.

Among many examples comparing the conductivities of glasses and crystals,[1,5,6,24,33,38,39] only one example is shown in Figure 2.[6] This figure shows a comparison of conductivities of glasses and crystals in the form of composition dependence of conductivities at room temperature for the AgI–Ag$_2$O–P$_2$O$_5$ system. The data for crystals are cited from Ref. 40. It is obvious that the conductivities are higher in the glasses than in the crystalline case for the whole range of glass formation. Firstly the glass forming regions, glass transition temperatures, structure and ionic conductivities of the newly found glasses will be described with a focus on borate systems in this review.

The absolute values of conductivity reach the order of 10^{-2} Scm^{-1} as seen in Figure 2, which are comparable to those of the 5% aqueous solution of NaCl or AgNO$_3$.[1] The conductivities increase with increasing AgI content. When the conductivities are extrapolated beyond the glass forming region up to 100% AgI, the values become very close to the conductivity of α-AgI.[13,33] It is well known that α-AgI exhibits a very high conductivity of about 1 Scm^{-1} at temperatures higher than 147°C. Unfortunately,

it transforms to the non-conductive β-form at 147°C and shows low conductivities below 147°C.

The author tried to expand the glass forming region by increasing the quenching rate of the melts in order to obtain the highest value of conductivities in the newly found glass forming systems. In the course of such trials, it happened that it was found that α-AgI was frozen and stabilized in the glass matrix.[41–56] In addition to the stabilization process by quenching the glass forming melts to freeze α-AgI in the glass matrix, stabilization of α-AgI was also obtained by heat treating or annealing the glasses.[57–67] Many, many investigations were undertaken to lower the transformation temperature below room temperature to find new materials with high conductivities. Several compounds were found to show high conductivities by adding another cation or anion to AgI. However no one succeeded in the stabilization of α-AgI itself at room temperature. Secondly in this review the stabilization process of α-AgI in borate glass matrix is discussed.

Finally in this review, tin borate and tin borophosphate glasses are considered as the negative electrode materials for Li$^+$ ion secondary batteries.[68–74] In recent years, significant attention has been paid to Li$^+$ ion secondary batteries, and tin borate and tin borophosphate glasses were found to show high capacities as negative electrodes in such batteries.[75]

2. Finding new glasses with very high ionic conductivities

2.1. Glass forming regions

The glass forming systems of the Ag$^+$ ion conducting glasses are summarized in Figure 3.[3,22] The fundamental combination of components for glass formation is AgI, Ag$_2$O and M$_x$O$_y$, where M$_x$O$_y$ is an oxide such as B$_2$O$_3$, SiO$_2$, GeO$_2$, P$_2$O$_5$, V$_2$O$_5$, As$_2$O$_5$, CrO$_3$, SeO$_3$ and MoO$_3$.[13] All the glass forming systems AgI–Ag$_2$O–M$_x$O$_y$ shown here constitute new types of

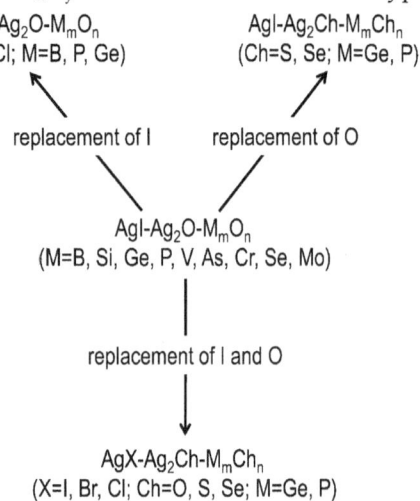

Figure 3. Variation of glass forming systems showing high ionic conductivities in the systems AgI–Ag$_2$O–M$_x$O$_y$

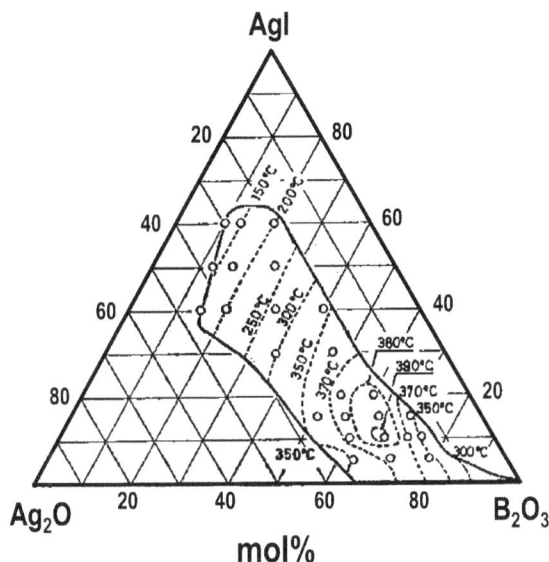

Figure 4. Contours of T_g of glasses in the system AgI–Ag$_2$O–B$_2$O$_3$

50 mol% B$_2$O$_3$, T_g is primarily determined by the B$_2$O$_3$ content and increases with increasing B$_2$O$_3$ content. In the high B$_2$O$_3$ content region, however, the variation with composition becomes complex and the so-called "boron anomaly" is observed. Another feature of T_g in the B$_2$O$_3$-based glasses is that T_g is relatively high compared with other ion conductive glasses;[4,6,14] the highest value of T_g is about 390°C for 10AgI.25Ag$_2$O.65B$_2$O$_3$ (mol%). The other glasses composed of P$_2$O$_5$, As$_2$O$_5$, SeO$_3$ and MoO$_3$ are built up from discrete, monomer ions, which are categorized as the "ionic glasses." The B$_2$O$_3$-based glasses have a condensed structure of BO$_3$ and BO$_4$ groups. The values of T_g, however, do not directly relate to the content of BO$_4$ groups,[17–19] and further detailed study is needed to explain the reasons for high T_g in B$_2$O$_3$-based glasses.

glasses. The development of these new glasses started from the accidental observation of glass formation in the system AgI–Ag$_2$SeO$_4$.[76] When M=B, Ge and P, the replacement of I and/or O with another halogen or oxygen group element is reported to produce new glasses.[1–3,9,18,19,21,22,77–81] Among these new glass forming systems, the B$_2$O$_3$-based system AgI–Ag$_2$O–B$_2$O$_3$ shows the widest glass forming region.[14] The halogen substituted systems AgBr–Ag$_2$O–B$_2$O$_3$ and AgCl–Ag$_2$O–B$_2$O$_3$ have similarly wide glass forming regions,[18,19] which will be shown later. Such wide glass forming regions are the striking features of the new B$_2$O$_3$-based glasses.

2.2. Glass transition temperatures

Figure 4 shows the iso-T_g curves in the AgI–Ag$_2$O–B$_2$O$_3$ system.[17–19] In the low B$_2$O$_3$ region, up to about

2.3. Structure

The structure of the new borate glasses was investigated by infrared absorption (IR) spectroscopy. We can make a glass balloon by blowing the glass forming melts. IR spectra were measured by two ways; by using such blown films (4–15 μm thick) from glass balloons and by the usual method of KBr pellets containing a small amount (1 wt%) of the powdered glass.

Figure 5 shows the wavenumbers of IR absorption maxima for BO$_3$ groups and BO$_4$ groups versus the Ag$_2$O content for many AgI–Ag$_2$O–B$_2$O$_3$ glasses for blown films (left hand figure), and for KBr pellets (right hand figure).[1,17–19] The wavenumbers for BO$_4$ groups shown by squares are located in nearly the same positions for all the samples, regardless of whether blown films or KBr pellets are measured. On the other hand, the wavenumbers for BO$_3$ groups shown by triangles are shifted to higher wavenumbers by about 100 cm^{-1} for the KBr pellets as compared

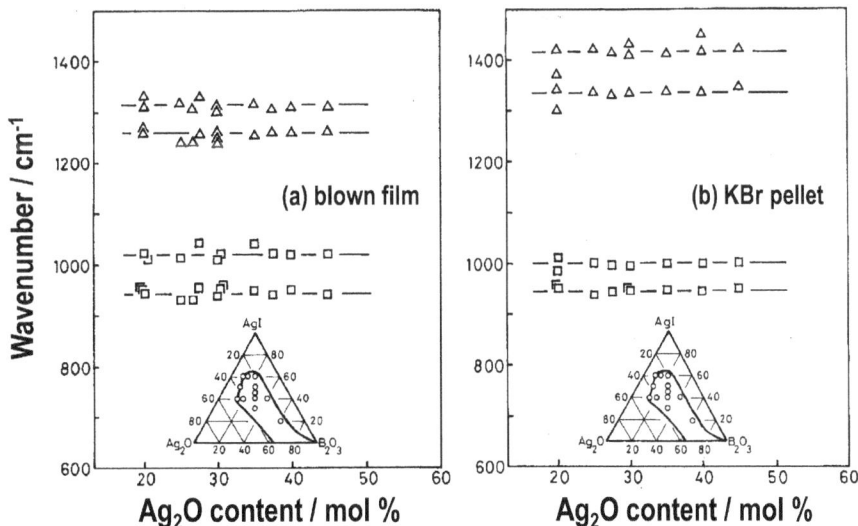

Figure 5. Positions of IR absorption bands for BO$_3$ (△) and BO$_4$ (□) groups in the AgI–Ag$_2$O–B$_2$O$_3$ glasses measured for blown films (left side) and KBr pellets (right side)

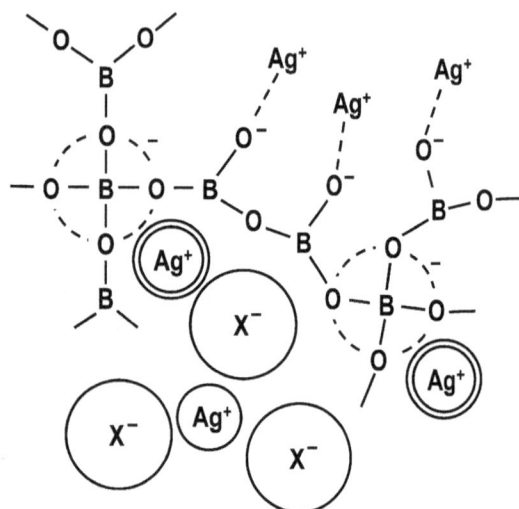

Figure 6. Schematic illustration of a structural model for the AgX–Ag_2O–B_2O_3 glasses where X=I, Br or Cl

to the blown films. These results indicate that K^+ ions substitute for Ag^+ ions located near the non-bridging oxygens, and thus the non-bridging oxygens are present only in the BO_3 groups. The shift of 100 cm^{-1} to higher wavenumbers also means that the partial covalency between Ag^+ ions and nonbridging oxygens is relatively strong.

The intensity of the BO_4 group absorption is reported to increase linearly with increasing Ag_2O content. The linear variation of BO_4 intensity with Ag_2O content passes through the origin, hence it is obvious that the formation of BO_4 groups is directly related to the addition of the Ag_2O.[19]

When AgI was replaced by AgBr or AgCl, the IR spectra and the intensity variation of BO_4 groups with composition did not change much compared with the original AgI–Ag_2O–B_2O_3 glasses.[1,17–19]

A structural model is proposed in Figure 6 on the basis of the IR spectra mentioned above.[1,18,19] The nonbridging oxygens are present in the BO_3 groups only. Three types of Ag^+ ions are illustrated; the first one is an Ag^+ ion bonded with nonbridging oxygens with strong partial covalency, the second one (circled) is an Ag^+ ion surrounded with X^- ions only, and the third one (double circled) is an Ag^+ ion interacting with BO_4 groups. The double circled Ag^+ ions occur only in borate glasses, which is a striking structural feature of the borate glasses. These double circled Ag^+ ions play an important role in the variation in conductivity with composition in the borate glasses.

2.4. Conductivities

2.4.1. Electrolytic properties

When we think whether the glasses are good for solid electrolytes or not, the electrolytic properties such as the magnitude of electronic conductivities (by electrons or holes) and the transport numbers of Ag^+ ions are very important as well as the magnitude of the total conductivities.

Electronic conductivities can be determined by the Wagner's DC polarization method.[82] In this method, the magnitude of electronic conductivities and also whether conduction is by electrons or holes are determined. The transport numbers can be measured by two ways: the Tubandt method[83] and the electromotive force (EMF) method.[40]

Table 1 shows the electrolytic properties of some borate glasses.[17–19] The total conductivities of these glasses are 10^{-3} to 10^{-4} Scm^{-1}. The electronic conduction was caused by electrons and the values are 10^{-8} to 10^{-9} Scm^{-1}, which are smaller by 5 to 6 orders of magnitude than the total conductivities. The transport numbers of Ag^+ ions are practically unity. These properties are the most important features for solid electrolytes. From these values shown in Table 1, we can conclude that these glasses are very good solid electrolytes.

2.4.2. Conduction mechanism

Figure 7 shows the iso-conductivity (25°C) curves for three systems of B_2O_3-based glasses containing AgI, AgBr or AgCl. There are two features in these B_2O_3-based glasses. First of all, the glass forming regions are wide and very similar for the three systems even if AgI is replaced by AgBr or AgCl. According to Figure 3, we expect that glasses can also be formed in the AgI–Ag_2O–GeO_2 and AgI–Ag_2O–P_2O_5 systems where AgI is replaced by AgBr or AgCl. In contrast to the B_2O_3-based system, the glass forming regions are different in those GeO_2- and P_2O_5-based systems containing different silver halides.[9,22] Secondly, the conductivities ranging from 10^{-2} to 10^{-7} Scm^{-1} vary quite similarly among these three systems. In the GeO_2- and P_2O_5-based glasses, however, the conductivities decrease in the order of AgI- to AgBr- to AgCl-containing glasses.[9,22]

It is reported that the Ag^+ ions move in the glass matrix accompanying the polarization of anions.[2,3,9,26,27,33] Elastic collisions should be avoided.

Table 1. Electrolytic properties of some borate glasses

Glass composition (mol%)	σ_{total} at 25°C (Scm^{-1})	σ_e at 25°C (Scm^{-1})	Transport number of Ag^+ ion	
			Tubandt method	EMF method
$60AgI.30Ag_2O.10B_2O_3$	$8·5\times10^{-3}$	$7·8\times10^{-9}$	$1·01\pm0·04$	$0·992$
$40AgI.30Ag_2O.30B_2O_3$	$2·6\times10^{-3}$	–	–	$0·993$
$50AgBr.25Ag_2O.25B_2O_3$	$2·7\times10^{-3}$	$7·2\times10^{-8}$	$0·97\pm0·03$	–
$40AgCl.30Ag_2O.30B_2O_3$	$6·3\times10^{-4}$	$2·4\times10^{-9}$	–	–

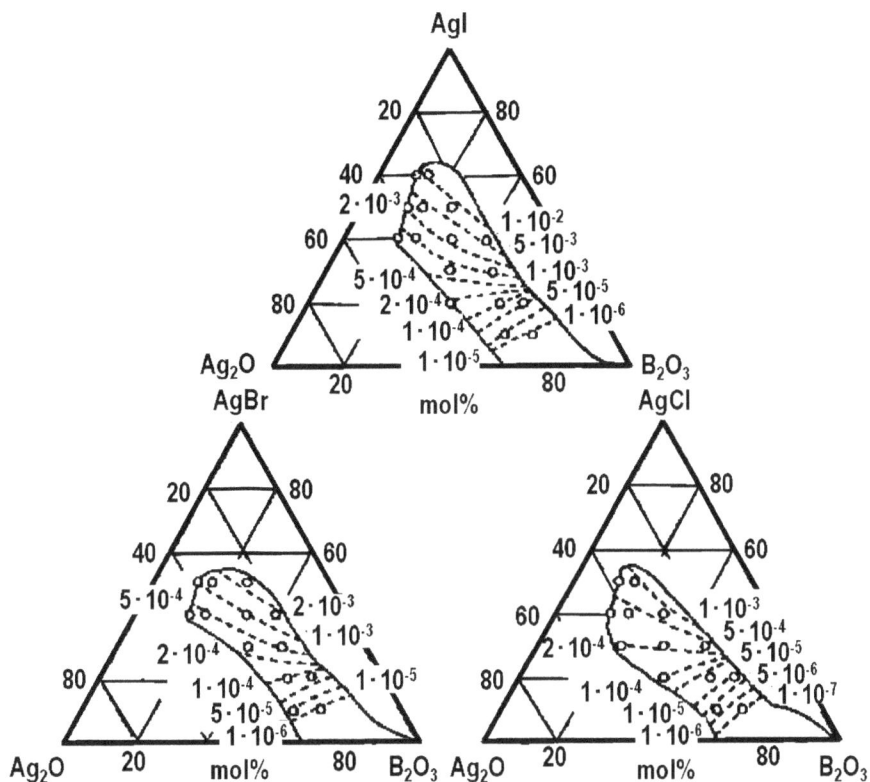

Figure 7. Contours of conductivity at 25°C for the AgX–Ag$_2$O–B$_2$O$_3$ glasses where X=I, Br or Cl

Thus larger anions with smaller charge are favourable for rapid motion of Ag$^+$ ions; halide ions and BO$_4$ groups must be such anions. It is strange in some senses that Ag$^+$ ions with a large ionic radius and a large atomic weight can move quickly in glasses. The reason for that can be explained on the basis of the Hard and Soft Acids and Bases theory (HSAB theory) proposed by Peason.[2,3,26,84] An Ag$^+$ ion has the d^{10} configuration and is classified as a soft acid, which can give high ionic movement in glasses. Cu$^+$ ion has the same d^{10} configuration and a smaller ionic radius than Ag$^+$ ion, and thus Cu$^+$ ion containing glasses like CuI–Cu$_2$O–MoO$_3$, CuI–Cu$_3$PO$_4$–Cu$_2$MoO$_4$, and CuBr–Cu$_3$PO$_4$–Cu$_2$MoO$_4$ also show higher ionic conductivities.[2,23–28,35,36] On the other hand, a Li$^+$ ion has a much smaller ionic radius than an Ag$^+$ ion and has the s^2 configuration. Thus Li$^+$ ions are classified as hard acids and such an ion cannot move quickly in an oxide glass matrix due to elastic collisions; an oxide ion is classified as a hard base. Rapid movement of Li$^+$ ions is realized in a soft base matrix such as a sulphide glasses, which will briefly be introduced in Section 3.2.

The composition dependence of conductivities was studied in quite some detail for the newly found ion conducting glasses; the most important conclusion is that all the Ag$^+$ ions do not contribute to conduction, but only Ag$^+$ ions brought into the glasses by the AgX component contribute to conduction.[1–3,5,6,9,13,18,19,22,33] In other words, Ag$^+$ ions surrounded by halide ions (X$^-$) can move rapidly and contribute to the conduc-

tion. The circled Ag$^+$ ions in Figure 6 correspond to such Ag$^+$ ions. The conductivities of the GeO$_2$- and P$_2$O$_5$-based glasses decrease with the replacement of AgI with AgBr and AgCl.[9,22] In the B$_2$O$_3$-based glasses, however, the conductivities do not change so much with the replacement of silver halides as shown in Figure 7, which means that the double circled Ag$^+$ ions interacting with BO$_4$ groups strongly contribute to the conduction, as well as the circled Ag$^+$ ions. The Ag$^+$ ions bonded with nonbridging oxygens contribute only very slightly to the conductivity.[17–19]

3. Stabilization of α-AgI in the glass matrix at room temperature

3.1. Stabilization of α-AgI by melt quenching

As already mentioned above, many investigations were undertaken to lower the transformation temperature of α-AgI to β-AgI below room temperature by partial replacement of Ag$^+$ ions or I$^-$ ions with other cations or anions. The most popular example of cationic substitution is RbAg$_4$I$_5$, which is a compound with 20% substitution of Ag$^+$ ions with Rb$^+$ ions, that is, 20RbI.80AgI.[85,86] There are many examples of anionic substitution; for example, Ag$_7$I$_4$PO$_4$ (=80AgI.20Ag$_3$PO$_4$), Ag$_{19}$I$_{15}$P$_2$O$_7$ (=94AgI.6Ag$_4$P$_2$O$_7$), Ag$_6$I$_4$WO$_4$ (=80AgI.20Ag$_2$WO$_4$) and Ag$_6$I$_4$SO$_4$ (=80AgI.20Ag$_2$SO$_4$).[40,87]

However, no one succeeded in the direct stabilization of α-AgI. We tried to expand the glass forming regions by increasing the melt quenching rate, and ac-

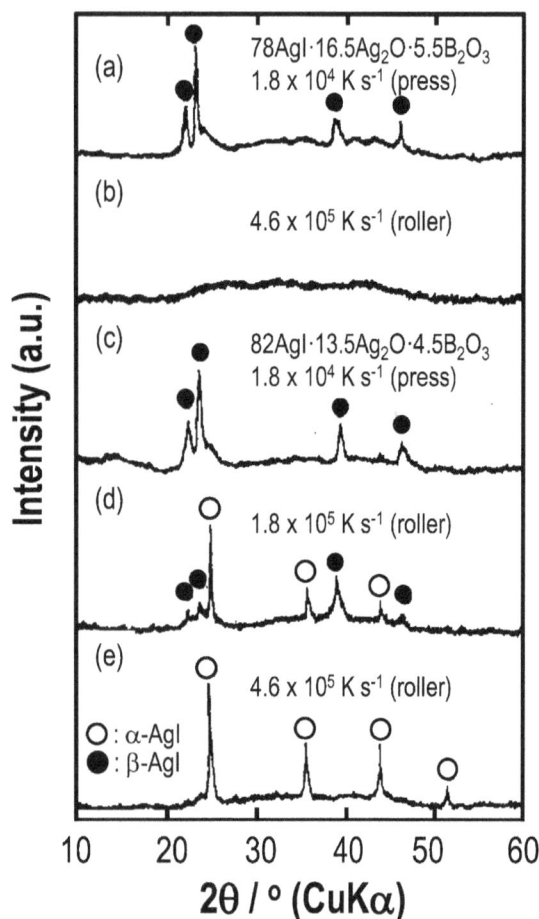

Figure 8. *X-ray diffraction patterns for rapidly quenched samples with different cooling rates and compositions in the system AgI–Ag$_2$O–B$_2$O$_3$. "Press" and "roller" indicate metal plate press quenching and twin roller quenching, respectively*

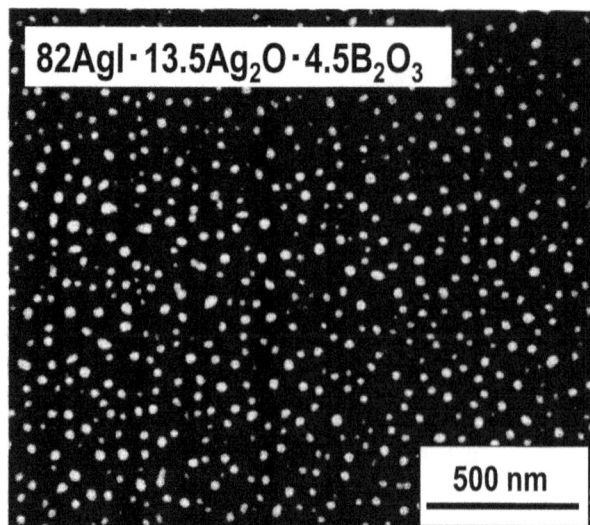

Figure 9. *FE-SEM photograph of the cross-section of the twin roller quenched 82AgI.13·5Ag$_2$O.4·5B$_2$O$_3$ (mol%) composite in which α-AgI was frozen at room temperature*

Ag$_2$O–P$_2$O$_5$, AgI–Ag$_2$O–V$_2$O$_5$ and AgI–Ag$_2$O–MoO$_3$. In all the cases the composition region, where α-AgI was frozen and stabilized in the glass matrix, was limited to a small area with higher AgI contents just outside of the glass forming region.[42,43,64]

Figure 9 shows a photograph of an FE-SEM observation for the cross section of quenched sample of 82AgI.13·5Ag$_2$O.4·5B$_2$O$_3$ (mol%), the composition of which is the same as that in Figure 8(e).[49] The white particles are frozen-in α-AgI and the continuous black part is the glass matrix. α-AgI crystals, the sizes of which are about 20–40 nm in diameter, are present as the dispersed phase in the composite.

3.2. Stabilization of α-AgI by annealing of the glasses

We thought at first that the rapid quenching of melts was the only way to stabilize α-AgI in the glass matrix at room temperature. However, we succeeded in stabilizing α-AgI in the glass matrix by annealing or heat treating the glasses.[57–66] The key point for the success was the careful observation of the fine structure of glasses by FE-SEM and x-ray diffraction.[48,52,53,58] Figure 10 shows a FE-SEM micrograph of the cross section of the rapidly quenched 78AgI.16·5Ag$_2$O.5·5B$_2$O$_3$ (mol%) glass which was then annealed at 110°C,[66] which is much lower than the α–β transformation temperature (147°C) of AgI. As shown in Figure 8(b), this composition is pure glass as judged by x-ray diffraction. Finely dispersed particles (20–30 nm in diameter) of α-AgI are observed in the glass matrix and the microstructure is very similar to Figure 9. This photograph clearly indicates that α-AgI can be stabilized at room temperature in the glass matrix by annealing or heat treating the glasses as well as by rapid quenching.

cidentally found that α-AgI was frozen and stabilized in the borate glass matrix.[41] Figure 8 shows the x-ray diffraction patterns for melt quenched samples with different cooling rates and different compositions in the AgI–Ag$_2$O–B$_2$O$_3$ system. Figure 8(a) is for the sample 78AgI.16·5Ag$_2$O.5·5B$_2$O$_3$ (mol%) cooled at a relatively rate. The peaks due to β-AgI are marked by closed circles. Figure 8(b) is for the same composition cooled at a higher rate and the crystalline peaks have disappeared and the sample is amorphous. Figure 8(c) is for the sample 82AgI.13·5Ag$_2$O.4·5B$_2$O$_3$ (mol%) containing a slightly higher AgI content than (a) and (b), and cooled at a relatively low rate. Only peaks due to β-AgI are seen. Figure 8(d) is for the same composition as (c) cooled at a relatively high rate. Peaks due to α-AgI are seen together with those due to β-AgI. Figure 8(e) is for the same composition cooled at the highest cooling rate, 4·6×10^5 K s^{-1}, we can achieve. In Figure 8(e) only diffraction peaks due to α-AgI are seen.

It was found that α-AgI can be frozen not only in the AgI–Ag$_2$O–B$_2$O$_3$ system but also in other glass forming systems such as AgI–Ag$_2$O–GeO$_2$, AgI–

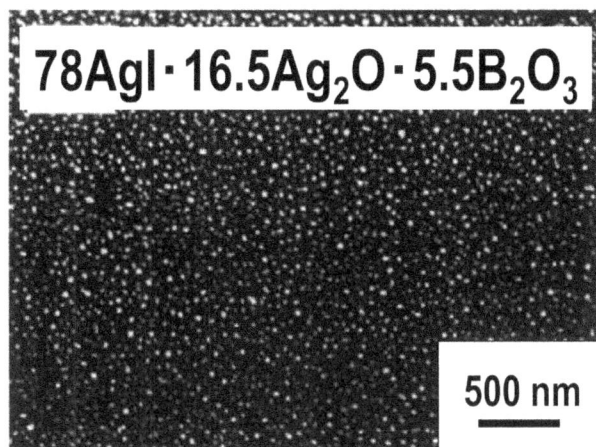

Figure 10. FE-SEM photograph of the cross-section of the twin roller quenched 78AgI.16·5Ag$_2$O.5·5B$_2$O$_3$ (mol%) glass after annealing at 110°C and cooling to room temperature

This technique for stabilizing a high temperature phase in glass matrix was exploited to develop high Li$^+$ ion conducting glass ceramics, which have very high potential for practical applications as solid electrolytes in the all solid state Li$^+$ ion secondary batteries. Sulphide and oxysulphide based glasses such as Li$_2$S–Li$_4$PS$_4$ and Li$_2$S–SiS$_2$–Li$_4$SiO$_4$ exhibit high Li$^+$ ion conductivities,[88-94] and the glasses heat treated to produce high temperature conductive crystalline phases show the conductivities of the order of 10^{-3} S cm^{-1} at room temperature.[97-99,103-108] These glass ceramics are expected to become solid electrolytes in practically all solid state Li$^+$ ion secondary batteries.[96,101,102]

3.3. Mechanism of stabilization of α-AgI by rapid quenching and annealing

In the last two sections, α-AgI was stabilized by rapid melt quenching and by annealing. The difference arises from a small difference in the chemical composition of glasses; an AgI-rich one stabilizes α-AgI by rapid quenching and AgI-poor one by annealing. Such a situation is summarized in Figure 11.[60] This figure is a different expression of the glass forming region of the system xAgI.(100−x)(0·75Ag$_2$O.0·25B$_2$O$_3$), that is, the system AgI–Ag$_3$BO$_3$. In regions A and B, homogeneous and phase separated glasses respectively are obtained with ordinary cooling rates (up to about 10^2 Ks^{-1}). Region C is usually beyond the glass forming limit. However, α-AgI is frozen and stabilized by rapid quenching at quenching rates greater than 10^5 Ks^{-1}. In region D, no glasses are obtained and β-AgI crystals are observed even if melts are rapidly quenched.

In region B, the separated phases are amorphous by x-ray diffraction, and the microstructure of the separated amorphous phases observed by FE-SEM was very similar to that shown in Figures 9 and

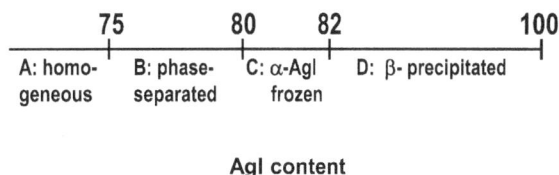

Figure 11. Summary of the α-AgI stabilization process in a glass matrix at room temperature for the xAgI. (100−x)(0·75Ag$_2$O.0·25B$_2$O$_3$) system

10.[48,52,53,58] From this kind of observation it is speculated that AgI rich separated phases contain α-AgI nuclei and aggregate on annealing to form island regions several hundred nanometres in size. The formation of α-AgI microcrystals 20–30 nm in diameter occurs in such island regions of the AgI rich phase. In region C, not only nucleation of α-AgI but also aggregation of the α-AgI rich phase and growth of α-AgI occurs during melt quenching.[57,60,64-66]

4. Negative electrode materials for Li$^+$ ion secondary batteries

As already mentioned above, tin borate and tin borophosphate glasses were reported to be good negative electrodes for Li$^+$ ion secondary batteries.[75] We introduced the mechanical milling technique instead of the usual melt quenching to prepare these glasses. The mechanical milling technique is now widely used for preparation of glass and glass ceramic solid electrolytes exhibiting Li$^+$ ion conduction.[93-109]

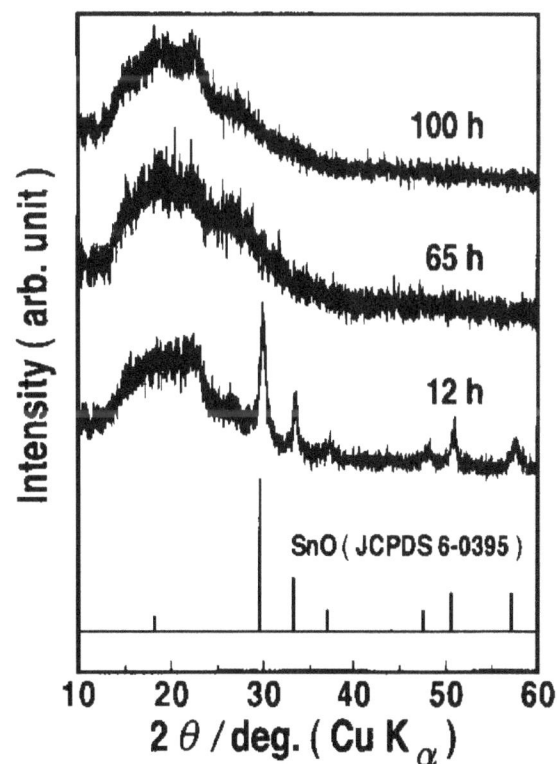

Figure 12. X-ray diffraction patterns of SnB$_{0.5}$P$_{0.5}$O$_3$ powders prepared by mechanical milling for different periods of time

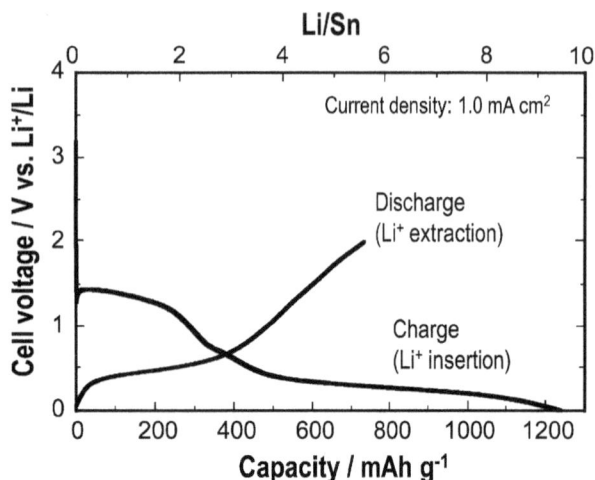

Figure 13. First charge-discharge curves for the electro-chemical cell using the SnB_2O_4 glass as a working electrode and lithium metal as a counter and reference electrode. The organic liquid electrolyte 1 M $LiPF_6$ in EC-DEC was used

Figure 12 shows the variation of the x-ray diffraction patterns with milling time for the tin borophosphate powders. The diffraction patterns become halos with increased milling time. A milling time over 65 h gives amorphous powders. The existence of glass transition temperatures were confirmed for these amorphous powders, which were thus concluded to be glasses.[68,73,74] The mechanical milling technique has several merits for the preparation of glassy electrodes namely (a) the process is simple, (b) the process can be carried out at room temperature, (c) powders are obtained for direct use as electrode and electrolyte materials and (d) new materials can be prepared, which could not be synthesized by the usual heating process .

Figure 13 shows the first charge discharge curves for a conventional rechargeable Li$^+$ ion cell using $50SnO.50B_2O_3$ (SnB_2O_4) glass as the negative electrode. The cell is composed of the electrolyte of ethylene carbonate and diethyl carbonate containing $LiPF_6$.[69,72] Charge and discharge means insertion and extraction respectively of Li$^+$ ions to/from the sample. The first charge capacity from the rest potential to 0 V at a constant current of 1·0 mA cm^{-2} achieves about 1200 mAh g^{-1} and a discharge capacity of 700 mAh g^{-1} is observed. It is concluded that tin borate glass is working as the negative electrode with a large capacity in Li$^+$ ion secondary batteries.

Two plateaus are observed during the first charge process. At the first plateau at 1·5 V, Sn (II) in the tin borate glass was reduced by the Li$^+$ insertion process and metallic tin Sn (0) particles embedded in lithium borate glass matrix were formed. This process is irreversible:

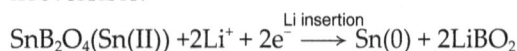

$$SnB_2O_4(Sn(II)) + 2Li^+ + 2e^- \xrightarrow{\text{Li insertion}} Sn(0) + 2LiBO_2$$

At the second plateau at 0·5 V, further lithiation transformed metallic tin Sn (0) into a Li–Sn alloy and

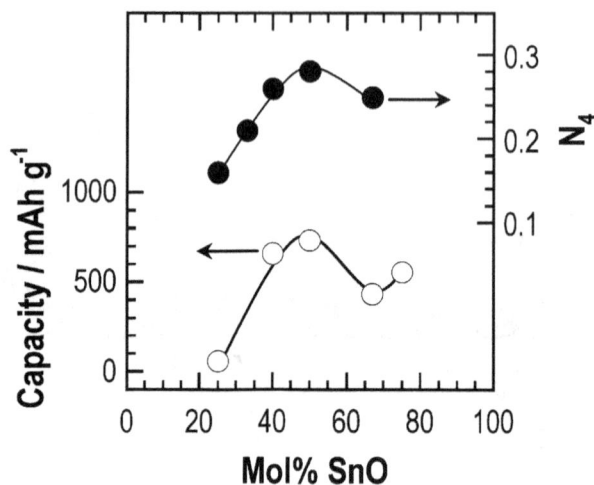

Figure 14. Relationship between N_4 and capacity of SnO–B_2O_3 glass electrodes

this process is basically reversible:

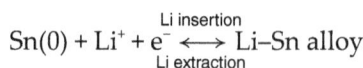

$$Sn(0) + Li^+ + e^- \underset{\text{Li extraction}}{\overset{\text{Li insertion}}{\longleftrightarrow}} Li\text{–}Sn \text{ alloy}$$

Figure 14 shows the composition dependence of the first discharge capacity of a SnO–B_2O_3 glass electrode together with the fraction of four-coordinated boron (N_4) in the glasses;[69,72] N_4 was determined by ^{11}B MAS-NMR.[70,71,73] The capacity has a maximum at about 50 mol% SnO, where N_4 also has a maximum. The glass transition temperatures also have a maximum at nearly the same composition.[70,73] From these results, we can say the capacity is enhanced by modifying and optimizing the local structure of tin borate glasses.

5. Conclusion

Many new glass forming systems were found for AgI–Ag_2O–M_xO_y, where M_xO_y is B_2O_3, SiO_2, GeO_2, P_2O_5, V_2O_5, As_2O_5, CrO_3, SeO_3 or MoO_3. The newly found glasses showed very high Ag$^+$ ion conductivities comparable to liquid electrolytes such as aqueous solutions of NaCl and $AgNO_3$ at room temperature. Among the glasses examined B_2O_3-based glasses had unique properties such as very wide glass forming regions, high glass transition temperatures, a network structure with BO_3 and BO_4, and replacement of AgI with AgBr or AgCl giving glasses with no appreciable difference in glass forming regions, structure and properties.

α-AgI, which is only stable at temperatures over 147°C, was stabilized at room temperature in silver borate glasses by rapid melt quenching and also by annealing of glasses. The stabilization mechanism was clarified.

Tin oxide containing borate and borophosphate glasses were proven to work as large capacity negative electrodes for Li$^+$ ion secondary batteries. The capacity was strongly related to the fraction of four-coordinated boron in the glasses.

As the most important conclusion, structural studies should always be undertaken to understand the reason for the excellent properties of glasses to be used in the field of solid state ionics.

Acknowledgments

The author wishes to thank Professor M. Tatsumisago and Professor A. Hayashi for their long period of co-operation. Without their help this review paper would not have been completed. He also gives sincere thanks to the many students and co-workers working together in his laboratory.

References

1. Minami, T. *J. Non-Cryst. Solids*, 1983, **56**, 15–26.
2. Minami, T. *J. Non-Cryst. Solids*, 1985, **73**, 273–284.
3. Minami, T. *J. Non-Cryst. Solids*, 1987, **95–96**, 107–118.
4. Minami, T., Nambu, H. & Tanaka, M. *J. Am. Ceram. Soc.*, 1977, **60**, 283–284.
5. Minami, T., Nambu, H. & Tanaka, M. *J. Am. Ceram. Soc.*, 1977, **60**, 467–469.
6. Minami, T., Takuma, Y. & Tanaka, M. *J. Electrochem. Soc.*, 1977, **124**, 1659–1662.
7. Minami, T., Katsuda, T. & Tanaka, M. *J. Non-Cryst. Solids*, 1978, **29**, 389–395.
8. Minami, T. Katsuda, T. & Tanaka, M. *J. Phys. Chem.*, 1979, **83**, 640–641.
9. Minami, T. & Tanaka, M. *Rev. Chim. Miner.*, 1979, **16**, 283–292.
10. Minami, T. & Tanaka, M. *J. Solid State Chem.*, 1980, **32**, 51–55.
11. Minami, T., Katsuda, T. & Tanaka, M. *J. Electrochem. Soc.*, 1980, **127**, 1308–1311.
12. Minami, T. & Tanaka, M. *J. Non-Cryst. Solids*, 1980, **38–39**, 289–294.
13. Minami, T., Imazawa, K. & Tanaka, M. *J. Non-Cryst. Solids*, 1980, **42**, 469–476.
14. Minami, T., Imazawa, K. & Tanaka, M. *J. Am. Ceram. Soc.*, 1980, **63**, 627–629.
15. Minami, T., Matsuda, K. & Tanaka, M. *J. Electrochem. Soc.*, 1981, **128**, 100–102.
16. Minami, T., Matsuda, K. & Tanaka, M. *Solid State Ionics*, 1981, **3–4**, 93–96.
17. Minami, T., Ikeda, Y. & Tanaka, M. *J. Chem. Soc. Jpn.*, 1981, 1617–1623. (In Japanese.)
18. Minami, T., Ikeda, Y. & Tanaka, M. *J. Non-Cryst. Solids*, 1982, **52**, 159–169.
19. Minami, T., Shimizu, T. & Tanaka, M. *Solid State Ionics*, 1983, **9–10**, 577–584.
20. Minami, T. & Kaneko, H. *Solid State Ionics*, 1985, **17**, 57–62.
21. Minami, T. & Shimizu, T. *J. Chem. Soc. Jpn.*, 1986, 420–427. (In Japanese.)
22. Minami, T. & Yamane, A. *Chem. Express*, 1986, **1**, 567–570.
23. Machida, N., Chusho, M. & Minami, T. *J. Non-Cryst. Solids*, 1988, **101**, 70–74.
24. Machida, N., & Minami, T. *J. Am. Ceram. Soc.*, 1988, **71**, 784–788.
25. Machida, N., Tsuboi, S. & Minami, T. *Proc. XV Int. Congr. on Glass*, 1989, **2b**, 211–214.
26. Minami, T. & Machida, N. *Mater. Chem. Phys.*, 1989, **23**, 63–74.
27. Machida, N., Shinkuma, Y. & Minami, T. *J. Ceram. Soc. Jpn.*, 1989, **97**, 1104–1108.
28. Machida, N., Shinkuma, Y. & Minami, T. *Solid State Ionics*, 1991, **45**, 123–129.
29. Minami, T. *Mater. Sci. Forum*, 1991, **67–68**, 575–586.
30. Shinkuma, Y., Machida, N. & Minami, T. *J. Am. Ceram. Soc.*, 1991, **74**, 3133–3135.
31. Machida, N., Tsuchida, S., Minami, T., Shigematsu, T. & Nakanishi, N. *J. Ceram. Soc. Jpn.*, 1992, **100**, 104–106.
32. Machida, N., Shigematsu, T, Nakanishi, N., Shinkuma, Y. & Minami, T. *Solid State Ionics*, 1992, **50**, 303–306.
33. Minami, T. & Machida, N., *Mater. Sci. Eng.*, 1992, **B13**, 203–208.
34. Machida, N., Shinkuma, Y., Minami, T., Shigematsu, T & Nakanishi, N. *J. Electrochem. Soc.*, 1992, **139**, 1380–1383.
35. Minami, T. & Machida, N. *Proc. XVI Int. Congr. on Glass*, 1992, **4**, 163–168.
36. Machida, N., Shigematsu, T., Nakanishi, N., Tsuchida, S. & Minami, T. *J. Chem. Soc. Faraday Trans.*, 1992, **88**, 3059–3062.
37. Hayashi, A., Ishida, T., Minami, T. & Tatsumisago, M. *Glass Technol.: Eur. J. Glass Sci. Tech. A*, 2007, **48**, 1–5.
38. Frischat, G. H. In *Mass Transport Phenomena in Ceramics*, Materials Science Research, Vol. 9, Plenum, New York, 1975, pp.285–295.
39. Tatsumisago, M. Hamada, A. Minami T. & Tanaka, M. *J. Am. Ceram. Soc.*, 1982, **65**, 575–577.
40. Takahashi, T., Ikeda, S. & Yamamoto, O. *J. Electrochem. Soc.*, 1972, **119**, 477–482.
41. Tatsumisago, M., Shinkuma, Y. & Minami, T. *Nature*, 1991, **354**, 217–218.
42. Tatsumisago, M., Shinkuma, Y., Saito, T. & Minami, T. *Solid State Ionics*, 1992, **50**, 273–279.
43. Tatsumisago, M., Taniguchi, A. & Minami, T. *J. Am. Ceram. Soc.*, 1993, **76**, 235–237.
44. Saito, T., Tatsumisago, M. & Minami, T. *Solid State Ionics*, 1993, **61**, 285–291.
45. Hanaya, M., Nakayama, M., Oguni, M., Tatsumisago, M., Saito, T. & Minami, T. *Solid State Commun.*, 1993, **87**, 585–588.
46. Tatsumisago, M., Saito, T., Minami, T., Hanaya, M. & Oguni, M. *J. Phys. Chem.*, 1994, **98**, 2005–2007.
47. Tatsumisago, M., Saito, T. & Minami, T. *Solid State Ionics*, 1994, **70–71**, 394–397.
48. Saito, T., Tatsumisago, M., Torata, N. & Minami, T. *Solid State Ionics*, 1995, **79**, 279–283.
49. Saito, T., Torata, N., Tatsumisago, M. & Minami, T. *J. Phys. Chem.*, 1995, **99**, 10691–10693.
50. Saito, T., Tatsumisago, M. & Minami, T. *J. Electrochem. Soc.*, 1996, **143**, 687–691.
51. Tatsumisago, M., Torata, N., Saito, T. & Minami, T. *J. Non-Cryst. Solids*, 1996, **196**, 193–198.
52. Tatsumisago, M., Saito, T. & Minami, T. *Thermochim. Acta*, 1996, **280**, 333–341.
53. Minami, T., Saito, T. & Tatsumisago, M. *Solid State Ionics*, 1996, **86-88**, 415–420.
54. Saito, T., Torata, N., Tatsumisago, M. & Minami, T. *Solid State Ionics*, 1996, **86–88**, 491–495.
55. Hayashi, A., Hama, S., Morimoto, H., Tatsumisago, M. & Minami, T. *J. Am. Ceram. Soc.*, 2001, **84**, 477–479.
56. Hayashi, A., Kitade, T., Ikeda, Y., Kohjiya, S., Matsuda, A., Tatsumisago, M. & Minami, T. *Chem. Lett.*, 2001, 814–815.
57. Taniguchi, A., Tatsumisago, M. & Minami, T. *J. Am. Ceram. Soc.*, 1995, **78**, 460–464.
58. Torata, N., Saito, T., Tatsumisago, M. & Minami, T. *J. Mater. Sci. Lett.*, 1997, **16**, 1012–1016.
59. Itakura, N. Tatsumisago, M. & Minami, T. *J. Am. Ceram. Soc.*, 1997, **80**, 3209–3212.
60. Tatsumisago, M., Itakura, N. & Minami, T. *J. Non-Cryst. Solids*, 1998, **232–234**, 267–273.
61. Tatsumisago, M. & Minami, T. *Solid State Ionics: Science & Technology* (Proc. 6th Asian Conference on Solid State Ionics), 1998, pp.81–90.
62. Tatsumisago, M., Okuda, K., Itakura N. & Minami, T. *Solid State Ionics*, 1999, **121**, 193–200.
63. Minami, T. & Tatsumisago, M. *Phys. Chem. Glasses*, 2000, **41**, 236–241.
64. Tatsumisago, M., Saito, T. & Minami, T. *Chem. Lett.*, 2001, 790–791.
65. Hayashi, A., Hama, S., Morimoto, H., Tatsumisago, M. & Minami, T. *Chem. Lett.*, 2001, 872–873.
66. I-io, K., Hayashi, A., Morimoto, H., Tatsumisago, M. & Minami, T. *Chem. Mater.*, 2002, **14**, 2444–2449.
67. Hayashi, A., Hama, S., Mizuno, F., Tadanaga, K., Minami, T. & Tatsumisago, M. *Solid State Ionics*, 2004, **175**, 683–686.
68. Morimoto, H., Nakai, M., Tatsumisago, M. & Minami, T. *J. Electrochem. Soc.*, 1999, **146**, 3970–3973.
69. Nakai, M., Hayashi, A., Morimoto, H., Tatsumisago, M. & Minami, T. *J. Ceram. Soc. Jpn.*, 2001, **109**, 1010–1016.
70. Hayashi, A., Nakai, M., Tatsumisago, M., Minami, T., Himei, Y., Miura, Y. & Katada, M. *J. Non-Cryst. Solids*, 2002, **306**, 227–237.
71. Hayashi, A., Nakai, M., Tatsumisago, M. & Minami, T. *Comptes Rendus Chimie*, 2002, **5**, 751–757.
72. Hayashi, A., Nakai, M., Tatsumisago, M., Minami, T. & Katada, M. *J. Electrochem. Soc.*, 2003, **150**, A582–A587.
73. Hayashi, A., Nakai, M., Morimoto, H., Minami, T. & Tatsumisago, M. *J. Mater. Sci.*, 2004, **39**, 5361–5364.
74. Hayashi, A., Fukuda, T. Morimoto, H., Minami, T. & Tatsumisago, M.

J. Mater. Sci., 2004, **39**, 5125–5127.

75. Idota, Y.., Kubota, T., Matsufuji, A., Maekawa, Y. & Miyasaka, T. *Science*, 1997, **276**, 1395–1397.

76. Kunze, D. In *Fast Ion Transport in Solids*, Ed. W. van Gool, North-Holland Pub., Amsterdam, 1973, pp.405–411.

77. Mallugani, J. P., Wasniewski, A., Doreau, M., Robert, G. & Al Rikabi, A. *Mater. Res. Bull.*, 1978, **13**, 427–433.

78. Maligani, J. P. & Robert, G. *Mater. Res. Bull.*, 1979, **15**, 715–720.

79. Chiodeli, G., Vigano, G. C., Flor, G., Magistris, A. & Villa, M. *Solid State Ionics*, 1983, **8**, 311–318.

80. Robinel, E., Carette, B. & Ribes, M. *J. Non-Crystalline Solids*, 1983, **57**, 49–58.

81. Malugani, J. P. & Mercier, R. *Solid State Ionics*, 1984, **13**, 293–299.

82. Wagner, C. *Proc. C.I.T.C.E.*, 1955, **VII**, 361.

83. Tubandt, C. *Handbuch der Experimetalphysik*, Vol. 12, Part 1, Eds., W. Wien & F. Harms, Akademische Verlagsgesellschaft, Leipniz, 1932, p.383.

84. Pearson, R. G. *J. Chem. Educ.*, 1968, **45**, 581–587.

85. Bradley, J. N. & Green, P. D. *Trans. Faraday Soc.*, 1967, **63**, 424–430.

86. Owens, B. B. & Argue, G. R. *Science*, 1967, **157**, 308–310.

87. Schiraldi, A., Chiodelli, G. & Magistris, A. *J. Appl. Electrochem.*, 1976, **6**, 251–255.

88. Tatsumisago, M., Hirai, K., Minami, T., Takada, K. & Kondo, S. *J. Ceram. Soc. Jpn.*, 1993, **101**, 1315–1317.

89. Hirai, K., Tatsumisago, M., Takahashi, M. & Minami, T. *J. Am. Ceram. Soc.*, 1996, **79**, 349–352.

90. Hayashi, A., Hirai, K., Takahashi, M., Tatsumisago, M. & Minami, T. *Solid State Ionics*, 1996, **86–88**, 539–542.

91. Tatsumisago, M., Hirai, K., Minami, T. & Takahashi, M. *Phys. Chem. Glasses*, 1997, **38**, 63–65.

92. Tatsumisago, M., Hayashi, A. & Minami, T. *Proc. XVII Int. Congr. on Glass*, 1998, **D7**, 118–123.

93. Morimoto, H., Yamashita, H., Tatsumisago, M. & Minami, T. *J. Am. Ceram. Soc.*, 1999, **82**, 1352–1354.

94. Hayashi, A., Tatsumisago, M. & Minami, T. *Phys. Chem. Glasses*, 1999, **40**, 333–338.

95. Tatsumisago, M., Saito, T. & Minami, T. *J. Non-Cryst. Solids*, 2001, **293–295**, 10–18.

96. Hayashi, A., Komiya, R., Tatsumisago, M. & Minami, T. *Solid State Ionics*, 2002, **152–153**, 285–290.

97. Tatsumisago, M., Hama, S., Hayashi, A., Morimoto, H. & Minami, T. *Solid State Ionics*, 2002, **154–155**, 635–640.

98. Hayashi, A., Ishikawa, Y., Hama, S., Minami, T. & Tatsumisago, M. *Electrochem. Solid State Lett.*, 2003, **6**, A47–A49.

99. Mizuno, F., Hayashi, A., Tadanaga, K., Minami, T. & Tatsumisago, M. *Electrochemistry*, 2003, **71**, 1196–1200.

100. Hayashi, A., Fukuda, T., Hama, S., Yamashita, H., Morimoto, H., Minami, T. & Tatsumisago, M. *J. Ceram. Soc. Jpn.*, 2004, **112**, S695–S699.

101. Hayashi, A., Konishi, T., Nakai, M., Morimoto, H., Tadanaga, K., Minami, T. & Tatsumisago, M. *J. Ceram. Soc. Jpn.*, 2004, **112**, S713–S716.

102. Hayashi, A., Konishi, T., Tadanaga, K., Minami, T. & Tatsumisago, M. *J. Power Sources*, 2005, **146**, 496–500.

103. Mizuno, F., Hayashi, A., Tadanaga, K. & Tatsumisago, M. *Adv. Mater.*, 2005, **17**, 918–921.

104. Minami, K., Hayashi, A., Ujiie, S. & Tatsumisago, M. *Solid State Ionics*, 2011, **192**, 122–125.

105. Minami, K., Hayashi, A. & Tatsumisago, M. *J. Non-Cryst. Solids*, 2010, **356**, 2666–2669.

106. Minami, K., Hayashi, A. & Tatsumisago, M. *J. Ceram. Soc. Jpn.*, 2010, **118**, 305–308.

107. Hayashi, A., Minami, K., Ujiie, S. & Tatsumisago, M. *J. Non-Cryst. Solids*, 2010, **356**, 2670–2673.

108. Minami, K. Hayashi A. & Tatsumisago, M. *J. Am. Ceram. Soc.*, 2011, **94**, 1779–1783.

Phys. Chem. Glasses: Eur. J. Glass Sci. Technol. B, April 2012, 53 (2), 52–60

The research history of Professor Tsutomu Minami

Akitoshi Hayashi*

Department of Applied Chemistry, Graduate School of Engineering, Osaka Prefecture University, 1-1 Gakuen-cho, Naka-ku, Sakai, Osaka 599-8531, Japan

Manuscript received 6 November 2011
Revised version received 28 March 2012
Accepted 11 April 2012

The Seventh International Conference on Borate Glasses, Crystals and Melts held in August 2011 in Halifax, Canada was dedicated to Dr Tsutomu Minami, Professor emeritus of Osaka Prefecture University. He started his research career in 1964 at Osaka Prefecture University and retired there in 2009. For over 45 years, he has devoted himself to the development of new glasses in the following fields: (1) vitreous semiconductors, (2) ionically conducting glasses, (3) rapid quenching for new glass preparation, (4) glass preparation by mechanical milling, and (5) coating film preparation by the sol-gel process. In this paper the author describes Professor Minami's research history in the field of glass science and technology. His scientific publications and awards are briefly summarised, and his activities at the University and in several scientific communities are reported.

1. Introduction

The Seventh International Conference on Borate Glasses, Crystals and Melts (Borate 7) took place on August 21–25, 2011 in Halifax, Canada, under the chairmanship of Professor Josef Zwanziger of Dalhousie University. The Borate Conference has always been held in honour of a scientist who has made an outstanding contribution to the study of borate materials. Historical details of the Borate Conference are summarised in an article written by Dr Alex Hannon.[1] The Borate 7 Conference was dedicated to Dr Tsutomu Minami, Professor emeritus of Osaka Prefecture University, who now works as Director general for the Technology Research Institute of Osaka Prefecture, Japan.

Professor Minami was the supervisor of the author when he was a student from bachelor to PhD, and before and after his graduation, Professor Minami has continuously given the author kind encouragements and insightful advices. The author would like to express his sincere thanks to Professor Minami for all of the guidance he has provided the author. It is thus a great pleasure for the author to have a chance to write the research history of Professor Minami. This article is based on the introductory talk at the Borate 7 Conference by the author.

Professor Minami started his research career in 1964 at Osaka Prefecture University and retired there in 2009. For over 45 years he has devoted himself to the development of new glasses. He has developed borate glasses both as superionic conducting glasses, and as electrode active materials for energy storage devices. The ion conduction mechanism in borate

Email hayashi@chem.osakafu-u.ac.jp
Original version presented at VII Int. Conf. on Borate Glasses, Crystals and Melts, Dalhousie University, Halifax, Nova Scotia, Canada, 21–25 August 2011

glass electrolytes was investigated from a structural point of view. In this paper, the author will widely describe Professor Minami's research history in the field of glass science and technology. Professor Minami's research on borate glass systems is summarised in his plenary article in the Borate 7 Proceedings,[2] and hence his investigations of other chalcogenide glasses and sol-gel derived coating materials will also be reported in this article.

2. Research history and research topics

Five main research topics of Professor Minami are shown in Table 1. He started his research career with vitreous chalcogenide semiconductors. At the time of the early 1960s, the word "semiconductor" was very attractive to him and he thus decided to study vitreous semiconductors. He had majored in applied chemistry and it was not so easy to do research on semiconductors for a chemist. He thus studied these materials with the advice of professors in the Department of Electrical Engineers at Osaka Prefecture University. From 1974 to 1975, he worked for Professor John D. Mackenzie as a postdoctoral researcher at the

Table 1. Rough sketch of the research history of Professor T. Minami

1. Vitreous Semiconductors: 1964–1989
    Development of *n*-type chalcogenide glasses
2. Ionically Conducting Glasses: 1977–2009
    Development of superionic conducting glasses
3. Rapid Quenching for New Glass Preparation: 1979–2009
    Stabilization of α-AgI at room temperature
4. Glass Preparation by Mechanical Milling: 1997–2009
    Preparation of high lithium ion conductors
    Application to all-solid-state lithium secondary batteries
5. Coating Film Preparation by the Sol-Gel Process: 1987–2009
    Superhydrophobic and superhydrophilic coatings

University of California, Los Angeles. After returning to Japan, he engaged in the development of ionically conducting glasses. In order to develop lithium ion conducting glasses, he invented a rapid quenching apparatus. By using this apparatus, he succeeded in stabilizing α-AgI at room temperature. His fourth research topic, the mechanical milling technique, has been very successful for the development of lithium ion conductors and all-solid-state lithium ion secondary batteries. His final research topic is the preparation of coating films by the sol-gel process. Superhydrophobic and superhydrophillic coatings have been developed, and several key developments are described here in detail.

2.1. Vitreous semiconductors

The first major achievement of Professor Minami's research life was the development of n-type semiconducting glasses in 1979. Until then, amorphous semiconductors had been only p-type, and the control of conduction type had been believed to be impossible in amorphous semiconductors. It was thus a significant advance to prepare n-type semiconducting glasses by doping Bi into Ge–Se chalcogenide glasses.[3,4]

Figure 1 shows the composition dependence of the Seebeck coefficient at 50°C, S_{50}, for melt quenched $Ge_{20}Bi_xSe_{80-x}$ and $Ge_{20}Bi_xSe_{70-x}Te_{10}$ glasses. The glasses containing 7·5 at% Bi are p-type semiconductors, similar to other melt quenched chalcogenide glasses.

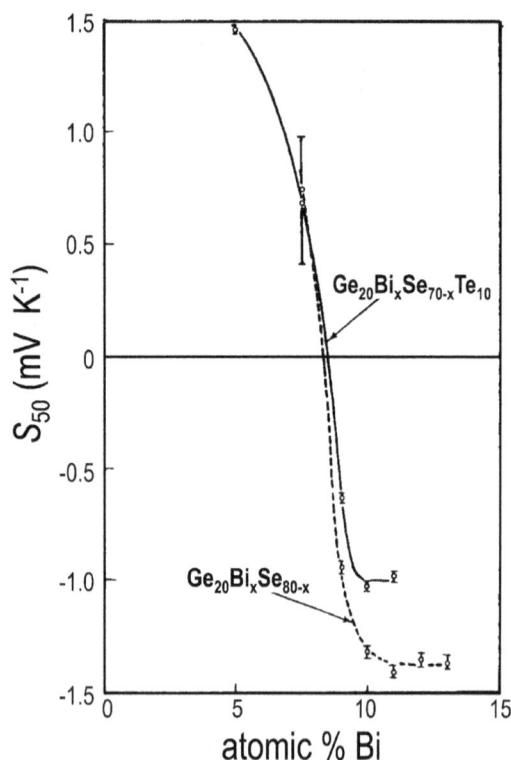

Figure 1. Composition dependence of the Seebeck coefficient at 50°C, S_{50}, for melt quenched $Ge_{20}Bi_xSe_{80-x}$ and $Ge_{20}Bi_xSe_{70-x}Te_{10}$ glasses

On the other hand, S_{50} changes from positive to negative values with an increase in the bismuth content to more than 9 at% Bi, and this is clear evidence that the glasses are n-type semiconductors. Bismuth thus plays an important role in the appearance of n-type conduction.

2.2. Ionically conducting glasses

Glass is advantageous as an ionic conducting material. Professor Minami discovered many glass forming systems with a high silver ion conductivity of the order of 10^{-2} S cm^{-1} at room temperature. The conductivity of these glasses is over ten orders of magnitude higher than that of conventional glasses, such as soda–lime–silica glasses, and is comparable to the conductivity of aqueous solution of sodium chloride or silver nitrate. He thus named these glasses as "superionic conducting glasses", and detailed results for Ag$^+$ ion conducting glasses are reported in his plenary paper.[2]

For application as all-solid-state batteries, the development of lithium ion conducting glasses is highly desirable. One of the important requirements for the preparation of highly conductive glasses is a high concentration of lithium ions in the glass. In order to increase the Li$^+$ ion concentration, a high cooling rate is essential. Hence a rapid quenching apparatus was invented by Professor Minami,[5] and this will be described in detail in the next section. Several papers were published on oxide glasses containing large amounts of Li$^+$ ions, prepared by rapid quenching.[6–10] For example, quasi-binary oxide glasses combining two lithium ortho-oxosalts, such as Li_3BO_3–Li_4SiO_4, showed the highest conductivity of 10^{-6} S cm^{-1} at room temperature.[7,9,10] However, it is difficult to further increase the Li$^+$ ion concentration in oxide glasses; there is a limit to the development of Li$^+$ ion conductors in oxide glasses. Professor Minami noticed that high Li$^+$ ion conductivity can be achieved by introducing highly polarizable anions into the glass, because then elastic collisions should be avoided during the movement of small sized Li$^+$ ions. He thus introduced sulphide components instead of oxide components.[11–18]

Several researchers have reported that sulphide glasses in the systems Li_2S–P_2S_5 and Li_2S–B_2S_3, prepared by melt quenching, show high conductivity from 10^{-5} to 10^{-4} S cm^{-1} at room temperature.[19–22] Further conductivity enhancement was achieved by doping highly polarizable iodide ions; a room temperature conductivity of 10^{-3} S cm^{-1} was reported for Li_2S–SiS_2–LiI glasses.[23,24] The doping of small amounts of lithium ortho-oxosalts also increased the conductivity of sulphide glasses.[11,15,25] Professor Minami prepared Li_2S–SiS_2–Li_4SiO_4 oxysulphide glasses with conductivity of 10^{-3} S cm^{-1} by a rapid quenching technique and analyzed their local structure by solid

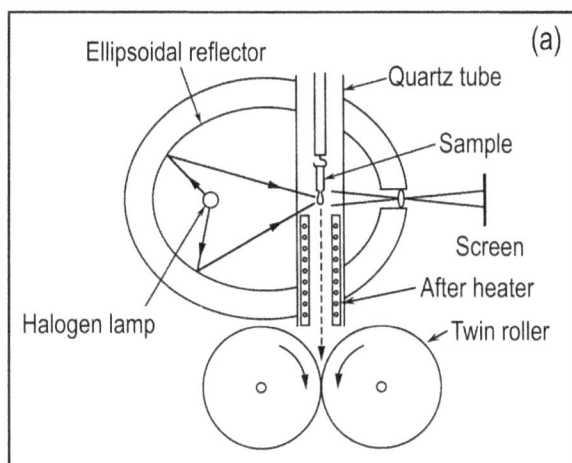

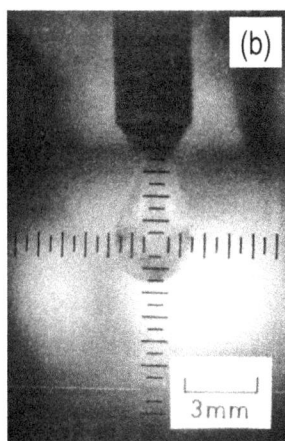

Figure 2. Schematic of rapid quenching system (a) and a view of a melting sample projected on a screen (b)

Figure 3. DTA-TG curves of a rapidly quenched BiCaSrCu₂Oₓ sample (a) and temperature dependence of resistivities for samples heat treated at 700 and 850°C (b)

state NMR and XPS techniques.[12,13,15] It was revealed that $Si_2OS_6^{6-}$ dimer ions, consisting of one bridging oxygen and six nonbridging sulphur atoms, were mainly present in the oxysulphide glass with 5 mol% Li_4SiO_4. He thus concluded that the introduction of bridging oxygen atoms into the sulphide matrix is a key to increase the conductivities of sulphide glasses.

2.3. Rapid quenching for new glass preparation

As mentioned above, Professor Minami invented an apparatus for rapid quenching of melts to prepare new glasses,[5] shown in Figure 2(a). The ellipsoidal reflector has two focal points. A halogen lamp is put on one of the focal points, and then the light beams gather on the other focal point, where the temperature becomes high. A sample put this second focal point can be melted by the light beams, and the molten part drops. The dropped molten part is then quenched by the rotating twin roller, which is made of steel. The quenching rate can be controlled by the rotation speed and the pressure of the two rollers, and the highest quenching rate is reported to be about 10^6 $K s^{-1}$. The molten state can be observed on the screen

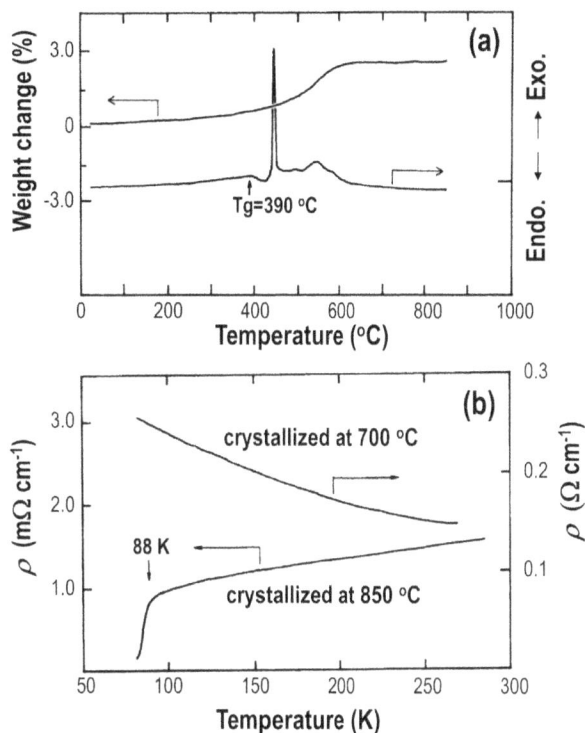

through a lens, as shown in Figure 2 (b).

One interesting example prepared by the rapid quenching apparatus is Bi–Ca–Sr–Cu–O glass, with the same composition as high temperature superconductor. Professor Minami successfully synthesized these glasses by rapid quenching.[26] Figure 3(a) shows DTA-TG curves of a rapidly quenched $BiCaSrCu_2O_x$ sample. The glass transition at 390°C and subsequent crystallisation are clearly observed on the DTA curve for the sample. The prepared sample also showed halo patterns in x-ray diffraction measurements, and thus it was concluded to be glass. Figure 3(b) shows the temperature dependence of resistivities for samples heat treated for 2 h at 700 and 850°C. The sample heat treated at 850°C shows onset of superconductivity at a transition temperature, T_c, of about 88 K, while the sample heat treated at 700°C is semiconducting. This is clear evidence that the glass crystallised at 850°C has become a superconductor. Another significant example is the stabilization of α-AgI by rapid quenching, which is described in Professor Minami's plenary paper.[2]

2.4. Glass preparation by mechanical milling

Professor Minami introduced a mechanical milling technique for glass preparation as well as a melt quenching technique. A high energy planetary ball mill apparatus was used for glass preparation. The base disc and zirconia vessels containing zirconia balls rotate independently in opposite directions,

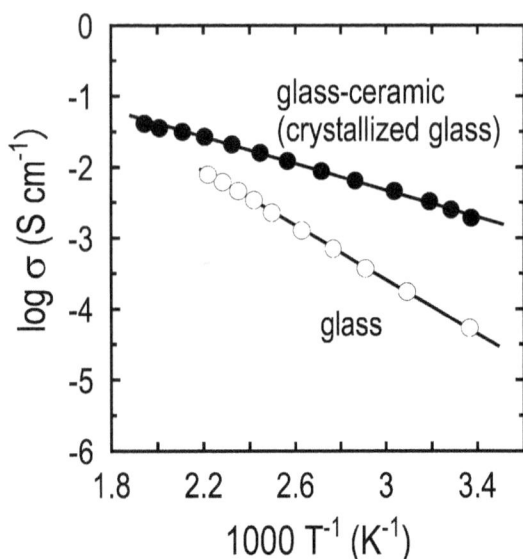

Figure 4. Temperature dependence of the conductivity of $70Li_2S.30P_2S_5$ (mol%) glass and glass-ceramic (crystallised glass)

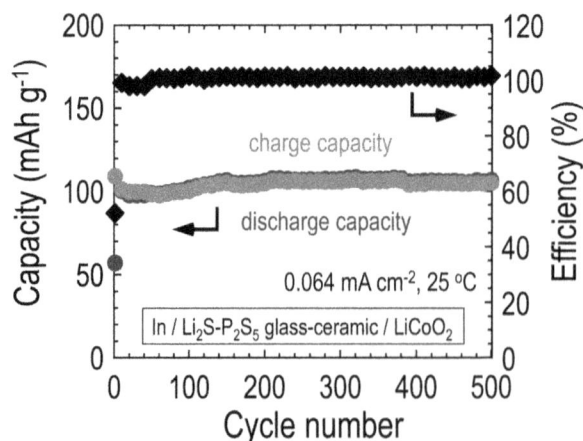

Figure 5. Charge–discharge cycle performance of an all-solid-state battery using an indium negative electrode, $80Li_2S.20P_2S_5$ glass-ceramic electrolyte, and a $LiCoO_2$ positive electrode

and sample powder placed inside the vessel is thus milled to become a glassy powder. This technique has several merits: the process can be done at room temperature, and sulphide glasses which have high vapour pressure at high temperatures are prepared at ambient pressure. The direct preparation of glass powders suitable for application in all-solid-state batteries is another advantage.

Professor Minami prepared a variety of sulphide, oxysulphide and oxide glasses by mechanical milling, and showed that the local structure of milled glasses is similar to that of melt-quenched glasses.[14,16,18,27,28] Glasses with high alkali contents can be prepared by mechanical milling. For example, lithium thio-orthophosphate glass, $75Li_2S.25P_2S_5$, which cannot be prepared by melt quenching, was prepared by milling.[16] In addition, he discovered that suitable crystallisation of Li_2S–P_2S_5 sulphide glasses drastically increases the ionic conductivity.[17] Figure 4 shows the temperature dependence of the conductivities of $70Li_2S.30P_2S_5$ (mol%) glass and glass-ceramic (crystallised glass). Heat treatment of the glass at temperatures above the first crystallisation temperature increases the conductivity by two orders of magnitude, and the obtained glass-ceramic shows a high lithium ion conductivity of over 10^{-3} Scm^{-1} at room temperature.[17,29–31] A superionic high temperature phase, $Li_7P_3S_{11}$,[32,33] was stabilized by suitable crystallisation of the glass. It is difficult to synthesize such a crystalline phase by conventional solid-state reaction. Glass electrolytes are thus very important as a precursor for precipitating superionic material in a high temperature or metastable phase. Similar enhancement in conductivity by crystallisation was observed for Li_2S–P_2S_5 glasses with a Li_2S content of 70 mol% Li_2S or greater.[30]

The prepared glass-ceramics can be used as a solid electrolyte in all-solid-state lithium ion secondary batteries. All-solid-state batteries are recognized as a promising innovative battery with high safety and high energy density, and have a big potential for replacing present lithium ion batteries which use an organic liquid electrolyte. Figure 5 shows the charge-discharge cycle performance of an all-solid-state battery using an indium negative electrode, $80Li_2S.20P_2S_5$ glass-ceramic electrolyte, and a $LiCoO_2$ positive electrode.[18,30,34] The battery test was carried out at 25°C under a constant current density of 0·064 $mA\,cm^{-2}$. The battery shows an excellent charge–discharge performance over 500 cycles, with a large capacity of $100\,mA\,h\,g^{-1}$ of $LiCoO_2$ active material, and 100% charge–discharge efficiency. It is noteworthy that the battery has continuously operated during two and a half years without any capacity fading. Professor Minami discovered several glass-based solid electrolytes suitable for all-solid-state battery technology[15,18] and his contribution has promoted research in the field of all-solid-state batteries.

2.5. Coating film preparation by the sol-gel process

Professor Minami developed the preparation of coating films by the sol-gel process. It is well known that various shapes of materials like bulk, fibre, coating film, and powder can be prepared by the sol-gel process. Among these, he focused his attention on the preparation of coating films on various substrates at relatively low temperatures because of their high potential for industrial applications. In this paper, superhydrophobic coating films are discussed as the most valuable example of coating films via the sol-gel process.[35–40]

In nature, superhydrophobic phenomena can be seen on the leaves of several plants. One popular

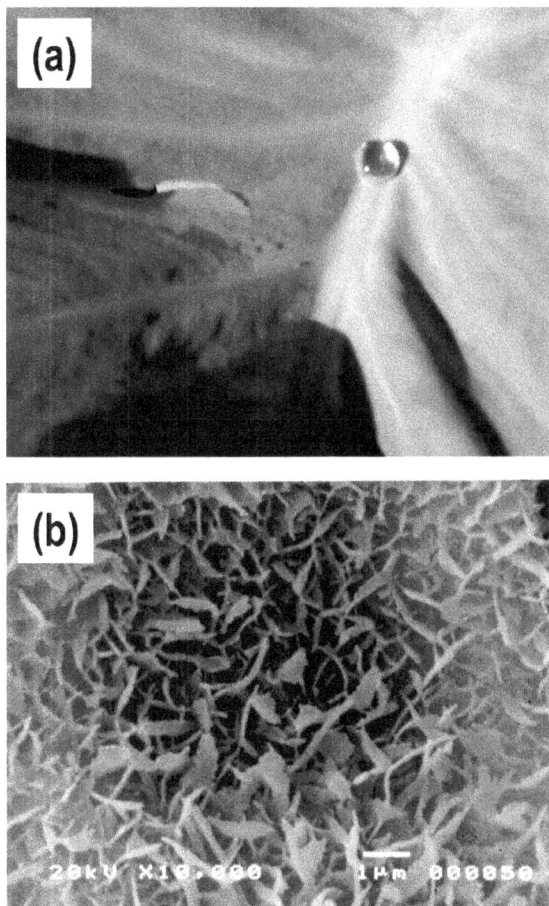

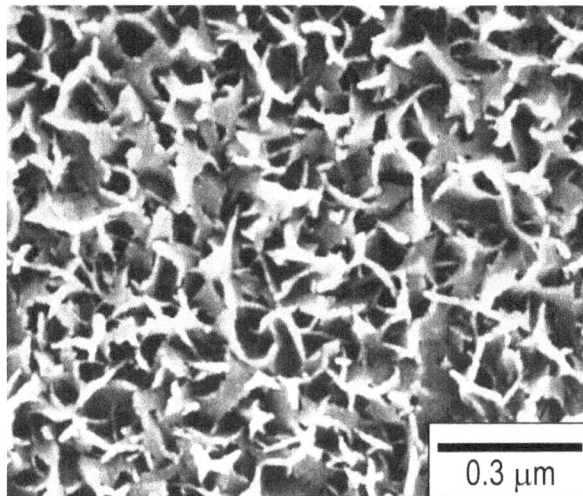

Figure 7. FE-SEM image of the surface of an alumina thin film prepared via the sol-gel process

Figure 6. Photograph of a water droplet on a taro leaf (a) and SEM image of the surface of a leaf (b)

agents, a superhydrophobic surface was obtained. A photograph of a water droplet of diameter about 1 mm on a FAS-coated flowerlike alumina film is shown in Figure 8. The surface is highly water-repellent with a contact angle for water of about 165°. Thus transparent flowerlike alumina on a glass plate is superhydrophobic.

Flowerlike alumina films on soda–lime–silica glass, non-alkali glass and quartz glass substrates also show excellent antireflective (AR) properties over a wide range of visible light wavelengths.[38–40,43] The transmittance spectra of the coated substrates

example is a water droplet on the leaf of taro, a kind of Japanese potato, as shown in Figure 6(a). Figure 6(b) shows a SEM image of the surface of a leaf. The surface is covered with microcrystallites of epicuticular wax. The surface roughness is thus formed by the crystallites, and the hydrophobic property of the wax is enhanced to give superhydrophobic properties to the leaf.[41]

Professor Minami succeeded in preparing thin films with a similar surface structure to the taro leaf.[35,36] Figure 7 shows a FE-SEM image of the surface of an alumina thin film. Films with a roughness of several tens of nanometres are formed by immersing porous alumina gel films prepared by the sol-gel process in hot water. Before the treatment, the alumina film had a very smooth surface. After the treatment, unevenness with a 20 to 50 nm petal structure is observed, referred to as a "flowerlike structure". This is due to the formation of pseudoboehmite nanocrystals by the dissolution–reprecipitation process due to the reaction between the sol-gel derived porous alumina film and water.[42] This flowerlike microstructure is very similar to that of the surface of the taro leaf as shown in Figure 6(b). After the surface of the flowerlike alumina was covered with fluoroalkylsilanes (FAS), which are known to be excellent water-repellent

Figure 8. Photograph of a water droplet of diameter about 1 mm on a FAS-coated flowerlike alumina film

Figure 9. A report about Professor Minami in the Japanese version of Weekly Playboy *in 1995*

Figure 10. Feature on Professor Minami in the book entitled 100 distinguished researchers referred to as Japanese brains *published in 1993 from Mita Publishing Co. Ltd*

show a high transparency of more than 99% in the visible light region, indicating that almost no light scattering occurred from the flowerlike alumina. For usual mono- or multilayered AR films, the AR properties show a large incident angle dependence. However, the AR effects based on the roughness of the flowerlike alumina have a very small incident angle dependence.

3. Records of scientific publications and awards

In this section, Professor Minami's scientific publications and awards are summarised. The number of his original papers is 488. The number of reviews is 84 and the number of articles in books is 45. The citation data for his publications from Web of Science (one of the most popular databases of scientific articles) are summarised in Table 2, where papers cited more than 100 times from 1974 to 2011 are listed. There are 12 papers; for example, the paper "fast ion conducting glasses" published in 1985 has been cited 244 times and the paper about "flowerlike alumina film" published in 1997 has been cited 179 times. It is noteworthy that the number of his papers registered on Web of Science is 386, and these papers have been cited 8164 times in total. The remainder of about 100

papers not registered was mostly published earlier than the start of the database.

Professor Minami has been identified as a unique researcher in both a popular magazine and a book. He was featured in the Japanese version of *Weekly Playboy* in 1995 as shown in Figure 9. This does not mean he is a real "playboy", but his creative research work was highlighted to give a scientific message to the young readers of this popular magazine. He was also selected as one of the top 100 researchers in a book entitled *100 distinguished researchers referred to as Japanese brains* published in 1993 from Mita Publishing Co. Ltd. as shown in Figure 10. The Japanese sentence in this figure means that his ion conducting glass is an innovative pioneering material in the research field of electronics and ionics.

Professor Minami has received many awards from a number of scientific communities, such as The Ceramic Society of Japan and The Chemical Society of Japan, including those which are summarised in Table 3. He received the W. H. Zachariasen Award on July 3, 1984, on the basis of his work on a series of electronically and ionically conducting glasses, published especially in the *Journal of Non-Crystalline Solids*. The award was established by the *Journal of Non-Crystalline Solids* to honour a young scientist for publications in a three year period preceding the year of the award. He is the first recipient of the award. The presentation ceremony was held at the banquet of the International Symposium on Glass Science and Technology, in honour of the 80th birthday of Professor Norbert Kreidl in Vienna. Professor Minami is very proud of this award, as important evidence that he was internationally recognized as a young scientist with potential. Figure 11 is a photograph showing Professor Minami receiving the award certificate from Professor John D. Mackenzie, the

Table 2. Citation data of Professor T Minami's publications from Web of Science

Paper title	Journal title	Times cited
Fast Ion Conducting Glasses **T Minami**	*J. Non-Cryst. Solids*, 1985, **73** (1–3), 273	244
Super-Water-Repellent Al_2O_3 Coating Films with High Transparency K. Tadanaga, N. Katata and **T. Minami**	*J. Am. Ceram. Soc.*, 1997, **80** (4), 1040	179
Superhydrophobic-Superhydrophilic Micropatterning on Flowerlike Alumina Coating Film by the Sol-Gel Method K. Tadanaga, Morinaga, A. Matsuda and **T. Minami**	*Chem. Mater.*, 2000, **12** (3), 590	176
Formation Region and Characterization of Superionic Conducting Glasses in the Systems $AgI-Ag_2O-M_xO_y$ **T. Minami**, K. Imazawa and M. Tanaka	*J. Non-Cryst. Solids*, 1980, **42** (1–3), 469	165
Formation Process of Super-Water-Repellent Al_2O_3 Coating Films with High Transparency by the Sol-Gel Method K. Tadanaga, N. Katata and **T. Minami**	*J. Am. Ceram. Soc.*, 1997, **80** (12), 3213	154
Electrical and Optical Properties of n-Type Semiconducting Chalcogenide Glasses in the System Ge–Bi–Se N. Tohge, **T. Minami**, Y. Yamamoto and M. Tanaka	*J. Appl. Phys.*, 1980, **51** (2), 1048	149
Superionic Conducting Glasses: Glass Formation and Ionic Conductivity in the $AgI-Ag_2O-P_2O_5$ System **T. Minami**, Y. Takuma and M. Tanaka	*J. Electrochem. Soc.*, 1977, **124** (11), 1659	141
Preparation and Properties of Superionic Conducting Glasses Based on Silver Halides **T. Minami**	*J. Non-Cryst. Solids*, 1983, **56** (1–3), 15	131
Preparation of $PbZrO_3-PbTiO_3$ Ferroelectric Thin Films by the Sol-Gel Process N. Tohge, S. Takahashi and **T. Minami**,	*J. Am. Ceram. Soc.*, 1991, **74** (1), 67	125
Stabilization of Superionic α-AgI at Room Temperature in a Glass Matrix M. Tatsumisago, Y. Shinkuma and **T. Minami**	*Nature*, 1991, **354** (6350), 217	114
Recent Progress in Superionic Conducting Glasses **T. Minami**	*J. Non-Cryst. Solids*, 1987, **95/96** (1), 107	109
Zirconia Coating on Stainless Steel Sheets from Organozirconium Compounds K. Izumi, M. Murakami, N. Tohge, **T. Minami** *et al*	*J. Am. Ceram. Soc.*, 1989, **72** (8), 1465	104

founder of the journal and the Editor in Chief at that time. The late Professor N. Kreidl can be seen on the right of the photo. Figure 12 is a photograph of Professor Minami proudly displaying the certificate. He is also a Fellow of The American Ceramic Society and The Chemical Society of Japan. On the basis of his several epoch-making findings in the sol-gel field, he has just received the Life Achievement Award at the 16th International Sol-Gel Conference held on August 28–September 2, 2011 in China.

These awards were given to Professor Minami on the basis of his major contributions mainly to two research fields, solid state ionics and sol-gel science. After he left the laboratory at Osaka Prefecture University, we have continued research inspired by him, and our core research projects are still in solid-state-ionics and sol-gel science.

4. Activities at Osaka Prefecture University and in the wider scientific community

Professor Minami's activities at Osaka Prefecture University are summarised here. He was Dean of the College of Engineering from 1994 to 1998, and President of the University from 2001 to 2009, and he

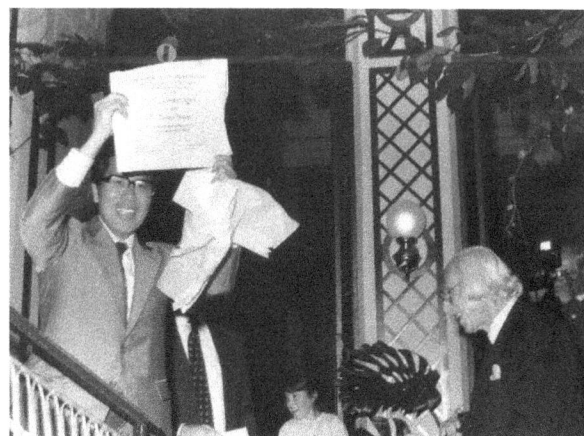

Figure 11. Photograph of Professor Minami receiving the certificate for the W. H. Zachariasen Award for 1983 from John D. Mackenzie, the Editor in Chief of the Journal of Non-Crystalline Solids, *at the presenting ceremony in Vienna on July 3, 1984. On the right is the late Professor Norbert Kreidl*

Figure 12. Photograph of Professor Minami proudly displaying the certificate for the W. H. Zachariasen Award

Table 3. History of winning awards for Professor T. Minami

1. Young Ceramist Award from The Ceramic Society of Japan (1975)
2. W. H. Zachariasen Award (1984)
3. Academic Achievements Award in Ceramic Science and Technology from The Ceramic Society of Japan (1987)
4. Osaka Science Prize (1988)
5. Fellow of The American Ceramic Society (1991)
6. The Chemical Society of Japan Award for Creative Work (1994)
7. Science and Technology Agency Award (2000)
8. Fellow of The Chemical Society of Japan (2009)
9. Life Achievement Award from The International Sol-Gel Society (2011) and so on

has devoted himself to education and governance at the University as well as scientific research for long periods of time.

Professor Minami organized a major research project in the field of solid state ionics, sponsored by the Ministry of Education, Culture, Sports, Science and Technology of Japan, for which he acted as project leader for five years from 1999 to 2003. He summarised the research outcomes in a book entitled *Solid State Ionics for Batteries*, published by Springer-Verlag, Tokyo, in 2005, for which he was Editor in Chief. The book cover is shown in Figure 13. He was also dedicated to developing the activities of The International Society for Solid State Ionics, acting as President and Vice President; he was elected as Vice President in 2003 and as President in 2005, and he

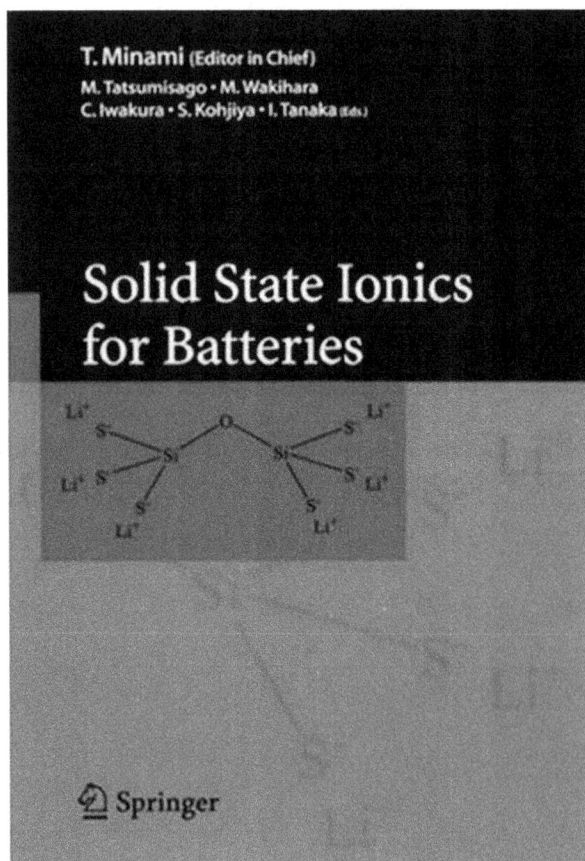

Figure 13. The cover of the book Solid State Ionics for Batteries published by Springer-Verlag, Tokyo, in 2005

performed these roles from 2003 to 2007. In the sol-gel field, he was also involved in the foundation of The Japanese Sol-Gel Society in 2003.[43]

5. Conclusions

Professor Tsutomu Minami has developed glass science and technology, especially in the field of Ag^+, Cu^+ and Li^+ ion conducting glasses for solid electrolytes in all-solid-state batteries and functional coating materials made via the sol-gel process. Borate glasses are one of the main glass-forming systems which he focused on as Ag^+ ion conductors and active materials for lithium ion secondary batteries, and his findings are summarised in his plenary article in the Proceedings of the Borate 7 Conference.[2] The author would like to finish this paper with Professor Minami's favourite phrase. He has often said to the author that students (young people) are the ultimate treasure. The author has learned that maintaining a good relationship between a supervisor and students is the key to enhance research progress in a university laboratory. The author appreciates and respects Professor Minami's superlative leadership.

References

1. Hannon, A. C. *Phys. Chem. Glasses: Eur. J. Glass Sci. Technol. B*, 2010, **51**, 40.
2. Minami, T., *Phys. Chem. Glasses: Eur. J. Glass Sci. Technol. B*, 2012, **53** (2), 17.
3. Tohge, N., Yamamoto, Y., Minami, T. & Tanaka, M. *Appl. Phys. Lett.*, 1979, **34**, 640.
4. Tohge, N., Minami, T. & Tanaka, M. *J. Non-Cryst. Solids*, 1980, **38–39**, 283.
5. Tatsumisago, M., Minami, T. & Tanaka, M. *J. Am. Ceram. Soc.*, 1981, **64**, C97.
6. Tatsumisago, M., Hamada, A., Minami, T. & Tanaka, M. *J. Am. Ceram. Soc.*, 1982, **65**, 575.
7. Tatsumisago, M., Minami, T. & Tanaka, M. *Glastech. Ber.*, 1983, **56K**, 945.
8. Tatsumisago, M., Narita, H., Minami, T. & Tanaka, M. *J. Am. Ceram. Soc.*, 1983, **66**, C210.
9. Tatsumisago, M. & Minami, T. *Mater. Chem. Phys.*, 1987, **18**, 1.
10. Tatsumisago, M., Yoneda, K., Machida, N. & Minami, T. *J. Non-Cryst. Solids*, 1987, **95/96**, 857.
11. Tatsumisago, M., Hirai, K., Minami, T., Takada, K., & Kondo, S. *J. Ceram. Soc. Jpn.*, 1993, **101**, 1315.
12. Hirai, K., Tatsumisago, M., Takahashi, M. & Minami, T. *J. Am. Ceram. Soc.*, 1996, **79**, 349.
13. Hayashi, A., Tatsumisago, M., Minami, T. & Miura, Y. *Phys. Chem. Glasses*, 1998, **39**, 145.
14. Morimoto, H., Yamashita, H., Tatsumisago, M. & Minami, T. *J. Am. Ceram. Soc.*, 1999, **82**, 1352.
15. Minami, T., Hayashi, A. & Tatsumisago, M. *Solid State Ionics*, 2000, **136–137**, 1015.
16. Hayashi, A., Hama, S., Morimoto, H., Tatsumisago, M & Minami, T. *J. Am. Ceram. Soc.*, 2001, **84**, 477.
17. Hayashi, A., Hama, S., Minami, T. & Tatsumisago, M. *Electrochem. Commun.*, 2003, **5**, 111.
18. Minami, T., Hayashi, A. & Tatsumisago, M. *Solid State Ionics*, 2006, **177**, 2715.
19. Souquet, J. L., Robinel, E., Barrau, B. & Ribes, M. *Solid State Ionics*, 1981, **3/4**, 317.
20. Menetrier, M., Estournes, C., Levasseur, A. & Rao, K. J. *Solid State Ionics*, 1992, **53–56**, 1208.
21. Mercier, R., Malugani, J. P., Fahys, B. & Robert, G. *Solid State Ionics*, 1981, **5**, 663.

22. Pradel, A. & Ribes, M. *Solid State Ionics*, 1986, **18-19**, 351.
23. Kennedy, J. H., & Yang, Y. *J. Electrochem. Soc.*, 1986, **133**, 2437.
24. Kennedy, J. H. *Mater. Chem. Phys.*, 1989, **23**, 29.
25. Aotani, N., Iwamoto, K., Takada, K. & Kondo, S. *Solid State Ionics*, 1994, **68**, 35.
26. Minami, T., Akamatsu, Y., Tatsumisago, M., Tohge N. & Kowada, Y. *Jpn. J. Appl. Phys.*,1988, **27**, L777.
27. Morimoto, H., Nakai, M., Tatsumisago, M. & Minami, T. *J. Electrochem. Soc.*, 1999, **146**, 3970.
28. Hayashi, A., Nakai, M., Morimoto, H., Minami, T. & Tatsumisago, M. *J. Mater. Sci.*, 2004, **39**, 5361.
29. Mizuno, F., Hayashi, A., Tadanaga, K. & Tatsumisago, M. *Adv. Mater.*, 2005, **17**, 918.
30. Mizuno, F., Hayashi, A., Tadanaga, K. & Tatsumisago, M. *Solid State Ionics*, 2006, **177**, 2721.
31. Hayashi, A., Minami, K., Ujiie, S. & Tatsumisago, M. *J. Non-Cryst. Solids*, 2010, **356**, 2670.
32. Yamane, H., Shibata, M., Shimane, Y., Junke, T., Seino, Y., Adams, S., Minami, K., Hayashi, A. & Tatsumisago, M. *Solid State Ionics*, 2007, **178**, 1163.
33. Minami, K., Hayashi, A. & Tatsumisago, M. *J. Ceram. Soc. Jpn.*, 2010, **118**, 305.
34. Tatsumisago, M. & Hayashi, A. *Funct. Mater. Lett.*, 2008, **1**, 31.
35. Tadanaga, K., Katata, N. & Minami, T. *J. Am. Ceram. Soc.*, 1997, **80**, 1040.
36. Tadanaga, K., Katata, N. & Minami, T. *J. Am. Ceram. Soc.*, 1997, **80**, 3213.
37. Matsuda, A., Matoda, T., Kogure, T., Tadanaga, K., Minami, T. & Tatsumisago, M. *Chem. Mater.*, 2005, **17**, 749.
38. Yamaguchi, Y., Tadanaga, K., Matsuda, A., Minami, T. & Tatsumisago, M. *J. Sol-Gel Sci. Tech.*, 2005, **33**, 175.
39. Yamaguchi, Y., Tadanaga, K., Matsuda, A., Minami, T. & Tatsumisago, M. *Surface Coating Tech.*, 2006, **201**, 3653.
40. Tadanaga, K., Yamaguchi, Y., Uraoka, Y., Matsuda, A., Minami, T. & Tatsumisago, M. *Thin Solid Films*, 2008, **516**, 4526.
41. Neinhuis, C. & Barthlott, W. *Ann. Botany*, 1997, **79**, 667.
42. Tadanaga, K., Kitamuro, K., Matsuda, A. & Minami, T. *J. Sol-Gel Sci. Tech.*, 2003, **26**, 705.
43. Minami, T. *J. Sol-Gel. Sci. Tech.* in press (DOI: 10.1007/s10971-011-2572-y).

Phys. Chem. Glasses: Eur. J. Glass Sci. Technol. B, February 2012, **53** *(1), 7–10*

Computational study of four-fold coordinate boron in borates: assignment of edge-shared structures

J. W. Zwanziger

Department of Chemistry and Institute for Research in Materials, Dalhousie University, Halifax, NS B3H 4J3 Canada

Manuscript received 30 October 2011
Revised version received 12 December 2011
Accepted 19 December 2011

Several issues raised by the recent discovery of compounds with four-fold coordinate boron in edge-shared pairs are addressed by means of first principles computations. First, site-specific assignments are made by computing the quadrupole coupling of each boron and comparing literature values. Then, the bonding contacts in this unusual arrangement are studied through the atoms-in-molecules approach. Finally, the unusual three-fold coordinate oxygen in another high pressure borate phase is studied in the same way. All computations were performed using the projector-augmented wave formalism.

1. Introduction

It is well-known that boron in oxide glass may be found in both three-fold and four-fold coordination, and the relative amount of each is important in describing the short and intermediate range structure of borate glass and its relationship to bulk properties. Typically boron-11 nuclear magnetic resonance (NMR) spectroscopy is used to characterize these sites, because the quadrupole coupling, which arises from the electric field gradient, is very different in the two coordination environments. The coupling in three-fold sites is typically about 2·7 MHz, and only around 0·5 MHz in four-fold coordinate sites due to their nearly tetrahedral symmetry. In modern high field NMR spectrometers these two kinds of sites are easily distinguished.

Recently there have been several reports of borate crystals containing fourfold coordinate sites in a rather unusual edge-shared configuration.[1,2] The NMR responses of these sites have been reported but were not fully assigned to the crystal structure.[2] Moreover, from the reported values alone it was not clear whether the boron–oxygen bonding in these sites shows any unusual features, and in particular whether due to the edge shared constraints, there is any boron–boron bonding.

In order to address the above questions we have carried out a first principles computational study of several borate crystals, including the new compounds that show edge shared units. We used the projector augmented wave formalism in a density functional theory approach to compute the electron density and

Email jzwanzig@dal.ca
Original version presented at VII Int. Conf. on Borate Glasses, Crystals and Melts, Dalhousie University, Halifax, Nova Scotia, Canada, 21–25 August 2011

electric field gradients. From the latter we were able to assign the NMR responses of the different sites, and by an atoms-in-molecules analysis of the density could investigate the bonding in these units.

2. Computational methods

The electric field gradient is computed as $-\nabla\nabla V(\mathbf{r})$, where V is the potential generated by the distribution of charges in the system (both electronic and ionic) and the derivatives are evaluated at the nuclear sites. The potential itself is determined from the charge density $n(\mathbf{r})$, which is determined via a density functional theory calculation. In particular the potential at nuclear position $\mathbf{R}$ is determined from Coulomb's law as

$$V(\mathbf{R}) = \int d^3\mathbf{r}\,\frac{n(\mathbf{r})}{|\mathbf{r}-\mathbf{R}|} \qquad (1)$$

After taking derivatives the resulting field gradient elements are denoted as V_{ij}, where

$$V_{ij} = -\left.\frac{\partial^2 V}{\partial x_i \partial x_j}\right|_* \qquad (2)$$

Principal axes x,y,z are chosen such that the V_{ij} matrix is diagonal. Note that these directions need not coincide with other coordinates of the system, such as those used to describe the unit cell or chemical bonds. Also, due to Laplace's equation the components satisfy $\sum V_{ii}=0$. It is conventional to choose the principal z axis to lie in the direction of the largest in magnitude principal value of V, and denote this term as $V_{zz}=eq$. The coupling constant C_q between the gradient eq and the nuclear quadrupole moment eQ is then denoted in frequency units as $C_q=e^2qQ/h$, where division is by Planck's constant h. The other directions are chosen such that

$$|V_{zz}| \geq |V_{yy}| \geq |V_{xx}| \qquad (3)$$

The asymmetry of the interaction is defined through

$$\eta = \frac{V_{xx} - V_{yy}}{V_{zz}} \qquad (4)$$

In order to obtain an accurate result for the gradient at the nuclear site it is necessary to obtain the potential very accurately in all regions of space. On the other hand, an all-electron code becomes highly time consuming for large simulation cells. A reasonable compromise is the projector augmented wave method (PAW).[3–5] The PAW method separates the electronic problem into spheres around atoms and interstitial regions, and also makes use of pseudopotentials.

Using the PAW method to compute the EFG is quite straightforward and amounts to a small post-processing calculation following a typical ground-state PAW determination of electronic structure.[6,7] The PAW method used here specifically is as outlined by Torrent *et al*, which closely parallels the treatment by Kresse & Joubert.[4,5] The charge distribution, which is the result of the PAW ground state calculation, is written as

$$n(\mathbf{r}) = \sum_i^N n_{Z,i}(\mathbf{r})\delta(\mathbf{r}-\mathbf{R}) + \tilde{n}(\mathbf{r}) + \hat{n}(\mathbf{r}) \qquad (5)$$
$$+ \left[n^1(\mathbf{r}) - \tilde{n}^1(\mathbf{r}) - \hat{n}^1(\mathbf{r}) \right]$$

The first term is the sum over the ionic charges, modelled as points, at the nuclear positions $\mathbf{R}_i$, summed over the N atoms in the unit cell. Other electronic terms are the pseudovalence density $\tilde{n}$, the all-electron density n^1 within spheres, and the pseudodensity within spheres $\tilde{n}^1$. The term $\hat{n}(\mathbf{r})$ is constructed so as to cancel all multipole moments of the on-site $(n^1-\tilde{n}^1)$ charge density originating within a sphere in regions outside that sphere. This construction ensures that the on-site charge densities

of different spheres do not interact. In a construction where the PAW atomic data sets are norm-conserving this term plays no role,[6] but does account for the case of soft pseudopotential construction. In the latter case the cut-off energy necessary for convergence may be significantly lower than in the norm-conserving case, so it is advantageous to use and account for the $\hat{n}$ term. To our knowledge it has not been included before in PAW treatments of the EFG.[6,7]

Once the charge density is constructed from the ground state density functional theory computation, the electric field gradient is computed as outlined previously.[6,7] The only difference is that the gradient due to the valence charge density now includes the $\hat{n}$ term, while the on-site contributions likewise include the $-\hat{n}$ portion. In the cases we have studied the cancellation between the two $\hat{n}$ contributions is almost always quite complete, the exception being oxygen in unusual bonding environments such as rutile TiO_2.

In order to estimate bonding contacts, the atoms-in-molecules (AIM) method was employed.[8] AIM is a technique used to analyze the ground state charge density distribution in terms of its interatomic connectivities, as judged by various types of saddle points between atoms. Specifically, two atoms are considered to show a bonding interaction when the Hessian matrix at a point between them contains two negative and one positive eigenvalue, indicating a minimum in two directions but non-zero density.

The above scheme for the electric field gradient, as well as AIM, were used in the open source density functional theory code ABINIT.[9] Extensive use was made of the PAW framework already existing in the code.[5] For the calculations shown, PAW atomic data sets were constructed using the ATOMPAW code[10] with an output filter to give files compatible with ABINIT. The PBE exchange and correlation functions were used,[11] and a planewave cutoff energy of 20 Ha. Experimental crystal structures were used without

Table 1. *Calculated and experimental values for the quadrupole coupling C_q, in MHz, and the asymmetry η, for the listed compounds. In each case the first reference is the crystal structure used in the calculations, and any additional references give experimental results for the listed parameters. The second column gives the boron site as three-fold or four-fold coordinate (B3, respectively, B4)*

Compound	Site	Calc C_q	η	Expt C_q	η
B_2O_3 [13, 14]	B3	3·085	0·183	2·690(5)	<0·05
		3·106	0·143		
$Li_2B_4O_7$ [15–17]	B3	2·964	0·200	2·59(2)	0·17(2)
	B4	0·613	0·533	<0·7, 0·276	
α-$Na_2B_4O_7$ [16,18]	B3	2·883	0·836	2·48(4)	0·65(2)
		2·947	0·123	2·59(4)	0·00(2)
		3·022	0·300	2·63(5)	0·20(3)
		3·061	0·281	2·63(4)	0·20(2)
		3·072	0·253	2·66(4)	0·20(2)
	B4	0·419	0·243	<0·3	
		0·483	0·320		
		0·601	0·754		
PbB_4O_7 [17,19]	B4	0·943	0·108	0·805(5)	0·09(3)
		1·100	0·784	0·960(5)	0·05(3)

further optimization or relaxation. Field gradients were converted to quadrupole coupling parameters using nuclear quadrupole moment data compiled by Pyykkö.[12]

3. Results

The primary results of this work are the computed quadrupole couplings and asymmetries at boron in a number of borate crystals relevant to glass. These results are summarized in Tables 1 and 2. Table 1 lists boron trioxide, some alkali borates, and lead borate. Table 2 lists values for the recently discovered compounds with edge-sharing four-fold coordinate boron.

4. Discussion

The accuracy of the PAW method for computing the electric field gradient can be assessed by comparison with both all-electron implementations of density-functional theory, and comparison with experiment. The former method is more rigorous as a test of the PAW approximations, in the sense that other variables (such as the use of single-particle density functional theory, choice of density functional model, etc.) can be held constant, while comparison with experiment is of course of more practical interest. The PAW method can reproduce all-electron density functional theory calculations to an accuracy of 1% or better, but this level of agreement with experimental data is rarely found, due presumably to the inability of the density functional models to mimic the very short-range electronic effects that dominate the electric field gradient, as well as the difficulty of generating accurately transferable PAW datasets. In fact it is common for the PAW method to overestimate the strength of the experimentally determined field gradient by about 10%, more or less systematically.[6] Nevertheless experience has shown that the method provides reliable orderings of field gradients when comparing different sites.

The results in Table 1 for the various B3 sites show both the overestimate noted above and also the precision in the method for obtaining relative orderings and hence site assignments. In both boron trioxide and the two alkali borates listed, the C_q values for the B3 sites are systematically overestimated. Note in the case of boron trioxide that only one resonance was resolved experimentally, while the crystal structure has two distinct but similar B3 sites.[13,14] The case of α-$Na_2B_4O_7$ is instructive in this regard. The crystal structure shows five different B3 sites, while Chen et al resolved four resonances in a 1:1:2:1 intensity distribution.[16,18] The change in C_q from site to site matches reasonably with the experimentally determined changes, and the asymmetry parameters match experiment to within the experimental

Table 2. Calculated and experimental values for C_q and η, as in Table 1, for the listed compounds

Compound	Site	Calc		Expt	
		C_q	η	C_q	η
$KZnB_3O_6$ [1,2]	B3	2·900	0·522	2·4(1)	0·3(2)
		2·986	0·290		
	B4	0·443	0·068	0·8(1)	0·4(1)
HP-KB_3O_5 [2]	B3	3·021	0·120	2·5(1)	0·17(3)
	B4	0·768	0·449	0·95(5)	0·55(4)
		1·019	0·823	1·2(1)	0·74(5)

uncertainty as well.

Accurate computation of the field gradients in the four-fold coordinate sites is more difficult than the three-fold sites, for the same reason that experimental determination of the parameters of these sites is difficult: the field gradients at these sites are simply very small and so subject to significant error. The data in Table 1 show that for the most part the computed C_q's for the B4 sites are in reasonable agreement with the experimentally determined ones, while the asymmetry parameters (when available from experiment) are less well reproduced.

Turning to the compounds showing edge-shared B4 sites (Table 2), we see again that the field gradient parameters determined by NMR are reasonably reproduced by the calculations, given the systematic overestimation. Concerning the B4 sites, the $KZnB_3O_6$ compound has only a single crystallographic site.[1] The high pressure phase HP-KB_3O_5 compound has two such sites, one edge-shared and one not, and they are resolved by the NMR experiment,[2] one with a smaller C_q and η and one with larger C_q and η. The calculations show the same pattern, with crystallographic site B(1) having the smaller coupling parameters and site B(3) the larger (crystallographic site B(2) is the three-fold coordinate site).[2] The B(3) site is the edge-shared site, and it has a larger spread of B–O bond lengths (1·41–1·55 Å) than does the B(1) site (1·45–1·52 Å). This larger spread indicates a less symmetric environment and hence is consistent with a larger C_q and η, as assigned.

In addition to assigning the spectrum to the crystallographic sites, it is of interest to investigate the bonding in these unusual coordination environments. This was done by using the AIM method to detect the so-called bond-critical points of the electron density distribution. These points indicate a bonding interaction in the sense that the electron density shows a saddle point between the two atoms, and the density at this point is a relative measure of bond strength. In the case of HP-KB_3O_5, we find the B3 site indeed shows three bond critical points to the three neighbouring oxygen, each with a density of about 0·2. In the case of the four-fold coordinate sites, we find four critical points for each boron, describing the B–O bonds, and the densities are somewhat lower (0·14–0·18) than the three-fold coordinate site. The lower densities are expected because the bonds are longer. Furthermore, the edge-shared site shows both a larger spread of critical point densities and a smaller minimum value, because it has a larger spread of bond lengths and the longest B–O bond of the two sites, as noted above. We do not see, in the edge-shared site, any evidence of B–B bonding character, at least as detected by the AIM formalism. The edge shared site in $KZnB_3O_6$ likewise shows four bond critical points, with similar densities as HP-KB_3O_5 and no B–B bond character.

Finally, we also investigated the bonding character around the unusual three-fold coordinate oxygen found in another high pressure borate phase, β-ZnB_4O_7.[20] Here the three-fold coordinate oxygen indeed exhibits three bond critical points, albeit with somewhat reduced densities as compared to the more typical two-fold coordinate sites ($0 \cdot 13$ as compared to $0 \cdot 17$). Thus here the close coordination distances do in fact correlate with bond formation.

5. Conclusions

In this work we have examined the electric field gradients at various boron sites in borate crystals with compositions relevant to glass formation. Key conclusions that can be drawn from this study are as follows. First, the PAW formalism provides a reasonable first-principles computational approach to the field gradients in these compounds, yielding broadly good agreement with experiment except for a systematic overestimation of the coupling strength. Second, based on the coupling strength and asymmetry parameters, the NMR resonances of the four-fold coordinate sites in the new HP-KB_3O_5 compound are assigned, with the edge-shared site exhibiting the larger field gradient and the other site the smaller gradient. This assignment is explained by the observation that the edge-shared site has both shorter bonds and a wider spread of bond lengths than does the other four-fold site. An AIM analysis of the electron density showed no particularly unusual bonding features, in particular no B–B bonding character in the edge-shared unit. Finally, the unusual three-fold coordinate oxygen in β-ZnB_4O_7 was also studied and the three close contacts were determined in fact to be chemical bonds, though somewhat weaker than the two-fold coordinate oxygen bonding in the same compound.

Acknowledgements

We thank Eric Walter, College of William and Mary, for helpful conversations on the computation of the electric field gradient, NSERC for funding, and ACEnet, the regional high performance computing consortium for universities in Atlantic Canada, for computational resources.

References

1. Jin, S., Cai, G., Wang, W., He, M., Wang, S. & Chen, X. Stable oxoborate with edge-sharing BO_4 tetrahedra synthesized under ambient pressure, *Angew. Chem. Int. Ed.*, 2010, **49**, 4967–70.
2. Neumair, S. C., Vanicek, S., Kaindl, R., Többens, D. M., Martineau, C., Taulelle, F., Senker, J. & Huppertz, H. HP-KB_3O_5 highlights the structural diversity of borates: Corner-sharing BO_3/BO_4 groups in combination with edge-sharing BO_4 tetrahedra, *Eur. J. Inorg. Chem.*, 2011, 4147–4152.
3. Blöchl, P. E. Projector augmented-wave method, *Phys. Rev. B*, 1994, **50**, 17953–17979.
4. Kresse, G. & Joubert, D. From ultrasoft pseudopotentials to the projector augmented-wave method, *Phys. Rev. B*, 1999, **59**, 1758–1775.
5. Torrent, M., Jollet, F., Bottin, F., Zárah, G. & Gonze, X. Implementation of the projector-augmented wave method in the abinit code: Application to the study of iron under pressure, *Comp. Mater. Sci.*, 2008, **42**, 337–351.
6. Profeta, M., Mauri, F. & Pickard, C. J. Accurate first principles prediction of ^{17}O NMR parameters in SiO_2: Assignment of the zeolite ferrierite spectrum, *J. Am. Chem. Soc.*, 2003, **125**, 541–548.
7. Zwanziger, J. W. & Torrent, M. First principles calculation of electric field gradients in metals, semiconductors, and insulators, *Appl. Magn. Reson.*, 2008, **33**, 477–456.
8. Bader, R. F. W. *Atoms in Molecules: A Quantum Theory*, Clarendon Press, Oxford, 1994.
9. Gonze, X., Amadon, B., Anglade, P.-M., Beuken, J.-M., Bottin, F., Boulanger, P., Bruneval, F., Caliste, D., Caracas, R., Cote, M., Deutsch, T., Genovese, L., Ghosez, P., Giantomassi, M., Goedecker, S., Hamann, D., Hermet, P., Jollet, F., Jomard, G., Leroux, S., Mancini, M., Mazevet, S., Oliveira, M. J. T., Onida, G., Pouillon, Y., Rangel, T., Rignanese, G.-M., Sangalli, D., Shaltaf, R., Torrent, M., Verstraete, M., Zerah, G. & Zwanziger, J. W. ABINIT: First-principles approach to material and nanosystem properties, *Comput. Phys. Commun.*, 2009, **180**, 2582–2615.
10. Holzwarth, N. A. W., Tackett, A. R. & Matthews, G. E. A projector augmented wave (PAW) code for electronic structure calculations, Part I: *atompaw* for generating atom-centered functions, *Comp. Phys. Commun.*, 2001, **135**, 329–347.
11. Perdew, J. P., Burke, K. & Ernzerhof, M. Generalized gradient approximation made simple, *Phys. Rev. Lett.*, 1996, **77**, 3865–3868.
12. Pyykkö, P. Year-2008 nuclear quadrupole moments, *Mol. Phys.*, 2008, **106**, 1965–1974.
13. Gurr, G. E., Montgomery, P. W., Knutson, C. D. & Gorres, B. T. The crystal structure of trigonal diboron trioxide, *Acta Cryst.*, 1970, **B26**, 906–915.
14. Kroeker, S. & Stebbins, J. F. Three-coordinated boron-11 chemical shifts in borates, *Inorg. Chem.*, 2001, **40**, 6239–6246.
15. Natarajan-Iyer, M., Faggiani, R. & Brown, I. D. Dilithium tetraborate, $Li_2B_4O_7$, *Cryst. Struct. Comm.*, 1979, **8**, 367–370.
16. Chen, B., Werner-Zwanziger, U., Nascimento, M. L. F., Ghussn, L., Zanotto, E. D. & Zwanziger, J. W. Structural similarity on multiple length scales and its relation to devitrification mechanism: A solid-state NMR study of alkali diborate glasses and crystals, *J. Phys. Chem.*, 2009, **C113**, 20725– 20732.
17. Bray, P. J. NMR and NQR studies of boron in vitreous and crystalline borates, *Inorg. Chim. Acta*, 1999, **289**, 158–173.
18. Krogh-Moe, J. The crystal structure of sodium diborate, $Na_2O(B_2O_3)_2$, *Acta Cryst.*, 1974, **B30**, 578–582.
19. Corker, D. L. & Glazer, A. M. Structure and optical non-linearity of $PbO.2B_2O_3$, *Acta Cryst.*, 1996, **B52**, 260–265.
20. Huppertz, H. & Heymann, G. Multianvil high-pressure/high-temperature preparation, crystal structure, and properties of the new oxoborate β-ZnB_4O_7, *Solid State Sci.*, 2003, **5**(2), 281–289.

Phys. Chem. Glasses: Eur. J. Glass Sci. Technol. B, June 2012, **53** (3), 121–127

SpectraFit: a new program to simulate and fit distributed ^{10}B powder patterns – application to symmetric trigonal borons

Victor Khristenko, Kevin Tholen, Nathan Barnes, Evan Troendle, David Crist, Mario Affatigato, Steve Feller*

Physics Department, Coe College Cedar Rapids, IA 52402, USA

Diane Holland, Thomas F. Kemp & Mark E. Smith

Physics Department, University of Warwick, UK CV4 7AL, UK

Manuscript received 3 November 2011
Revised version received 22 December 2011
Accepted 2 March 2012

We have developed a new program in C++ to simulate ^{10}B NMR powder patterns. The program takes into account the quadrupolar interaction of ^{10}B (I=3) and it allows for the use of a Gaussian distribution of the quadrupolar interaction parameters to fit solid amorphous and crystalline materials. A key aspect of this program is that it has been automated to find the best-fit parameters of the quadrupole coupling constant (C_Q) and the asymmetry parameter (η) as well as the widths (σ_{CQ} and σ_η) of their distributions. Initially, this program has been extensively used, with the aid of the earlier and non-automatic QuadFit developed by Tom Kemp at the University of Warwick (UK), to find the best fit parameters of ^{10}B NMR spectra in amorphous B_2O_3 prepared at different cooling rates as well as polycrystalline lithium orthoborate, Li_3BO_3. The technique is sensitive enough to clearly see differences in the three-coordinated boron quadrupole parameters.

I. Introduction

The intermediate range order (IRO) in glass, at the level up to several angstroms or so, is currently of great interest. Nuclear magnetic resonance (NMR) in certain circumstances may be suited to the study of the IRO. Normally, we examine the short range order (SRO) with NMR since it uses the nucleus to probe its local and perhaps nearest neighbor bonding environments. However, modern NMR techniques such as dynamic angle spinning (DAS) or double rotation (DOR), or the examination of certain nuclei with especially strong interactions are sometimes able to reach out a bit further and probe the IRO in terms of superstructural borate groups for example.

^{10}B NMR is sensitive to this level of order within the glass since it has a very strong quadrupole interaction, with coupling constants (C_Q) that exceed 5 MHz for the three-coordinate borons. It also has a spin of 3 which produces six transitions which provides exquisite detail to exploit. The resulting experimental powder patterns are also extremely sensitive to the asymmetry parameter, η, as well as to distributions of C_Q and η. The standard deviations of the respective distributions are denoted σ_{CQ} and σ_η. The result is an ability to probe further into the glass structure, even beyond the SRO.

Over the past few years we have begun to systematically prepare borate glasses and crystals, enriched in ^{10}B at levels that exceed 95 atomic%, of differing superstructural groupings, to determine the above four quadrupole parameters that characterize short range borate structures in intermediate range structures. However, until now, we have not had the means to obtain the true best fit parameters.

We report on fitting ^{10}B NMR powder patterns using a new automated fitting procedure that we have named *SpectraFit*. This program is capable of determining these best four fit parameters for a given site within borate glasses and crystals. This approach allows us to determine the four quadrupole parameters *individually* to a high precision. Thus, we have the chance to observe borons in differing environments, even within different superstructual groups, by systematically changing boron environments in different crystals and glasses.

In this paper we used the new technique, with its added precision, to specifically examine polycrystalline lithium orthoborate and vitreous boron oxide prepared using three cooling rates spanning a range of 10^8 K/s. Each of these materials has a symmetric SRO consisting of a boron triangle with either all bridging oxygens (vitreous B_2O_3) or nonbridging oxygens (crystalline $Li_3B_3O_3$).

II. Procedures

Experimental and simulations
1. Sample preparation
Glassy boron oxide was prepared three ways using enriched ^{10}B (greater than 95 atomic%) boric acid.

---

*Corresponding author. Email sfeller@coe.edu
Original version presented at VII Int. Conf. on Borate Glasses, Crystals and Melts, Dalhousie University, Halifax, Nova Scotia, Canada, 21–25 August 2011

(a)

Slow-Cooled B$_2$O$_3$: NRP = 0.0112711

(b)

Medium-Cooled B$_2$O$_3$: NRP = 0.0124779

(c)

Fast-Cooled B$_2$O$_3$: NRP = 0.0141819

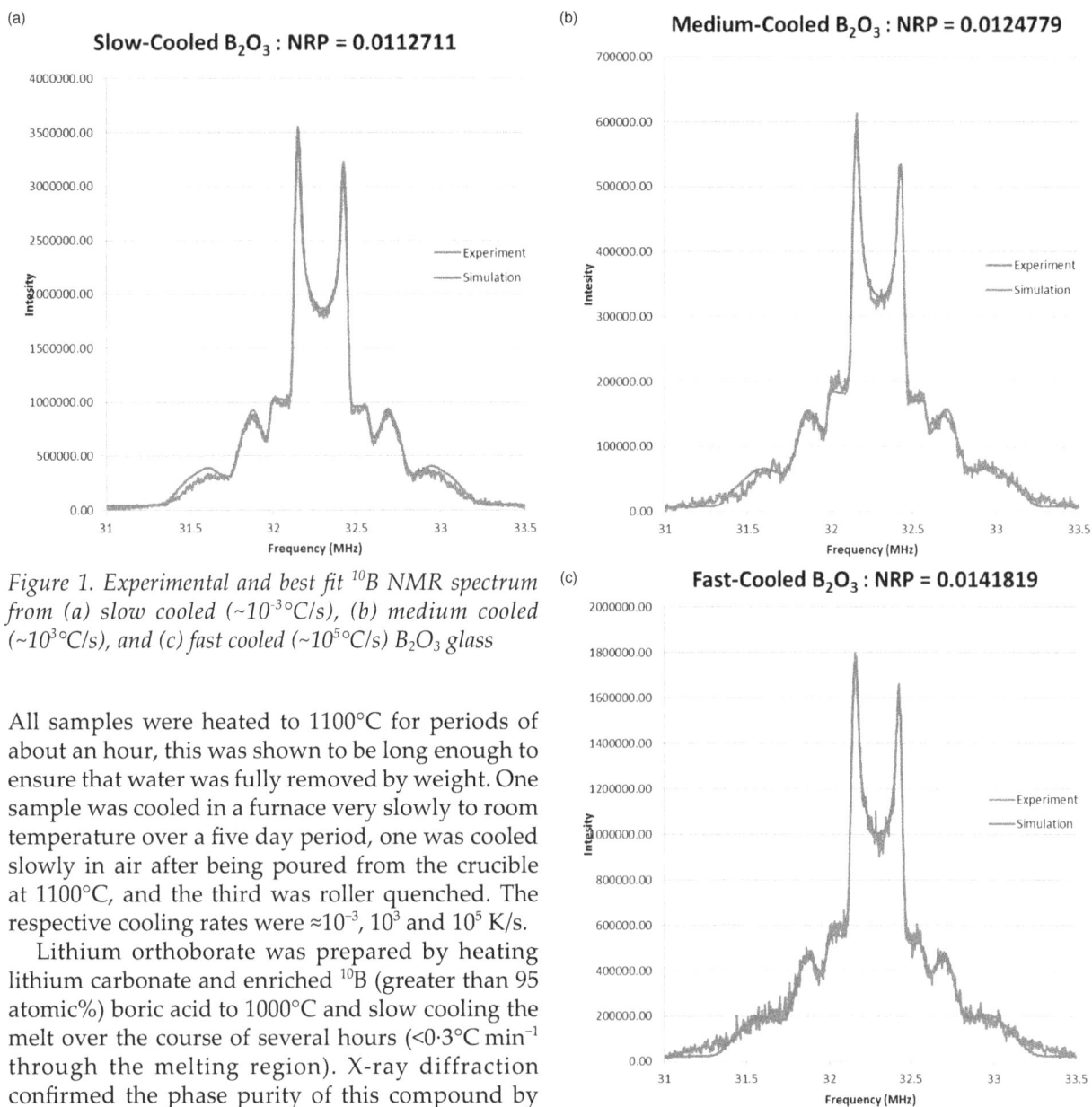

Figure 1. Experimental and best fit ^{10}B NMR spectrum from (a) slow cooled (~10^{-3}°C/s), (b) medium cooled (~10^3°C/s), and (c) fast cooled (~10^5°C/s) B$_2$O$_3$ glass

All samples were heated to 1100°C for periods of about an hour, this was shown to be long enough to ensure that water was fully removed by weight. One sample was cooled in a furnace very slowly to room temperature over a five day period, one was cooled slowly in air after being poured from the crucible at 1100°C, and the third was roller quenched. The respective cooling rates were $\approx 10^{-3}$, 10^3 and 10^5 K/s.

Lithium orthoborate was prepared by heating lithium carbonate and enriched ^{10}B (greater than 95 atomic%) boric acid to 1000°C and slow cooling the melt over the course of several hours (<0·3°C min^{-1} through the melting region). X-ray diffraction confirmed the phase purity of this compound by comparison with the pattern calculated from the single crystal study of Stewner.[1] Furthermore, Raman spectra were obtained from the sample that confirmed the presence of the crystalline phase when compared with the data of Kamitsos et al.[2]

2. NMR

Standard static pulse-echo ($\pi/2 \rightarrow \pi$) experiments were performed at 7·0 T at an operating frequency of 31·49 MHz with the following conditions: τ=3 ms and $\tau/2$=1·5 ms; 10–30 s pulse delay; 1 MHz spectrum width at each frequency. A novel aspect of the present work was the acquisition of spectra at several different magnetic fields through the use of a field-step unit. This was necessary because of the extreme width of the powder pattern. Typically a frequency range of 1·44 MHz was covered in 57·6 kHz steps, with 1500 to 2500 acquisitions taken at each frequency to give an acceptable signal-to-noise

ratio. The final experimental powder pattern is the summation of the spectra obtained at each magnetic field. It was found that a significant number of steps (on the order of 25) were needed to avoid artifacts from the step size appearing in the spectrum.

3A. Simulating ^{10}B powder patterns:

The powder patterns were calculated from an original automated fitting program that used the third order perturbation theory developed by Jellison, Feller & Bray[3] and the calculational ideas developed by Kemp,[4] including an Alderman Grant interpolation.[5] Quadrupolar perturbed NMR frequencies were thus calculated, using six transitions weighted properly in terms of the quantum mechanical probabilities.[3] This resulted in an NMR powder pattern. The effects of the distribution of sites in a glass were achieved by assuming a Gaussian distribution for

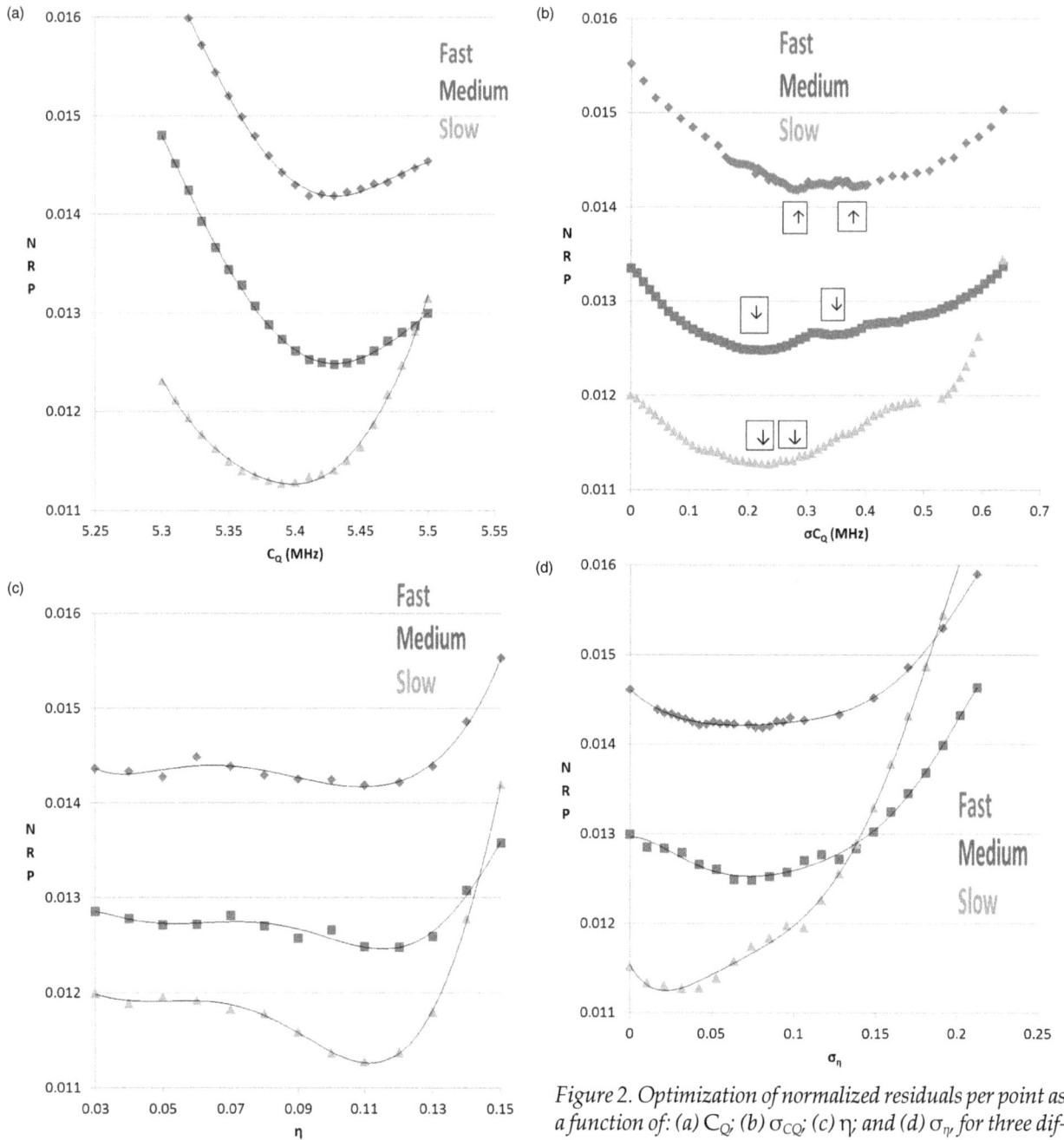

Figure 2. Optimization of normalized residuals per point as a function of: (a) C_Q; (b) σ_{CQ}; (c) η; and (d) σ_η for three differently cooled B_2O_3 glasses Any curve is a guide to the eye

values of C_Q and η, whose standard deviations are σ_η, and σ_{CQ}, and summing the spectra simulated by the Gaussian function with up to 81 individual spectra to create a distributed spectrum.[6] The distributions were sampled out to +/− two standard deviations from the mean.

3B. SpectraFit fitting procedures

The distributed powder patterns were then calculated and compared to the experimental spectrum by the use of the *normalized residuals per point (NRP)*. The NRPs were minimized by searching systematically in the four-parameter space (C_Q, η, σ_η, and σ_{CQ}). This was done until plots of the NRP were made against

optimized values for each of these four parameters. Minima in these plots represented best fit parameters from which the powder patterns were calculated. In some cases two minima were determined and these, we believe, represent two boron environments as will be discussed below.

III. Results

A. Vitreous B_2O_3

Experimental and best-fit simulated powder patterns were overlaid for the three differently cooled samples of ^{10}B enriched glassy boron oxide and these are shown in Figures 1(a, b and c). Figures 2(a)–(d) as well as Table 1 depict the results from the

Table 1. Quadrupole parameters from variously cooled boron oxide glasses (v-B$_2$O$_3$) and crystalline lithium or-thoborate (c-Li$_3$BO$_3$)

Sample (B Site)	Cooling rate (K/s)	C_Q (MHz)	σ_{CQ} (MHz)	η	σ_η	NRP
v-B$_2$O$_3$ ring	10^5	5·43	0·382	0·11	0·042	0·0142125
v-B$_2$O$_3$ non-ring	10^5	5·43	0·285	0·11	0·081	0·0141819
v-B$_2$O$_3$ ring	10^3	5·44	0·340	0·11	0·053	0·0126468
v-B$_2$O$_3$ non-ring	10^3	5·43	0·223	0·12	0·074	0·0124779
v-B$_2$O$_3$ ring	10^{-3}	5·39	0·265	0·11	0·021	0·0113059
v-B$_2$O$_3$ non-ring	10^{-3}	5·39	0·234	0·11	0·032	0·0112711
c-Li$_3$BO$_3$		5·55	0·159	0·05	0·011	0·0152082

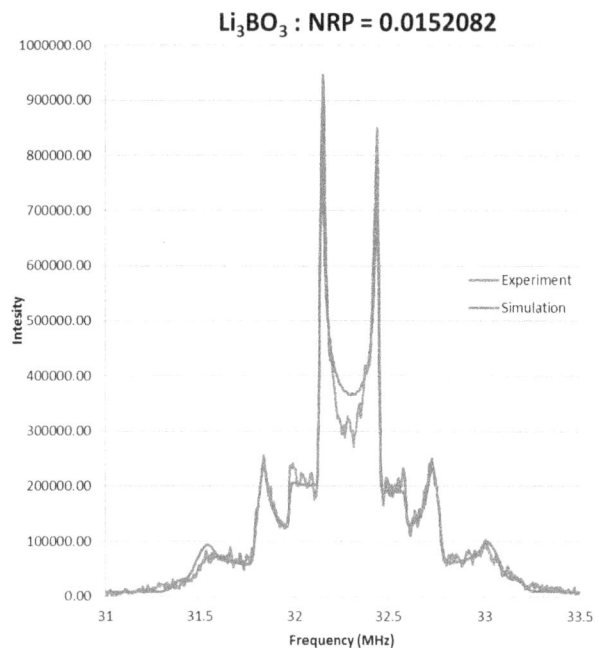

Figure 3. Experimental and best fit ^{10}B NMR spectrum from crystalline lithium orthoborate

minimization of the NRP for each of the quadrupole parameters: C_Q, σ_{CQ}, η, and σ_η. We see in Figure 2(b) the plots of the quadrupole distribution width, σ_{CQ}, evidence for two boron sites. The resulting distribution width in η was used to distinguish the two boron sites, ring and non-ring borons. We assume the ring boron has a smaller σ_η than the non-ring boron since the asymmetry parameter is especially sensitive to deviations from axial symmetry as it is a measure of planar deviations of the electric field gradient in the principal axis system. Thus, we presume the rings are composed of rigid and planar borate triangles with less variation in this respect than the non-ring borons, Ref. 9 gives the mean ring internal B–O–B angle of 120·0± 0·7° with a small standard deviation of 3·2±0·4. For the external B(ring)–O–B(non-ring) the values are 135·1° and 6·7±0·4.

We note that σ_{CQ} is greater in the ring whereas η is the same or slightly smaller in the ring. Also, the smaller value of η in the ring for the medium cooled case is consistent with the site identification since this implies a more cylindrically symmetric site.

B. Polycrystalline Li$_3$BO$_3$

Experimental and simulated powder patterns were overlaid for polycrystalline Li$_3$BO$_3$ as shown in Figure 3. Figures 4(a)–(d) as well as Table 1 depict the results from the minimization of the NRP for each of the quadrupole parameters: C_Q, η, σ_η, and σ_{CQ}.

IV. Discussion

A. Boron oxide glass
Examination of Figures 2(a)–(d) indicates several interesting features. Figure 2(a) shows the NRP versus C_Q for the three boron oxide samples. The slow cooled sample displays a slightly lower C_Q than the two faster cooled samples, with a difference of only about 0·7%. Figure 2(b) for NRP versus σ_{CQ} shows the interesting presence of multiple minima as discussed above. Also present is the clear trend of σ_{CQ} decreasing in the region of the minima in NRP as we proceed from fast to slow cooled boron oxide glass. Further, Figures 2(b and d) indicate that the NRP trend versus distributions widths σ_{CQ} and σ_η become

more sharply defined as the cooling rate is lowered, this is especially true of the slow cooled sample in σ_η. Corroborating this result is the trend seen for η in Figure 2(c). While η hardly changes, the minimum in the NRP becomes much more clearly defined. All of these trends indicate more ordering taking place in the glass as the cooling rate is lowered by a factor of 10^8.

Numerous authors, using several techniques, have described the structure of boron oxide glass as being a composite of boroxol ring (B$_R$) and non-ring (B$_{NR}$) boron atoms with a ratio of approximately 3 or 4 to 1.[7] As indicated above on this basis we attributed the site with the lower asymmetry parameter distributions to the ring boron atoms and the site with the larger distributions to the non-ring boron atoms.

The literature contains several results that we can compare our data and simulations to. In 1978 Jellison, Panek & Bray[8] were able to observe two of the six ^{10}B NMR transitions from plate quenched (medium cooled) B$_2$O$_3$. Using the third order perturbation theory of Jellison, Feller & Bray[3] and by fitting an aspect of the resulting derivative powder pattern that was sensitive to distributions in the asymmetry parameter they determined a single site fit with:

$$C_Q = 5 \cdot 51 \text{ MHz} \tag{1a}$$

$$\sigma_{CQ} = 0 \cdot 21 \text{ MHz} \tag{1b}$$

$$\eta = 0 \cdot 12 \tag{1c}$$

$$\sigma_\eta = 0 \cdot 05 \tag{1d}$$

Taking the average of our two sites for our medium

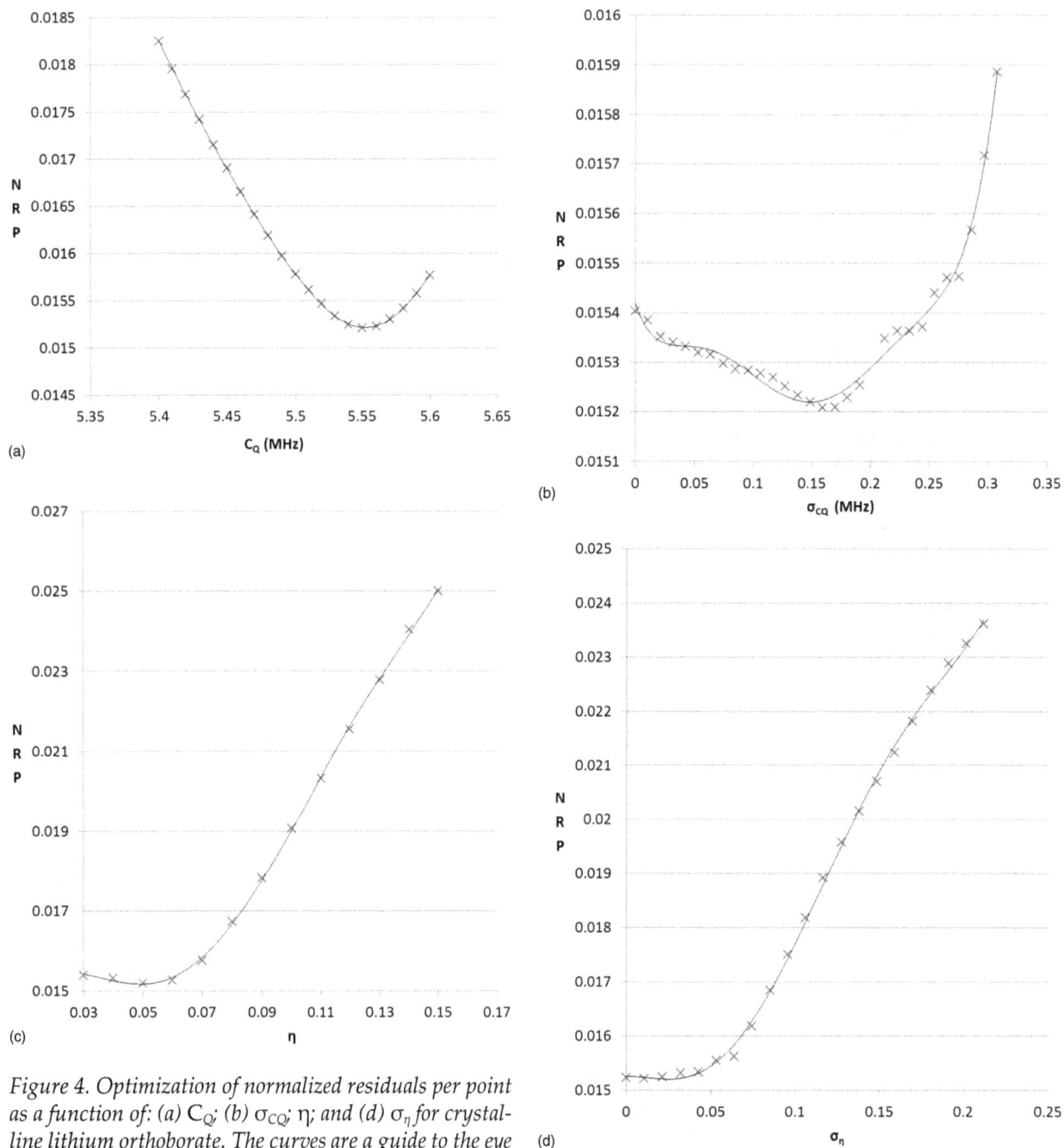

Figure 4. Optimization of normalized residuals per point as a function of: (a) C_Q; (b) σ_{CQ}; η; and (d) σ_η for crystalline lithium orthoborate. The curves are a guide to the eye

cooled sample gives:

$$C_Q = 5 \cdot 44 \text{ MHz} \tag{2a}$$

$$\sigma_{CQ} = 0 \cdot 28 \text{ MHz} \tag{2b}$$

$$\eta = 0 \cdot 12 \tag{2c}$$

$$\sigma_\eta = 0 \cdot 064 \tag{2d}$$

The agreement is quite reasonable.

The quadrupole interaction parameter, P_Q, is defined by

$$P_Q = C_Q (1 + \eta^2/3)^{0 \cdot 5} \tag{3}$$

and was recently measured by double rotation (DOR)

^{11}B NMR experiments by Dupree & Holland.[9] They reported values for P_Q and σP_Q in their paper using a medium cooled sample and found:

$$P_Q(B_R) = 2 \cdot 67 \text{ MHz} \tag{4a}$$

$$P_Q(B_{NR}) = 2 \cdot 61 \text{ MHz} \tag{4b}$$

$$\sigma P_Q(B_R) = 0 \cdot 08 \text{ MHz} \tag{4c}$$

$$\sigma P_Q(B_{NR}) = 0 \cdot 12 \text{ MHz} \tag{4d}$$

Using the ratios of the quadrupole moments of ^{10}B and ^{11}B we can find $C_Q(^{11}\text{B})$ from $C_Q(^{10}\text{B})$:

$$C_Q(^{11}\text{B}) = (Q^{B11}/Q^{B10}) * C_Q(^{10}\text{B}) = (1/2 \cdot 084) * C_Q(^{10}\text{B}) \tag{5}$$

Using Table 1 and Equations (1) and (4) we deduced the following values for the ^{11}B P_Q and σP_Q from our ^{10}B NMR results for our medium cooled sample:

$$P_Q(B_R) \text{ (from }^{10}\text{B)}=2\cdot62 \text{ MHz} \tag{6a}$$

$$P_Q(B_{NR}) \text{ (from }^{10}\text{B)}=2\cdot61 \text{ MHz} \tag{6b}$$

$$\sigma P_Q(B_R) \text{ (from }^{10}\text{B)}=0\cdot17 \text{ MHz} \tag{6c}$$

$$\sigma P_Q(B_{NR}) \text{ (from }^{10}\text{B)}=0\cdot12 \text{ MHz} \tag{6d}$$

The agreement is seen to be good for the frequencies and fairly close for the distribution values. The assignment of the two boron sites from ^{10}B is consistent with the ^{11}B analysis in that the two techniques both observed the higher C_Q for the ring borons.

Finally, Bray *et al* performed nuclear quadrupole resonance measurements (NQR) on boron oxide glass and found two peaks.[10] The quadrupole frequencies, ν_{BR} and ν_{BNR} as well as $\sigma\nu_{BR}$ and $\sigma\nu_{BNR}$ for the ^{11}B NQR responses in the ring and non-ring borons were found to be:

$$\nu_{BR}=1\cdot357 \text{ MHz} \tag{7a}$$

$$\nu_{BNR}=1\cdot305 \text{ MHz} \tag{7b}$$

and

$$\sigma\nu_{BR}=0\cdot00837 \text{ MHz} \tag{7c}$$

$$\sigma\nu_{BNR}=0\cdot00972 \text{ MHz} \tag{7d}$$

The quadrupole resonance frequency is $\frac{1}{2}P_Q$ or

$$\nu_Q = \frac{C_Q}{2}\left(\frac{1+\eta^2}{3}\right)^{0\cdot5} \tag{8}$$

Use of our ^{10}B results yields the following comparisons to the NQR data:

$$\nu_{BR} \text{ (from }^{10}\text{B)}=1\cdot308 \text{ MHz} \tag{9a}$$

$$\nu_{BNR} \text{ (from }^{10}\text{B)}=1\cdot306 \text{ MHz} \tag{9b}$$

and from our distributions in the quadrupole parameters

$$\sigma\nu_{BR} \text{ (from }^{10}\text{B)}=0\cdot085 \text{ MHz} \tag{9c}$$

$$\sigma\nu_{BNR} \text{ (from }^{10}\text{B)}=0\cdot059 \text{ MHz} \tag{9d}$$

The frequencies agree reasonably well whereas the distributions do not. The distributions are lower in the NQR data when compared with the ^{10}B NMR results.

B. Polycrystalline Li$_3$BO$_3$

On the whole it is still a very reasonable fit although overall the NRP for polycrystalline Li$_3$BO$_3$ is somewhat worse compared to those of boron oxide (0·01521 versus 0·01264). We are confident in the purity of the sample as it was checked both by x-ray diffraction and Raman spectroscopy, as discussed above. In particular, the central feature shows a poorer fit compared to the rest of the spectrum. One likely explanation is the possible preferred orientation of the crystalline powder in the NMR probe. Also, the crystal reasonably exhibits smaller distributions in the quadrupole parameters leading to inherently sharper features indicative of smaller distribution widths. A single minimum was found for the four quadrupole parameters as given in Table 1. This is consistent with the published crystal structure of Stewner[1] which indicates only one crystallographically distinct B site. The value of $\eta=0\cdot05$ makes sense for a short range unit consisting of isolated $(BO_3)^{3-}$ triangles whose oxygens are all nonbridging. Indeed, η is smaller for this case compared to v-B_2O_3 where the boron triangles in the rings are bonded to both ring and non-ring borons and all borons have a larger variation in bond angles. In Figures 4(a)–(d) the minima are sharper than in the case of boron oxide presumably because of the less variability in bond angles and lengths in the crystal. The sharpness of the NRP minimum for η and σ_η near 0·05 and 0·01, respectively, is especially evident in Figures 4(c) and (d) and is indicative of the highly symmetric $(BO_3)^{3-}$ isolated borate triangles with nonbridging oxygens.

A few reports of quadrupolar parameters from orthoborate anions in lithium borate glasses, have been given in the literature. In 1978 Yun reported[11] the C_Q and η for the orthoborate anion from ^{11}B NMR in $((BO_3)^{3-}+3Li^+)$ to be (average):

$$C_Q(^{11}\text{B})=2\cdot66 \text{ MHz} \tag{10a}$$

$$\eta=0\cdot08 \tag{10b}$$

The corresponding present values from ^{10}B NMR and converted to ^{11}B values using Equation (5) are:

$$C_Q \text{ (}^{11}\text{B converted from }^{10}\text{B)}=2\cdot66 \text{ MHz} \tag{11a}$$

$$\eta=0\cdot05 \tag{11b}$$

The agreement is good.

In 1982 Feller, Dell & Bray[12] were able to observe two of the six ^{10}B NMR transitions from plate quenched (medium cooled) lithium borate glasses of high lithia content. They determined for the orthoborate anion fit parameters of:

$$C_Q=5\cdot75 \text{ MHz} \tag{12a}$$

$$\sigma_{CQ}=0\cdot22 \text{ MHz} \tag{12b}$$

$$\eta=0\cdot00 \tag{12c}$$

$$\sigma_\eta=0\cdot05 \tag{12d}$$

The comparison values for crystalline Li$_3$BO$_3$ are

found in Table 1:

$$C_Q = 5 \cdot 55 \text{ MHz} \tag{13a}$$

$$\sigma_{CQ} = 0 \cdot 159 \text{ MHz} \tag{13b}$$

$$\eta = 0 \cdot 05 \tag{13c}$$

$$\sigma_\eta = 0 \cdot 011 \tag{13d}$$

C_Q and η are each a little different in the crystal compared with the glass parameters.

The distribution parameters, σ_{CQ} and σ_η, are each lower in the crystal than found in the glass; this is a sensible result.

V. Conclusions

We have demonstrated that an automatic fitting routine was effectively used in the fitting of ^{10}B NMR spectra from borate glasses and crystals. The fitting is able to examine distributions of the quadrupole parameters. This allowed us to find two boron sites in each of the glassy boron oxide samples as well a single boron site in polycrystalline lithium orthoborate. The two sites in v-B_2O_3 are attributed to ring and non-ring borons. Importantly, we saw increased ordering in the slow cooled boron oxide glass compared with the two faster cooled samples through a close examina-tion of its quadrupole parameters. In particular the distribution widths narrowed as the cooling rate changed by a factor of 10^8.

VI. Acknowledgements.

The National Science Foundation is acknowledged for support of this work through grant NSF-DMR-0904615.

VII. References

1. Stewner, F. *Acta Cryst.*, 1971, **B27**, 904.
2. Chryssikos, G. D., Kamitsos, E. I., Patsis, A. P., Bitsis, M. S. & Karakas-sides, M. A. *J. Non-Cryst. Solids*, 1990, **196**, 42.
3. Jellison, Jr, G. E., Feller, S. A. & Bray, P. J. *J. Mag. Res.*, 1977, **27**, 121–132.
4. Kemp, T. F. & Smith, M. E. *Solid State Nucl. Magn. Resonance*, 2009, **35** (4), 243–252.
5. Alderman, D. W., Solum, M. S. & Grant, D. M. *J. Chem. Phys.*, 1986, **84** (7), 3717.
6. In some instances, when η was near zero and the distribution would have required non-physical (-η) values, fewer than 81 spectra were used.
7. See summary in Wright, A. C. *Phys. Chem. Glasses: European J. Glass Science Technol. B*, 2010, **51** (1), 1–39.
8. Jellison Jr. G. E., Panek, L. W., Bray, P. J. & Rouse Jr. G. B. *J. Chem. Phys.*, 1977, **66**, 802.
9. Hung, I., Howes, A. P., Parkinson, B. G, Anupold, T., Samoson, A., Brown, S. P., Harrison, P. F., Holland, D. & Dupree, R. *J. Solid State Chem.*, 2009, **182**, 2402–2408.
10. Bray, P. J., Emerson, J. F., Lee, D. H., Feller, S., Bain, D. L. & Feil, D. A. *J. Non-Cryst. Solids*, 1991, **129**, 240–248.
11. Yun, Y. H. PhD Thesis, Brown University, 1978.
12. Feller, S., Dell, W. J. & Bray, P. J. *J. Non-Cryst. Solids*, 1982, **51**, 21–30.

Phys. Chem. Glasses: Eur. J. Glass Sci. Technol. B, June 2012, 53 (3), 128–131

Mechanochemical synthesis of BaO–B$_2$O$_3$ glass and glass-ceramic phosphor powders containing europium ions

*Atsuko Shinomiya, Akitoshi Hayashi, Kiyoharu Tadanaga & Masahiro Tatsumisago**

Department of Applied Chemistry, Graduate School of Engineering, Osaka Prefecture University, 1-1 Gakuencho, Naka-ku, Sakai, Osaka 599-8531, Japan

Manuscript received 24 October 2011
Revised version received 12 December 2011
Accepted 23 March 2012

Glasses in the system Eu$_2$O$_3$–BaO–B$_2$O$_3$ were prepared by a mechanochemical process using B$_2$O$_3$, BaO, and Eu$_2$O$_3$ as starting materials. X-ray diffraction patterns indicate that amorphous 99·5(0·2BaO.0·8B$_2$O$_3$).0·5Eu$_2$O$_3$ (mol%) powder was obtained using a planetary ball mill. From infrared and Raman spectra, the structural units observed in the milled glass with milling times longer than 5 h were the same as ones in the melt quenched glass. In photoluminescence measurements, the milled 99·5(0·2BaO.0·8B$_2$O$_3$).0·5Eu$_2$O$_3$ glasses showed red emission due to Eu^{3+} ions. By heating the milled glass in air at temperatures higher than the crystallization temperature, 99·5(0·2BaO.0·8B$_2$O$_3$).0·5Eu$_2$O$_3$ glass-ceramic was also prepared. The glass-ceramic showed broad emission in the range 380-450 nm from Eu^{2+} ions. Crystallization of the milled glass induced reduction of Eu^{3+} to Eu^{2+} in air.

Introduction

Mechanochemical processing using a milling apparatus is a way to prepare glassy materials in which the introduced mechanical energy leads to chemical reaction as well as pulverization. For example, we have reported that glasses in the system Li$_2$S–P$_2$S$_5$ have been prepared by mechanical milling, and can be used as the electrolytes for all solid state lithium secondary batteries.[1] We have also mechanochemically synthesized SnO–B$_2$O$_3$ glasses which can be used as negative electrodes with large capacity for lithium secondary batteries.[2] In this process, fine glass powders are obtained at room temperature, and thus mechanochemical synthesis is suitable for preparation of materials for applications like electrolytes and electrodes which require powders.[1,2] Since phosphors are also generally used as powders, the glassy phosphor powder prepared through mechanochemical process can be directly applied.

Europium ions (Eu^{2+} and Eu^{3+}) have been widely used as activators in phosphors.[3,4] Eu^{2+} ions show higher emission efficiency than Eu^{3+} ions, but reduction of the Eu^{3+} ion, which is stable in ambient atmosphere, needs a reductive atmosphere. On the other hand, it has been reported that Eu^{3+} in certain matrices can be reduced to Eu^{2+} in an oxidizing atmosphere,[5,6] and Machida *et al* have reported that crystallization of barium borate glasses in air induced the reduction of Eu^{3+} to Eu^{2+}.[7]

In the present study, we synthesized barium borate phosphor powders containing Eu^{3+} by mechanical milling, and prepared glass-ceramic phosphor powder containing Eu^{2+} by heating the milled glass in air.

Experimental

Glasses in the Eu$_2$O$_3$–BaO–B$_2$O$_3$ system were prepared by mechanical milling and melt quenching. In the mechanical milling, reagent grade BaO (Wako Pure Chemical Industries), B$_2$O$_3$ (Aldrich) and Eu$_2$O$_3$ (Wako Pure Chemical Industries) powders were used as starting materials. The B$_2$O$_3$ was dried *in vacuo* at 150°C for more than 12 h before the all experiments. The mechanochemical treatment was carried out for 1 g batches of the mixed materials at the composition 99·5(0·2BaO.0·8B$_2$O$_3$).0·5Eu$_2$O$_3$ (mol%) in a ZrO$_2$ pot (volume 45 ml) with 160 ZrO$_2$ balls (5 mm in diameter), using a high energy planetary ball mill apparatus (Pulverisette 7, Fritsch GmbH). The rotation speed was 370 rpm. The obtained glassy powder was heated at temperatures higher than the crystallization temperatures to yield Eu$_2$O$_3$–BaO–B$_2$O$_3$ glass-ceramics.

The glasses were also prepared by melt quenching. A mixture of BaCO$_3$ (Wako Pure Chemical Industries), H$_3$BO$_3$ (Wako Pure Chemical Industries) and Eu$_2$O$_3$ (Wako Pure Chemical Industries) was put into a platinum crucible and melted at 1200°C for 15 min in air. The melt was poured onto a stainless steel plate and quenched by a conventional press technique.

X-ray diffraction (XRD) measurements (Cu K$_a$) were performed using a diffractometer (XRD-6000, Shimadzu Co.). Differential thermal analyses (DTA) were carried out using a thermal analyser (Thermo-Plus 8120, Rigaku Co.) on obtained powders in a Pt

* Corresponding author. Email tadanaga@chem.osakafu-u.ac.jp
Original version presented at VII Int. Conf. on Borate Glasses, Crystals and Melts, Dalhousie University, Halifax, Nova Scotia, Canada, 21–25 August 2011

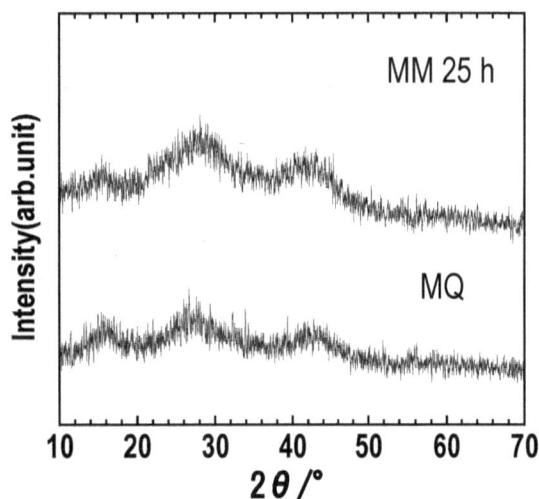

Figure 1. XRD patterns of 99·5(0·2BaO.0·8B$_2$O$_3$).0·5Eu$_2$O$_3$ (mol%) glasses prepared by mechanical milling (MM) for 25 h and melt quenching (MQ)

pan. The heating rate was 10°C min^{-1}. Fourier transform infrared (FT-IR) spectra of the prepared samples were collected using a FT-IR spectrometer (Spectrum GX, Parkin Elmer Co.). Raman spectra of the glasses were measured with Raman spectrophotometer (LabRAM HR800, Horiba Jobin Yvon) using the 325 nm line of a He–Cd laser. Photoluminescence (PL) spectra were measured at room temperature with a fluorescence spectrometer (FP-6500, JASCO Co.).

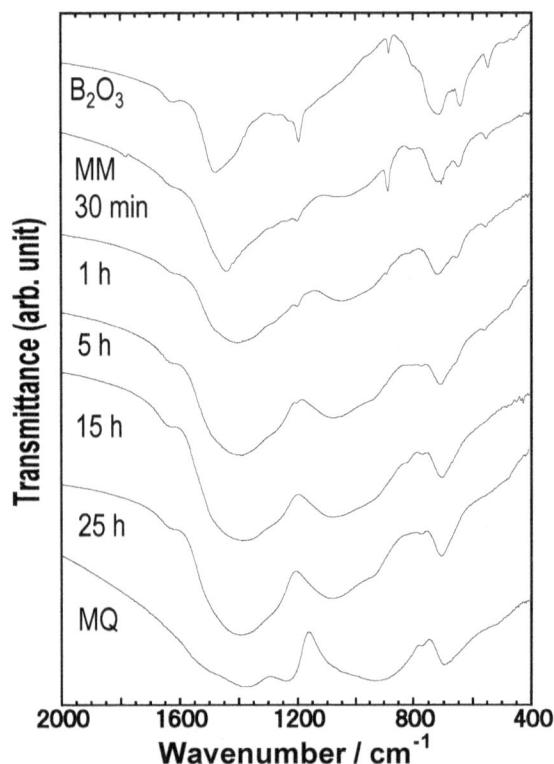

Figure 2. FT-IR spectra of 99·5(0·2BaO.0·8B$_2$O$_3$).0·5Eu$_2$O$_3$ (mol%) sample after different mechanical milling (MM) times, and as prepared by melt quenching (MQ). The spectrum of the dried regent grade B$_2$O$_3$ is also shown for comparison

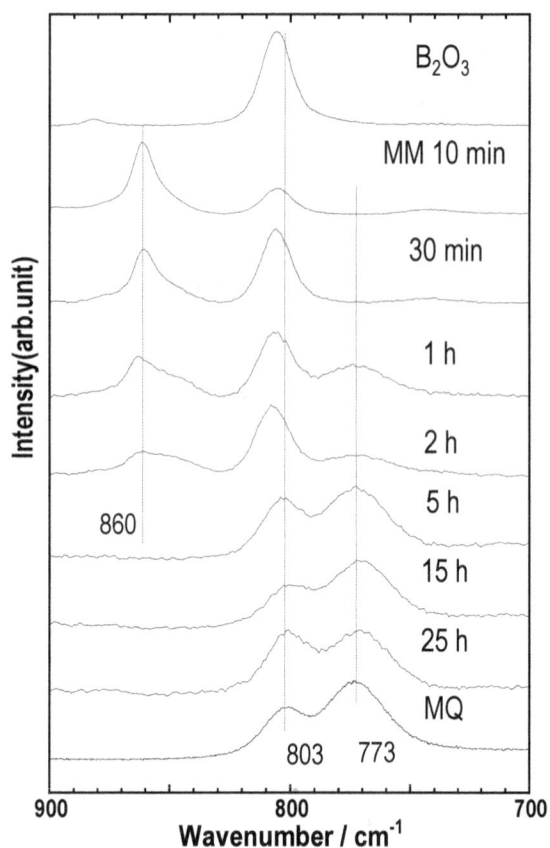

Figure 3. Raman spectra of 20BaO.80B$_2$O$_3$ (mol%) samples after different mechanical milling (MM) times, and as prepared by melt quenching (MQ). The spectrum of the dried regent grade B$_2$O$_3$ is also shown for comparison

Results and discussion

Figure 1 shows XRD patterns of 99·5(0·2BaO.0·8B$_2$O$_3$).0·5Eu$_2$O$_3$ (mol%) powders prepared by mechanical milling for 25 h and melt quenching. Halo patterns were observed, indicating the milled and melt quenched samples were amorphous. In DTA measurements for the milled sample, an exothermic peak due to crystallization was observed at 620°C and an endothermic change due to a glass transition appeared at about 560°C. Therefore, the powder prepared by mechanical milling is in the glassy state.

Figure 2 shows the FT-IR spectra of 99·5(0·2BaO. 0·8B$_2$O$_3$).0·5Eu$_2$O$_3$ (mol%) glasses prepared by mechanical milling and melt quenching. The spectrum of the dried regent grade B$_2$O$_3$ is also shown for comparison. The bands at 800–1100 cm^{-1} and 1200–1500 cm^{-1} are assigned to BO$_4$ and BO$_3$ units, respectively.[8] With increasing milling time, the intensity ratio of the two bands became similar to that of two bands observed for the melt quenched glass. This suggests that the abundance ratio of BO$_4$ and BO$_3$ units in the milled glass became similar to the ratio for the melt quenched glass.

Figure 3 shows Raman spectra of 20BaO.80B$_2$O$_3$ (mol%) glasses prepared by mechanical milling and melt quenching. The spectrum of dried regent

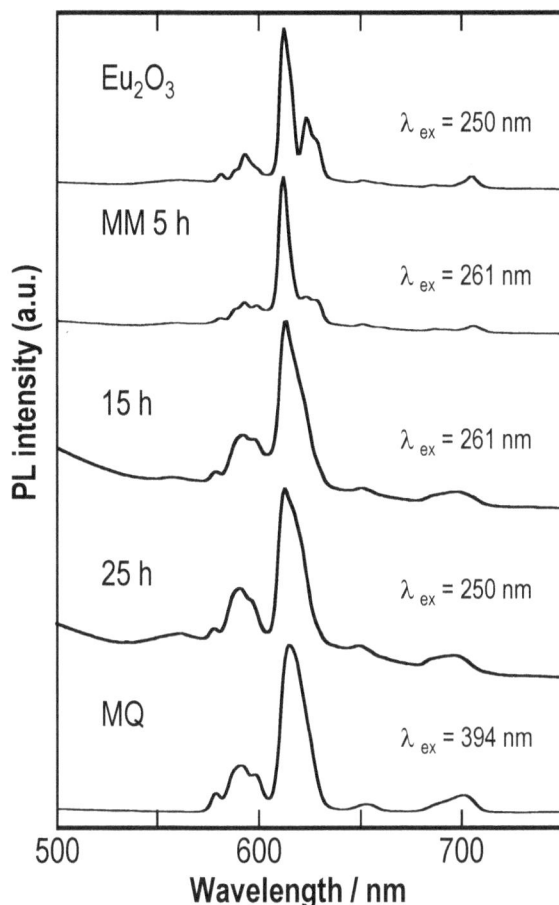

Figure 4. Photoluminescence spectra of $99.5(0.2BaO.$ $0.8B_2O_3).0.5Eu_2O_3$ (mol%) samples after different mechanical milling (MM) times, and as prepared by melt quenching (MQ). The spectrum of pure crystalline Eu_2O_3 powder is also shown for comparison

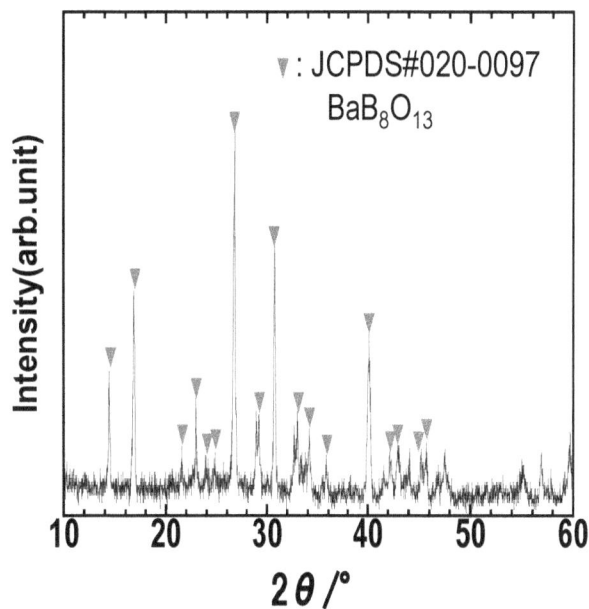

Figure 5. The XRD pattern of $99.5(0.2BaO.0.8B_2O_3).$ $0.5Eu_2O_3$ (mol%) glass-ceramic

grade B_2O_3 is also shown for comparison. Up to 2 h of milling, a peak was observed at 860 cm^{-1}, which is assigned to the pyroborate ($B_2ØO_4^{4-}$) unit.[9] After milling for 5 h, two peaks at 803 cm^{-1} and 773 cm^{-1} were observed, and the spectrum of the milled glass is similar to that of the melt quenched glass. These results suggest that at the initial stage of milling, pyroborate units existed in the milled sample. It seems that a BaO rich phase was firstly formed on the particle surface and structural units which have more nonbridging oxygens were formed. After milling for 5 h, main structural units were the boroxol ring (803 cm^{-1}) and six-membered borate rings with BO$_4$ units (773 cm^{-1}),[10] and the local structure around boron atoms of the milled glass was almost the same as that of the melt quenched glass.

Figure 4 shows PL spectra of $99.5(0.2BaO.0.8B_2O_3).$ $0.5Eu_2O_3$ (mol%) glasses prepared by mechanical milling and melt quenching. The PL spectrum of pure crystalline Eu_2O_3 powder is also shown for comparison. The peak at 613 nm was observed for the milled and melt quenched glasses, and is assigned to the $^5D_0 \rightarrow {}^7F_2$ transition of Eu^{3+} ions. For the milled sample, the IR and Raman spectra showed that a

change of glass structure was not observed with milling time longer than 5 h, but in the emission spectra, emission peaks due to Eu^{3+} around 613 nm became broader with increasing milling time. It seems that the reaction of europium oxide with the barium borate matrix took longer than the formation of the matrix. Since B_2O_3 is an acidic oxide, it reacts preferentially with BaO, which is a basic oxide, to form the barium borate matrix, and Eu_2O_3 should then react with the barium borate matrix.

The glass prepared by milling for 25 h was heated at 700°C for 1 h to make glass-ceramic $99.5(0.2BaO.$ $0.8B_2O_3).0.5Eu_2O_3$ (mol%), and Figure 5 shows the XRD pattern of this glass-ceramic. The diffraction pattern is

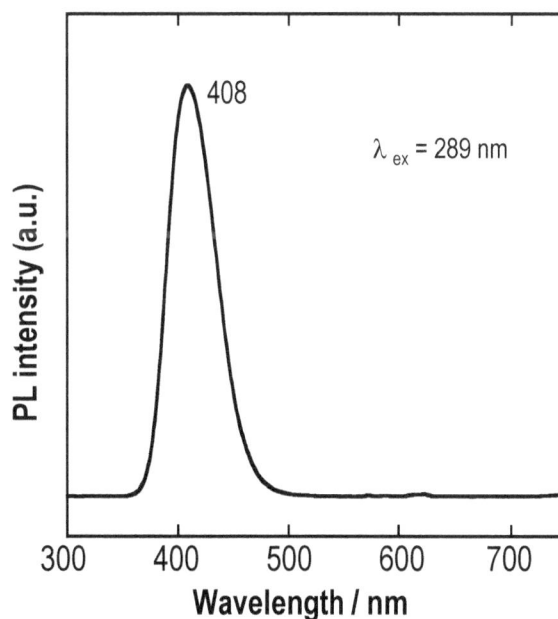

Figure 6. The photoluminescence spectrum of $99.5(0.2BaO.$ $0.8B_2O_3).0.5Eu_2O_3$ (mol%) glass-ceramic

attributable to crystalline BaB_8O_{13}. Figure 6 shows the PL spectrum of $99 \cdot 5(0 \cdot 2BaO.0 \cdot 8B_2O_3).0 \cdot 5Eu_2O_3$ (mol%) glass-ceramic. The broad emission band peaking at 408 nm was assigned to the $4f^65d \rightarrow 4f^7$ transition of Eu^{2+} ions. Therefore, crystallization of the milled glass induced reduction of Eu^{3+} to Eu^{2+}, as reported for the melt-quenched glass.[7] In the spectrum, very weak emission from Eu^{3+} was also observed at around 620 nm, indicating that the crystallization of the milled glass was not completed, and a trace of Eu^{3+} still remained in the glass phase.

Glass powders prepared by mechanical milling have large surface area compared to bulk glasses, and thus this is advantageous for crystallization because it usually starts from the particle surface. Therefore the mechanochemical process is a promising procedure for the preparation of glass and glass-ceramic phosphor powders.

Conclusions

BaO–B_2O_3 phosphor powders containing Eu^{3+} were successfully prepared by a mechanochemical process.

After milling for 5 h, structural units were the same in the milled samples as in a melt quenched sample. Photoluminescence emission peaks from Eu^{3+} became broader with increasing milling time. Glass-ceramic phosphor powders were prepared by heating the milled sample, and it showed broad and strong photoluminescent emission from Eu^{2+}.

References

1. Hayashi, A., Hama, S., Morimoto, H., Tatsumisago, M. & Minami, T. *J. Am. Ceram. Soc.*, 2001, **84**, 477.
2. Hayashi, A., Konishi, T., Nakai, M., Morimoto, H., Tadanaga, K., Minami, T. & Tatsumisago, M. *J. Ceram. Soc. Jpn.*, 2004, **112**, S713.
3. Chang, N. C. *J. Appl. Phys.*, 1963, **34**, 3500.
4. Xie, R. J., Hirosaki, N., Sakuma, K., Yamamoto, Y. & Mitomo, M. *Appl. Phys. Lett.*, 2004, **84**, 5404.
5. Pei, Z., Su, Q. & Zhang, J. *J. Alloy. Compd.*, 1993, **198**, 51.
6. Lian, Z., Wang, J., Lu, Y., Wang, S. & Su, Q. *J. Alloy. Compd.*, 2007, **430**, 257.
7. Machida, K., Ueda, D., Inoue, S. & Adachi, G. *Electrochem. Solid State Lett.*, 1999, **2**, 597.
8. Wong, J. & Angel, C. A. *Glass structure by spectroscopy*, Marcel Dekker, New York, 1976, p429.
9. Konijnendijk, W. L. *Philips Res. Rep. Suppl.*, 1975, No. 1.
10. Yiannopoulos, Y. D., Chryssikos G. D. & Kamitsos, E. I. *Phys. Chem. Glasses*, 2001, **42**, 164.

Phys. Chem. Glasses: Eur. J. Glass Sci. Technol. B, June 2012, **53** (3), 132–140

Double rotation ^{11}B NMR applied to polycrystalline barium borates

Oliver L. G. Alderman,* Dinu Iuga, Andrew P. Howes, Diane Holland & Ray Dupree

Department of Physics, University of Warwick, CV4 7AL, UK

Manuscript received 17 November 2011
Revised version received 12 January 2012
Accepted 28 February 2012

The potential of ^{11}B double rotation nuclear magnetic resonance to provide detailed structural information is demonstrated through the application of the technique to polycrystalline materials obtained by devitrification of barium diborate glasses. Isotropic chemical shift values and quadrupole interaction parameters for the four three-coordinated sites in α-BaB$_4$O$_7$ are extracted by exploiting the field dependency of the quadrupolar interaction. The technique is applied to the polymorph β-BaB$_4$O$_7$, and two new barium borate materials of undetermined structure, and its role as a powerful structural probe is demonstrated and discussed.

Introduction

The magic-angle spinning (MAS) technique is routinely applied in the field of solid state NMR to obtain high resolution powder spectra via the averaging to zero of various angle-dependent terms of the nuclear spin Hamiltonian. In the case of dilute, spin $I=\frac{1}{2}$ nuclei, MAS results in narrow resonance lineshapes free from anisotropic broadenings. However, in the case of quadrupolar (spin $I>\frac{1}{2}$) nuclei, MAS does not completely remove the second order quadrupolar broadening due to the different angular dependence of this interaction, which results in anisotropic broadening of spectra when significantly large electric field gradients (EFGs) are present. This is the case in numerous systems containing commonly-studied nuclei such as ^{27}Al and ^{17}O (both $I=5/2$) and ^{11}B ($I=3/2$). In particular, in borate materials, the trigonal planar, 3-fold coordination (B^{III}) of boron to oxygen results in large EFGs at the boron nuclei and quadrupole coupling constants, C_Q, of typically 2·4 to 2·9 MHz.[1] This is in contrast to values of $\leq$0·5 MHz for tetrahedral, 4-fold coordinated (B^{IV}) sites.[2] At sufficiently high field, under MAS, the residual second order quadrupolar broadening of the spectral lineshape arising from B^{III} sites is sufficiently reduced that the B^{III} resonance is separated from any spectral contribution from B^{IV} sites (Figure 1) since the differing shift ranges for the two coordination states, $-4 \lesssim \delta_{iso} \lesssim 2$ ppm and $12 \lesssim \delta_{iso} \lesssim 19$ ppm for B^{IV} and B^{III} sites, respectively,[2] is greater than the quadrupolar broadening. However, borate systems at ambient pressure commonly contain more than one distinct B^{III} site, and their spectral contributions will inevitably overlap. This tendency can be seen from comparison of the typical shift range of 7 ppm, with typical second order

quadrupolar widths which scale as v_Q^2/v_0,[2] where $v_Q=3C_Q/[2I(2I-1)]$ is the quadrupolar frequency and v_0 is the Larmor frequency. Even at a large field of 20 T, these widths are typically 4 to 9 ppm.

The technique of double rotation (DOR) NMR[3] can be used to remove the second order quadrupolar broadening and hence eliminate (or very much reduce) this undesired overlap of B^{III} site resonances. DOR involves *simultaneous* spinning of the sample about both the magic angle (54·74°) and a second axis chosen to average to zero second order effects. Rapid rotation about the former (MAS) axis allows averaging, to null, those terms in the Hamiltonian that are modulated by the second order Legendre polynomial $P_2(\cos\theta)=\frac{1}{2}(3\cos^2\theta-1)$, such as those arising from the dipolar interaction and the first order quadrupolar interaction. Rotation about the latter axis similarly removes broadening from terms modulated by the fourth order Legendre polynomial, namely the second order quadrupolar interaction. The result is a powder spectrum composed of simple, narrow resonances. The high resolution and quantitative nature of DOR NMR have been successfully applied to the study of the ^{11}B nucleus in glasses such as vitreous B$_2$O$_3$[4] and, more recently, to Pyrex®.[5] In these cases, despite the intrinsic structural disorder of the glassy state, multiple resonances from distinct B^{III} environments are resolved. Surprisingly, ^{11}B DOR NMR has not previously been applied to polycrystalline materials, in which one expects intrinsically much narrower resonances than in glasses, although ^{17}O DOR has, for instance, been exploited in the assignment of crystallographically unique oxygen sites in ferrierite[6] and monosodium glutamate.[7,8] ^{27}Al DOR has also been applied to the crystalline aluminoborate 9Al$_2$O$_3$.2B$_2$O$_3$,[9] in which the tetrahedral, two pentahedral and one octahedral aluminium sites are all fully resolved. Furthermore, the 2D spin

*Corresponding author. Email o.alderman@warwick.ac.uk
Original version presented at VII Int. Conf. on Borate Glasses, Crystals and Melts, Dalhousie University, Halifax, Nova Scotia, Canada, 21–25 August 2011

diffusion ^{27}Al DOR experiment allowed assignment of the pentahedral resonances to the known crystallographic sites.

The aim of this contribution is to demonstrate how relatively simple ^{11}B DOR NMR experiments on polycrystalline borates can be used to access structurally related information, not easily obtainable by other means.

A case study is made of the barium diborate, $BaO.2B_2O_3$, system which has previously received attention owing to its observed polymorphism and thermal expansion behaviour.[10–15] The structure of the α-BaB_4O_7 polymorph is known[16] and consists of a continuous network of ditriborate ($B_3O_3\emptyset_5^{2-}$) and dipentaborate ($B_5O_6\emptyset_5^{2-}$) superstructural units, with eight crystallographically distinct boron sites, and is unique among the known diborate crystal structures,[17] including the recently discovered β-SrB_4O_7.[18] The reported β-BaB_4O_7[13] has been characterised only by its powder diffraction pattern and thermal properties, with no space-group or lattice parameters determined. There exists also a high pressure form of barium diborate,[19] also termed β-BaB_4O_7, however, unless explicitly stated, we shall here reserve this nomenclature for the ambient pressure polymorph.

DOR NMR

For a half integer quadrupolar nucleus, such as ^{11}B, the dominant higher order term in the nuclear spin Hamiltonian is the second order quadrupolar interaction. As mentioned earlier MAS removes first order broadenings and this term is only partially averaged, with the residual lineshape containing both an isotropic and an anisotropic contribution, giving rise to the complicated lineshapes observed for B^{III} resonances, see Ref. 1 for example, and Figure 1. Under DOR this anisotropic term is time averaged and an isotropic line is observed, such that the DOR resonance occurs at a shift of

$$\delta_{DOR} = \delta_{iso} + \delta_{QIS} \qquad (1)$$

where δ_{QIS} is the quadrupolar induced shift, which for the central, $\frac{1}{2}$ to $-\frac{1}{2}$, transition (that is usually the only transition observed), and for spin $I=3/2$ nuclei, such as ^{11}B, is given by[2]

$$\delta_{QIS} = -\frac{1}{40}\frac{C_Q^2}{\nu_0^2}\left(1+\frac{\eta^2}{3}\right) = -\frac{1}{40}\frac{P_Q^2}{\nu_0^2} \qquad (2)$$

In Equation (2) P_Q is the quadrupole coupling parameter and, as can be seen, is a function of the quadrupole coupling constant C_Q and the asymmetry parameter η. The latter two variables are defined as

$$C_Q = \frac{e^2 q Q}{h} \qquad (3)$$

$$\eta = \frac{V_{xx} - V_{yy}}{V_{zz}} \qquad (4)$$

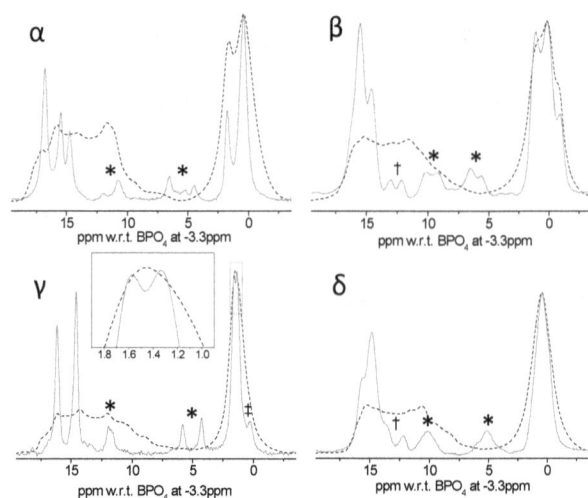

Figure 1. ^{11}B NMR spectra for the four barium borate polymorphs. Solid (red) lines: DOR spectra taken at 20 T. Dashed (black) lines: MAS spectra taken at 14·1 T. Spectra are arbitrarily scaled such that the largest B^{IV} peaks are the same height for both DOR and MAS spectrum of each material. Asterisks denote the second order spinning sidebands from the outer DOR rotor. The outer rotor spinning frequencies were 1400, 1200, 1400 and 1300 Hz for the α to δ phases respectively. † denote peaks due to barium metaborate impurity and ‡ an unknown impurity. Inset highlights the resolution gained under DOR for the B^{IV} manifold of the γ-phase

where e is the electron charge, Q the electric quadrupole moment of the nucleus, h Planck's constant and $eq=V_{zz}$ the largest component of the diagonal EFG tensor in its principal axis system. This parameterisation is conventional with $|V_{zz}| \geq |V_{yy}| \geq |V_{xx}|$. Satisfaction of Laplace's equation requires that $\Sigma V_{ii}=0$ such that Equations (3) and (4) uniquely define the EFG tensor at the nucleus. Note however that only the magnitude of C_Q is determined experimentally from a powder spectrum, and by definition $0 \leq \eta \leq 1$, with $\eta=0$ corresponding to axial symmetry.

Experimental details

Crystallisation of barium diborate glass

Polycrystalline barium borates were all prepared by devitrification of nominally stoichiometric barium diborate, $BaO.2B_2O_3$, glasses. A number of routes to devitrification, as well as to vitrification, were explored:

- α-BaB_4O_7 was obtained as follows. Initially a melt, batched to yield 10 g of glass, was created by heating from room temperature, at 600°C/h, 6·728 g $BaCO_3$ (Alfa Aesar 99·0–101·0%) and 4·775 g B_2O_3 (99·62 at% ^{11}B enriched, Eagle Picher) glass. The vitreous B_2O_3 had previously been treated to remove absorbed water. This process consisted of melting at 1000°C, in a large 9Pt–Rh crucible, approximately 50 g of material, quenching the base of the crucible into water and immediately storing the glass obtained in a desiccator, after breaking it into large chunks, as required. The barium borate melt was quenched in a similar manner, from 1050°C, but was not removed from the crucible prior to devitrification at 715°C for 20 h.

- β-BaB$_4$O$_7$ was obtained in a similar manner to that described by Polyakova & Pevzner.[13] BaCO$_3$ (Sigma Aldrich 99+%) and B$_2$O$_3$ (Alfa Aesar 99·98%) were combined and mixed in sufficient quantity to yield 20 g of Ba diborate glass. An additional 0·1 mol% Fe$_2$O$_3$ (Aldrich 99+%) was added. The mixture was heated at 600°C/h to 1250°C and held for 1 h before being poured into a steel mould to yield a glass block of approximate dimensions 15×15×20 mm. The block was subjected to twice repeated crystallisation-melting cycles with the intention of removing any dissolved gases (CO$_2$). The first consisted of 900°C/h heating to 715°C, holding for 1·5 h, 900°C/h heating to 1250°C, holding for 0·5 h, and quenching the base into water. The second was the same, except that the latter hold was extended to 54 min and the melt was poured into a mould. The resulting glass block was then annealed by placing in a furnace at 550°C (approx. 50°C below the glass transition temperature) for 5·1 h and cooled to room temperature at 300°C/h. Devitrification was performed by a two stage nucleation–growth process, heating at 300°C/h to 610°C, holding for 17 h and then heating again at 300°C/h to 710°C, holding for 4 h before finally cooling to room temperature at 300°C/h. The resultant polycrystalline material was visibly not phase pure. The bulk consisted of dark material (due to the presence of iron), whilst on the surface a material white in appearance was obtained. These were manually separated and the surface forming phase was found to have a distinct powder diffraction pattern (Figure 2), and shall be referred to for convenience as the γ-phase.

- The same γ-phase was obtained in relatively pure form by following the synthesis route of the α-BaB$_4$O$_7$ with the following differences. Natural isotopic abundance B$_2$O$_3$ (Alfa Aesar 99·98%) was used and the barium diborate melt was allowed to cool naturally upon removal from the furnace, as opposed to being quenched in water. The glass thus obtained in the base of the crucible was heated to 715°C at 300°C/h and held for 48 h before cooling in the furnace at 100°C/h. The resulting polycrystalline material is that studied in the experiments described in this paper.

- A further material, distinguished by its powder diffraction pattern (Figure 2) was produced by devitrification of finely powdered glass and

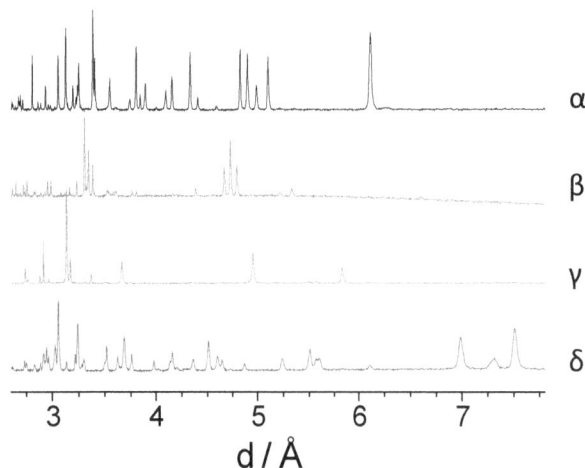

Figure 2. X-ray diffraction powder patterns for the four barium borate polymorphs as a function of planar spacing d. Top to bottom, α- to δ-phases

shall be referred to as the δ-phase. The synthesis route was again similar to that for the α-BaB$_4$O$_7$ with the following differences. 0·1 mol% Fe$_2$O$_3$ (Aldrich 99+%) was added to the batch, and the barium diborate melt was obtained by heating at 900°C/h to 1250°C. The glass was knocked out of the crucible and ground in an agate pestle and mortar to a fine powder before being returned to the crucible and devitrified by heating to 815°C at 300°C/h, holding for 20 h and cooling at 300°C/h to room temperature.

NMR experiments

^{11}B MAS and DOR NMR experiments were performed at three fields using the following spectrometers: a Brucker Avance III 500 MHz (11·75 T); a Brucker Avance II+ 600 MHz (14·1 T); and a Brucker Avance III 850 MHz (20·0 T), employing either dedicated MAS or DOR probes, the latter constructed in Tallinn, Estonia. MAS and DOR probes used at 850 MHz gave a boron background signal (significant for the Brucker 4mm MAS) which was subtracted from the spectra. No boron background signal was present when using a Varian T3 3·2 mm probe for MAS at 600 MHz. The rotors were spun around the magic angle at frequencies of 12 or 12·5 kHz for MAS experiments at 14·1 and 20·0 T, respectively. DOR rotors consisted of an inner ~3·2 mm rotor set at 30·56° to the axis of a ~9·4 mm outer rotor, itself set at the magic angle to the applied field. Outer rotor spinning frequencies ranged between 1200 and 1600 Hz, chosen to minimise overlap of second order spinning sidebands and the centrebands. First order sideband suppression was employed,[20] which relies upon synchronisation of the signal acquisition with the inner rotor position. The inner rotor frequency was a factor of 4·5 to 5 times that of the outer, this is a function of the moments of inertia of the inner rotor[21] as well as how well the respective centres of mass of the two rotors were aligned. Short (1–4 μs), central transition selective ('soft') radio frequency pulses were used throughout, along with relaxation delays deemed long enough to give quantitative spectra, from 1 to 128 s. [11]B chemical shifts of the MAS spectra collected at 14·1 T are referenced to BTE (chemical formula $BF_3.O(CH_2CH_3)_2$) via the secondary reference BPO_4 which was spun under MAS conditions and taken to have δ_{iso}=−3·3ppm.[22] All other spectra are internally referenced to the B^{IV} resonances in the MAS spectra, this being deemed to result in smaller uncertainties as compared to independent referencing. In the case of DOR spectra, Equation (2) shows that there will be a quadrupole induced shift of the B^{IV} resonances, estimated to be $\delta_{QIS} \leq 0·3$ ppm at 11·75 T, $\leq 0·2$ ppm at 14·1 T and $\leq 0·1$ ppm at 20·0 T (C_Q=0·5 MHz, η=1), such that the systematic uncertainty is smallest at higher field.

X-ray powder diffraction patterns were collected using Cu $K_{\alpha 1}$ radiation and a Panalytical X-Pert Pro MPD, with samples spun to remove in-plane preferred orientation effects.

Results

Figure 2 shows the distinct x-ray powder diffraction patterns obtained for the samples of α-BaB_4O_7, β-BaB_4O_7 and the two new compounds, a γ-phase and δ-phase. MAS NMR spectra taken at 14·1 T are shown overlaid with the DOR NMR spectra taken at 20·0 T in Figure 1. DOR NMR experiments on α-BaB_4O_7 at two additional fields (14·1 and 11·75 T) were performed to allow extraction of the quadrupolar parameter P_Q and isotropic chemical shift δ_{iso} for the B^{III} sites. Figure 3 shows δ_{DOR} as a function of $10^{16}/\nu_0^2$ which, from Equations (1) and (2), can be seen to allow extraction of P_Q and δ_{iso}, from the gradient and infinite field ($1/\nu_0^2$=0) extrapolation, respectively, of a least squares linear fit to the data. The results from this analysis are presented in the first three columns of Table 1, and were used to constrain the fits of second order quadrupolar powder lineshapes to MAS NMR spectra taken at 20·0 and 14·1 T, Figure 4, with results also shown in Table 1. Note that, of the four B^{III} sites

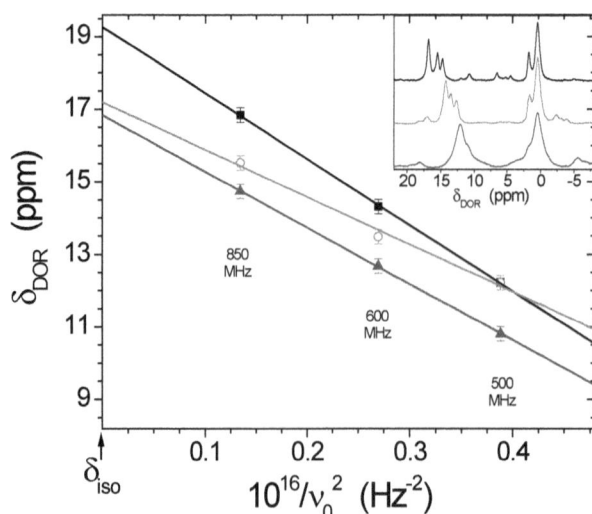

Figure 3. Field dependency of the three resolved B^{III} resonance DOR shifts for α-BaB_4O_7. The three δ_{DOR} values show linear correlation with $1/\nu_0^2$, solid lines are least-square fits to the data. Inset: DOR spectra taken at the three fields, from which the points in the main plot were extracted by pseudo-Voigt peak fitting of the B^{III} region. Outer rotor frequencies were 1400, 1600 and 1400 Hz for the spectra taken on 850, 600 and 500 MHz spectrometers, respectively

known to exist in α-BaB_4O_7, two of their resonances remain unresolved at all three fields under DOR. This indicates that they not only have very similar δ_{iso}, but quadrupolar parameters as well. Therefore, a single powder lineshape was used to simulate these two sites in the fits. The γ-phase material has two clearly resolved B^{III} resonances under DOR, Figure 1. This relative simplicity allowed fitting of the 20·0 and 14·1 T MAS NMR spectra without further constraint on the quadrupolar parameters. The results of such fits are shown in Figure 5 and Table 2. Also shown in Table 2 is the predicted δ_{DOR} at 20·0 T using the fit results along with Equations (1) and (2), which is compared with the measured values in the final column.

The fraction N_4, of B^{IV} to total boron (B^{IV}+B^{III}) can be extracted by integration of the MAS spectra in Figure 1. All four spectra have $0·50 \leq N_4 \leq 0·54$ and the corresponding crystal structures are therefore all thought to contain equal numbers of B^{III} and B^{IV} sites. The integrated intensities should ideally be corrected

Table 1. Extracted NMR parameters for α-BaB_4O_7. The results from fitting to the MAS spectra at two fields, Figure 4, were the same within experimental uncertainty. Quadrupole parameters for the B^{IV} sites are small and were not determined

Multifield DOR			DOR informed fit to MAS at 850 MHz			
Site	δ_{iso} (ppm)	P_Q (MHz)	δ_{iso} (ppm)	P_Q (MHz)	C_Q (MHz)	η
	±0·3	±0·2	±0·2	±0·1	±0·1	±0·1
B_{III}(1)	19·3	2·7	19·2	2·8	2·8	0·1
B_{III}(2)	17·2	2·3	17·4	2·5	2·5	0·0
B_{III}(3)	16·8	2·5	16·4	2·5	2·5	0·3
B_{IV}(1)	1·7	n.d.	1·7	≤0·5	≤0·5	n.d.
B_{IV}(2)	0·5	n.d.	0·4	≤0·5	≤0·5	n.d.

Table 2. Extracted NMR parameters for B^{III} sites in γ-BaB_4O_7. The results from fitting to the MAS spectra at two fields, Figure 5, were the same within experimental uncertainty. The penultimate column gives the predicted DOR shift using the δ_{iso} and P_Q values from the fits. The final column gives the measured DOR shift at 20·0 T

MAS fitting parameters				Calculated	Measured
Site	δ_{iso} (ppm)	P_Q (MHz)	C_Q (MHz)	η	δ_{DOR} (ppm)	δ_{DOR} (ppm)
	±0·2	±0·1	±0·1	±0·1	±0·3	±0·2
B_{III}(1)	18·6	2·7	2·7	0·2	16·2	16·2
B_{III}(2)	17·0	2·8	2·7	0·3	14·5	14·6

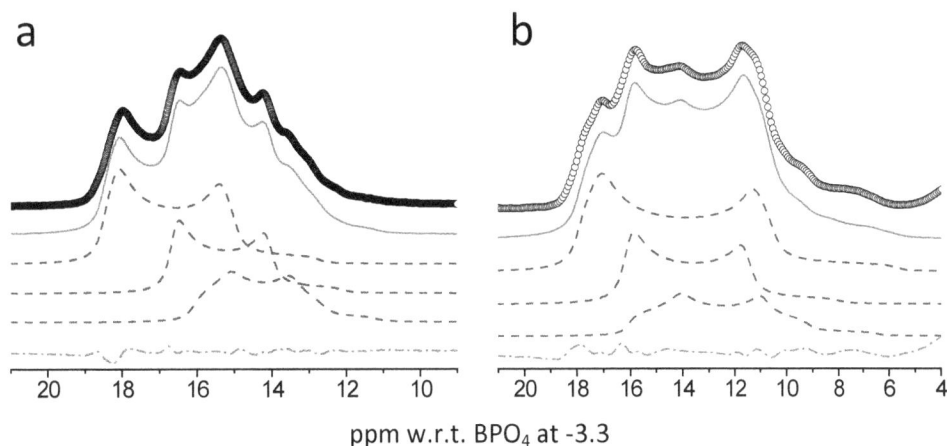

Figure 4. MAS spectra (black open circles, upper) of α-BaB_4O_7 at (a) 20·0 T and (b) 14·1 T, showing the B^{III} regions only. Solid (red) lines offset below the data are the results of least-squares fitting using second order quadrupolar broadened powder lineshapes. Three peaks, shown offset for clarity as dashed (blue) lines, were used to simulate the experimental spectra, using the fact that the two sites unresolved under DOR have both similar δ_{iso} and P_Q. The fits were otherwise unconstrained. Dash-dot lines (magenta) are residuals

(Massiot *et al*[23]) to account for the spinning sideband intensities, however, this requires knowledge of the quadrupole parameters. Due to the fact these are not known in all cases, and the presence of impurity phases, this correction is not attempted, although, using the experimental Larmor and MAS frequencies, typical quadrupole parameters for B^{III} and B^{IV} sites would yield a reduction in N_4 of about 0·03.

Discussion

Polymorphism of barium diborate

The preparation of polycrystalline materials by the crystallisation of barium diborate glass has previously been reported to be extremely sensitive to the presence of dissolved gases and surface preparation of the glass.[13] In the present study it has been found that four distinct polycrystalline materials can be ob-

tained, in contrast to the two obtained by Polyakova & Pevzner.[13] The differences probably arise from the particular impurities present, and possible changes in glass composition due to preferential volatilisation from the melt, as well as absorbed water in the B_2O_3 feedstock. Note also that the different phases were obtained by a number of routes, not just those described above for the particular samples used in the NMR experiments. It was found that particular devitrification routes did not always result in the same crystalline product, due to the sensitivity of the system to parameters not under experimental control. However, the following general trends were observed:

- α-BaB_4O_7 could be obtained from either monolithic or powdered glass upon holding at 715°C, with the latter resulting in a much smaller average crystallite size.

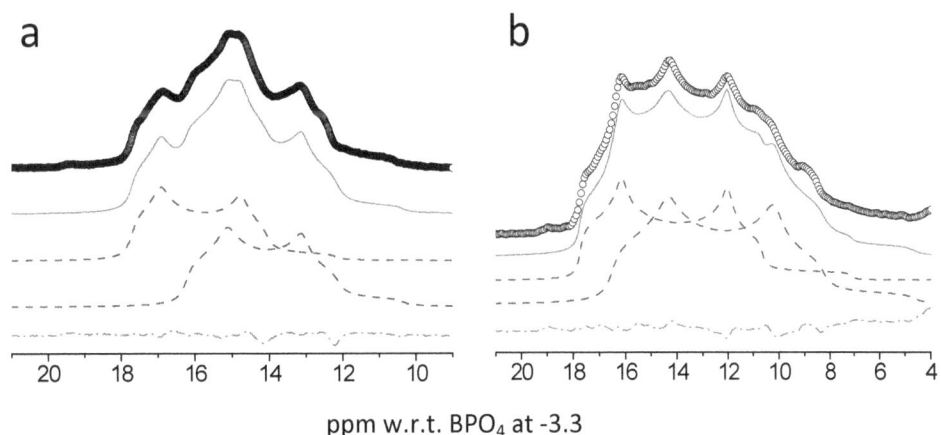

Figure 5. MAS spectra (black open circles, upper) of γ-phase barium borate at (a) 20·0 T and (b) 14·1 T, showing the B^{III} regions only. Solid (red) lines offset below the data are the results of least-squares fitting using second order quadrupolar broadened powder lineshapes. Individual peaks are shown offset for clarity as dashed (blue) lines. Dash-dot lines (magenta) are residuals

- β-BaB$_4$O$_7$ was only ever obtained from monolithic glass blocks. The nucleation period at 610°C is thought to be essential, with its omission resulting in formation of other phases. Differential thermal analysis (DTA) of the glass, once melted into a DTA crucible, cooled through the glass transition and reheated would result in crystallisation of the β-BaB$_4$O$_7$, as found by Polyakova & Pevzner.[13] However, we have also found that the crystallisation event observed in DTA is much larger when an additional 0·1 mol% Fe$_2$O$_3$ was present, indicating that the iron impurities act as nucleation sites for the β-BaB$_4$O$_7$.

- The γ-phase was obtained from the as-quenched glass in the base of the crucible, or, more often, in coexistence with either the β-BaB$_4$O$_7$, i.e. on the surface of a monolithic sample, or with the δ-phase, in a devitrified powder.

- The δ-phase was usually observed to form in powdered glass samples, held at either 715°C or 815°C, presumably requiring surface defects for nucleation. On one occasion the δ-phase was formed in a monolithic glass block containing 0·1 mol% Fe$_2$O$_3$. Glass batches using boric acid precursor, as opposed to vitreous B$_2$O$_3$ were powdered and devitrified at both 715°C and 815°C and yielded δ-phase material, although these typically contained a higher proportion of γ-phase, as identified by x-ray powder diffraction.

It is clear that the system is sensitive to many variables often assumed extraneous, including the glass powder grain size. The question arises as to whether the γ- and δ-phases are barium diborate polymorphs, or if their stoichiometry is in fact different. Were the stoichiometry different, it would probably be barium-rich since B$_2$O$_3$ is more often preferentially volatilised from the melt and, being hygroscopic, any water present would lead to a deficit in boron oxide in the pre-melt mixture. The most recent equilibrium phase diagram[24] shows that the adjacent congruently melting compounds to barium diborate are the metaborate (BaB$_2$O$_4$) and the dibarium pentaborate (Ba$_2$B$_{10}$O$_{17}$). Bragg peaks corresponding to the Ba$_2$B$_{10}$O$_{17}$ phase[25] were never observed in any of the samples devitrified from barium diborate glass, although the material Ba$_2$B$_{10}$O$_{17}$ was easily obtained upon devitrification of its stoichiometric glass powder at 815°C. β-BaB$_2$O$_4$[26] was identified as a minor phase in the powder diffraction of the β-BaB$_4$O$_7$ sample. The structure[26] contains two boron sites, both B^{III}, and these are apparent in the ^{11}B DOR NMR spectra for the β-BaB$_4$O$_7$, *as well as* for the δ-phase in Figure 1 at δ_{iso}=13·0 and 12·2 ppm. This supports the supposition that glasses were marginally barium-rich. However, considering that the metaborate (50 mol% BaO) has 16·67 mol% BaO more than the diborate, and that both α- and β-BaB$_4$O$_7$ were obtained by crystallisation of glasses, it follows

that the γ- and δ-phases are likely barium diborate polymorphs or else have slightly Ba-rich stoichiometries. This is supported by DTA measurements on powdered glass samples, which did not reveal any endothermic feature corresponding to the liquidus of the metaborate field which rises sharply from the eutectic point at approximately 36 mol% BaO.[24] Finally, there is little evidence for residual glass in any of the diffraction, DTA or NMR measurements made on crystalline samples. This question is not discussed further in this work, and requires further measurements either to determine the stoichiometries directly, or to elucidate the inter-relationships of the various phases, for example using thermal analysis and high temperature x-ray diffraction. However, it is worth noting that the peritectically melting 4SrO.7B$_2$O$_3$ compound (36·36 mol% SrO) has recently been found to exist in the SrO–B$_2$O$_3$ system,[27] although no compound of this stoichiometry has thus far been reported for the barium borate system.

NMR results

The primary purpose of this contribution is to demonstrate the applicability of the DOR NMR technique to ^{11}B in crystalline borates. The huge increase in resolution that can be gained under DOR is clearly evident in Figure 1 for the B^{III} resonances in the $10 \lesssim \delta_{iso} \lesssim 19$ ppm regions of the spectra. It is known that α-BaB$_4$O$_7$ contains four crystallographically distinct B^{III} sites, and their associated nuclear magnetic resonances are resolved into three spectral peaks under DOR, with two sites remaining unresolved. The four B^{IV} sites are split into two spectral peaks with 1:3 intensity ratio, both under MAS and DOR. It can be noted that the resolution, even of the lower C_Q B^{IV} sites, is better under DOR, which is true also for MAS and DOR spectra obtained at the same field. This point will be discussed below. Figure 1 also shows DOR spectra for the other three polymorphs, all of which show marked resolution enhancements. Peak widths are typically 0·5 ppm for the α-BaB$_4$O$_7$ and as low as 0·37 ppm for the γ-phase B^{III} sites, but are noticeably broader for the β-BaB$_4$O$_7$ and δ-phase due to the presence of paramagnetic Fe impurities. The B^{III} peak separation is smaller for these materials (at 20·0 T). Again there is an apparent resolution enhancement in moving from MAS to DOR even for the B^{IV} sites, where the B^{IV} manifold for the γ-phase, which appears as a single feature under MAS, is split into two contributions under DOR, with peak separation of only 0·25 ppm (68 Hz). This phenomenon may be attributable to the small but finite EFGs at the B^{IV} site nuclei, which, at 20·0 T, could lead to broadening of up to $(7/24) \times (\nu_Q^2 / \nu_0)$=0·2 ppm ($C_Q$=0·5 MHz, η=1),[2] consistent with the observed difference between MAS and DOR B^{IV} peak widths. The observed narrowing can be used to estimate an upper bound on

B^{IV} site quadrupole parameters, but not to accurately measure them due to the possible presence of other broadening mechanisms including: multiple strong non-commuting homonuclear dipolar couplings[28] between ^{11}B nuclei,[29] anisotropy of the bulk magnetic susceptibility, which is only partially removed by MAS[30] and dipole–quadrupole cross-term interactions.[31,32]

The δ_{QIS} for the B^{IV} sites are smaller than the uncertainty in the chemical shift scale in this study, and hence do not provide any constraint on the B^{IV} $P_{Q}s$. On the other hand, the δ_{QIS} for the B^{III} sites are much larger, typically of order 3 to 4 ppm at 20·0 T and 5 to 10 ppm at 11·75 T. ^{11}B DOR spectra of α-BaB$_4$O$_7$ taken at three fields are shown in the inset of Figure 3 and, from Equations (1) and (2), the expected linear correlation between δ_{DOR} and $1/\nu_0^2$ has been used in the main part of the figure to allow extraction of P_Q from the gradient of the fitted function and δ_{iso} from the infinite field extrapolation. The extracted values are given in Table 1. It is clear, from both the figure and tabulated values, that the P_Q parameters are different for the individual sites, with the site of intermediate δ_{iso} having the smallest $P_Q=2·3\pm0·2$ MHz and the two unresolved sites with largest δ_{iso} also having the largest $P_Q=2·7\pm0·2$ MHz. This results in a crossing point at around 11·75 T ($10^{16}/\nu_0^2=0·3885$ Hz^{-2}) such that a DOR spectrum at lower field would show their peak positions in reversed order. The information extracted from the ^{11}B DOR NMR has been used to constrain the fitting of the MAS NMR spectra, as shown in Figure 4. Clearly such a fit would have been impossible without the δ_{iso} and P_Q values from DOR which were used to initiate the fit. Furthermore, the fact that two sites have *both* similar δ_{iso} and P_Q values could not have been known *a priori* without the DOR experiments, and one would have had to fit not three, but four lineshapes to the MAS spectrum. Of course we have not allowed for different $[C_Q,\eta]$ possibilities for the two sites as the difference is likely to be small with $P_Q(\eta=1)/P_Q(\eta=0)\approx1·15$ for a given C_Q.

The ^{11}B DOR NMR spectrum for the γ-phase, shown in Figure 1, displays two narrow B^{III} resonances, as well as two closely separated B^{IV} resonances, indicating that there are a minimum of four crystallographically distinct boron sites in the unit cell. This relative simplicity with respect to α-BaB$_4$O$_7$ allows an attempt to be made to fit the MAS spectra, as shown in Figure 5, without the use of multiple experiments at different fields. The results are given in Table 2 and, despite the presence of a small amount of an unidentified impurity phase leading to a less good fit than in the α-BaB$_4$O$_7$ case, indicate that the two B^{III} sites have similar quadrupole parameters C_Q and η. Equations (1) and (2) have been used to calculate δ_{DOR} at 20·0 T from the fit results (Table 2, penultimate column) and these agree, within experimental uncertainty, with the measured δ_{DOR} (Table 2, final column). One can

speculate upon the network structure of the γ-phase barium borate, where the existence of only four boron sites limits greatly the possible borate superstructural units present, assuming that any are present, as is the case in the vast majority of M^+ and M^{2+} binary crystalline borates.[17] Possible solutions are the diborate ($B_4O_5\text{Ø}_4^{2-}$) unit alone, as found in Li,[33] Mg,[34] Zn[35] and Cd[36] diborates. In the case of a single diborate unit, the high symmetry means that the four sites of the diborate superstructural unit comprise only two crystallographically distinct sites. The latter three cases are isostructural with each other, but do not appear to be isostructural with the γ-phase Ba borate, judging by the powder diffraction patterns. The remaining options are a triborate unit ($B_3O_3\text{Ø}_4^-$) accompanied by an isolated tetrahedron ($B\text{Ø}_4^-$) as for the β-SrB$_4$O$_7$[18] compound or, a ditriborate unit ($B_3O_3\text{Ø}_5^{2-}$) accompanied by a neutral isolated $B\text{Ø}_3$ triangle, which has not been observed in any M_2O–B_2O_3 or MO–B_2O_3 crystal structure to date. β-SrB$_4$O$_7$ is not found to be isostructural with the γ-phase barium borate. Further experiment is required in order to extract more information, and, in the absence of suitable single crystals for diffraction, DOR NMR can be further exploited toward this goal. In particular, 2D homonuclear correlation experiments, performed under DOR, could provide information regarding the relative proximities of individual boron sites, as demonstrated for aluminium sites by ^{27}Al spin-diffusion DOR NMR in 9Al$_2$O$_3$.2B$_2$O$_3$.[9] A more difficult, but potentially more rewarding experiment involves double quantum homonuclear dipolar recoupling (DQDOR), and differs from the spin-diffusion experiment in that not only dipolar coupling between different sites can be observed, but also that between spatially proximate sites that are crystallographically identical. DQDOR had been demonstrated for ^{23}Na ($I=3/2$) and ^{27}Al,[37] and under MAS for ^{11}B in lithium diborate[38] where it was clearly demonstrated that the B^{III} sites are surrounded by B^{IV} sites alone, which is also the case in the cadmium diborate structure class.

The amount of information available from the 1D ^{11}B DOR NMR experiments on β-BaB$_4$O$_7$ and the δ-phase (Figure 1) is less than for the other two materials. This is due to the lower resolution (paramagnetic impurity), presence of impurity phases, and small separation of the resonances. This situation could be improved by synthesising samples without the addition of paramagnetic ions. Furthermore, the collection of ^{11}B DOR NMR spectra at other field strengths could not only shed light on the quadrupole parameters, but also on the number of sites present. The latter point would only be applicable if sites with different P_Q were present, which would lead to splitting of any unresolved (at 20·0 T) B^{III} resonances at other fields. Finally, the application of 2D homonuclear correlation experiments, performed under DOR, could allow assignment of the α-BaB$_4$O$_7$ resonances to the known structural sites, and elucidate the connectivity of the boron sites in the other three phases.

Although a detailed comparison is outside of the scope of this paper, it should be noted that the popular and more readily accessible multiple-quantum MAS (MQMAS) technique[39,40] can be used to yield much the same information as that extracted using the combination of DOR and MAS employed here, albeit in an intrinsically two-dimensional, and therefore time-consuming, experiment. However, the combined DOR and MAS approach is fully quantitative and has been shown to give a much better signal-to-noise than MQMAS.[41] Of course, the choice of method, in the end, will always depend upon the experimental questions posed and the hardware available.

It should be further noted that MQDOR, which in a single experiment separates out δ_{iso} and δ_{QIS}, is feasible (albeit with loss of sensitivity and quantification) and has successfully been applied, for example, to ^{11}B in vitreous B_2O_3.[42]

Summary and conclusions

^{11}B DOR NMR has been shown to yield extremely high resolution spectra in polycrystalline borates owing to the time-averaging of second order effects including the second order quadrupolar interaction. Quadrupole parameters for the four B^{III} sites in α-BaB_4O_7 were extracted by performing DOR NMR experiments at three fields. The high resolution and quantitative nature of the technique allows the number of sites to be determined in materials of undetermined structure, as demonstrated here for the γ-phase barium borate. As such, ^{11}B DOR NMR may be used as a tool to help answer structural questions in borate crystal chemistry, such as those posed by Wright[17] at the previous borates conference,[43] with the great potential of 2D ^{11}B homonuclear correlation experiments under DOR yet to be exploited.

Acknowledgements

This work was funded by the STFC Centre for Materials Physics and Chemistry under Grant CMPC09105 and the EPSRC. The UK 850 MHz solid-state NMR Facility used in this research was funded by EPSRC and BBSRC, as well as the University of Warwick including via part funding through Birmingham Science City Advanced Materials Projects 1 and 2 supported by Advantage West Midlands (AWM) and the European Regional Development Fund (ERDF). J. V. Hanna is thanked for access to the 500 and 600 MHz spectrometers.

References

1. Kroeker, S. & Stebbins, J. F. Three-coordinated boron-11 chemical shifts in borates. *Inorg. Chem.*, 2001, **40**(24), 6239–6246.
2. Mackenzie, K. J. D. & Smith, M. E. *Multinuclear solid state NMR of inorganic materials*. Pergamon Materials Series, Ed. R. W. Cahn. Vol. 6. 2002.
3. Samoson, A., Lippmaa, E. & Pines, A. High-Resolution Solid-State NMR Averaging of 2nd-Order Effects by Means of a Double-Rotor. *Mol. Phys.*, 1988, **65**(4), 1013–1018.
4. Hung, I., Howes, A. P., Parkinson, B. G., Anupõld, T., Samoson, A., Brown, S. P., Harrison, P. F., Holland, D. & Dupree, R. Determination of the bond-angle distribution in vitreous B_2O_3 by B-11 double rotation (DOR) NMR spectroscopy. *J. Solid State Chem.*, 2009, **182**(9), 2402–2408.
5. Howes, A. P., Vedishcheva, N. M., Samoson, A., Hanna, J. V., Smith, M. E., Holland, D. & Dupree, R. Boron environments in Pyrex (R) glass-a high resolution, Double-Rotation NMR and thermodynamic modelling study. *Phys. Chem. Chem. Phys.*, 2011, **13**(25), 11919–11928.
6. Bull, L. M., Bussemer, B., Anupõld, T., Reinhold, A., Samoson, A., Sauer, J., Cheetham, A. K. & Dupree, R. A high-resolution O-17 and Si-29 NMR study of zeolite siliceous ferrierite and ab initio calculations of NMR parameters. *J. Am. Chem. Soc.*, 2000, **122**(20), 4948–4958.
7. Wong, A., Howes, A. P., Pike, K. J., Lemaître, V., Watts, A., Anupõld, T., Past, J., Samoson, A., Dupree, R. & Smith, M. E. New limits for solid-state O-17 NMR spectroscopy: Complete resolution of multiple oxygen sites in a simple biomolecule. *J. Am. Chem. Soc.*, 2006, **128**(24), 7744–7745.
8. Wong, A., Howes, A. P., Yates, J. R., Watts, A., Anupõld, T., Past, J., Samoson, A., Dupree, R. & Smith, M. E. Ultra-high resolution ^{17}O solid-state NMR spectroscopy of biomolecules: A comprehensive spectral analysis of monosodium L-glutamate.monohydrate. *Phys. Chem. Chem. Phys.*, 2011, **13**(26), 12213–12224.
9. Hung, I., Howes, A. P., Anupõld, T., Samoson, A., Massiot, D., Smith, M. E., Brown, S. P. & Dupree, R. Al-27 double rotation two-dimensional spin diffusion NMR: Complete unambiguous assignment of aluminium sites in $9Al_2O_3.2B_2O_3$. *Chem. Phys. Lett.*, 2006, **432**(1–3), 152–156.
10. Klyuev, V. P. & Pevzner, B. Z. Thermal expansion of some borate and silicate compounds in glassy and crystalline states. *Phys. Chem. Glasses*, 2004, **45**(2), 146–148.
11. Klyuev, V. P. & Pevzner, B. Z. Thermal expansion of $BaO.2B_2O_3$ in the vitreous and crystalline states: II. Thermal expansion of polycrystalline barium diborate prepared by crystallization of glass powder of the same composition. *Glass Phys. Chem.*, 2005, **31**(5), 630–636.
12. Pevzner, B. Z., Klyuev, V. P. & Polyakova, I. G. Thermal expansion of $BaO.B_2O_3$ in the vitreous and crystalline states: I. Thermal expansion and density of barium diborate prepared from monolithic glass of the same composition. *Glass Phys. Chem.*, 2005, **31**(5), 621–629.
13. Polyakova, I. G. & Pevzner, B. Z. Crystallization of barium diborate glass. *Glass Phys. Chem.*, 2005, **31**(2), 138–144.
14. Filatov, S. K., Nikolaeva, N. V., Bubnova, R. S. & Polyakova I. G. Thermal expansion of beta-BaB_2O_4 and BaB_4O_7 borates. *Glass Phys. Chem.*, 2006, **32**(4), 471–478.
15. Pevzner, B.Z., Klyuev, V. P., Polyakova, I. G. & Borodzyulya, V. F. Peculiar properties of barium diborate polycrystals. *Phys. Chem. Glasses: Eur. J. Glass Sci. Technol. B*, 2006, **47**(4), 534–537.
16. Block, S. & Perloff, A. The Crystal Structure of Barium Tetraborate, $BaO.2B_2O_3$. *Acta Crystallogr.*, 1965. **19**, 297–300.
17. Wright, A. C. Borate structures: crystalline and vitreous. *Phys. Chem. Glasses: Eur. J. Glass Sci. Technol. B*, 2010, **51**(1), 1–39.
18. Vasiliev, A. D., Cherepakhin, A. V. & Zaitsev, A. I. The trigonal polymorph of strontium tetraborate, beta-SrB_4O_7. *Acta Crystallogr. E*, 2010, **66**, I48–U132.
19. Knyrim, J.S., Römer, S. R., Schnick, W. & Huppertz, H. High-pressure synthesis and characterization of the alkaline earth borate beta-BaB_4O_7. *Solid State Sci.*, 2009, **11**, 336–342.
20. Samoson, A. & Lippmaa, E. Synchronized Double-Rotation NMR-Spectroscopy. *J. Magn. Reson.*, 1989, **84**(2), 410–416.
21. Dupree, R. Double Rotation NMR, In: *Encyclopedia of Magnetic Resonance* DOI: 10.1002/9780470034590.emrstm1203, Eds R. K. Harris & R. E. Wasylishen, John Wiley: Chichester, 2011.
22. Turner, G. L., *et al* B-11 Nuclear-Magnetic-Resonance Spectroscopic Study of Borate and Borosilicate Minerals and a Borosilicate Glass. *J. Magn. Reson.*, 1986, **67**(3), 544–550.
23. Massiot, D., *et al* A Quantitative Study of Al-27 MAS NMR in Crystalline YAG. *J. Magn. Reson.*, 1990, **90**(2), 231–242.
24. Hovhannisyan, R. M. Binary alkaline earth borates: phase diagram correction and low thermal expansion of crystallised stoichiometric glass compositions. *Phys. Chem. Glasses: Eur. J. Glass Sci. Technol. B*, 2006, **47**(4), 460–464.
25. Hubner, K.-H. Ueber die Borate $2BaO.5B_2O_3$, tief-$BaO.B_2O_3$, $2BaO.B_2O_3$ und $4BaO.B_2O_3$. *Neues Jahrb. Mineral.*, 1969, 335–343.

26. Lu, S. F., Huang, Z. X. & Huang, J. L. Meta-barium borate, II-BaB$_2$O$_4$, at 163 and 293 K. *Acta Cryst. C*, 2006, **62**, 173–175.

27. Polyakova, I. G. & Litovchik, E. O. Crystallization of glasses in the SrO-B$_2$O$_3$ system. *Glass Phys. Chem.*, 2008, **34**(4), 369–380.

28. Zorin, V. E., Brown, S. P. & Hodgkinson, P. Origins of linewidth in H-1 magic-angle spinning NMR. *J. Chem. Phys.*, 2006, **125**(14), 144508.

29. Barrow, N. S., Yates, J. R., Feller, S. A., Holland, D., Ashbrook, S. E., Hodgkinson, P. & Brown, S. P. Towards homonuclear J solid-state NMR correlation experiments for half-integer quadrupolar nuclei: experimental and simulated ^{11}B MAS spin-echo dephasing and calculated ^{2}J$_{BB}$ coupling constants for lithium diborate. *Phys. Chem. Chem. Phys.*, 2011, **13**(13), 5778–5789.

30. Alla, M. & Lippmaa, E. Resolution Limits in Magic-Angle Rotation NMR-Spectra of Polycrystalline Solids. *Chem. Phys. Lett.*, 1982, **87**(1), 30–33.

31. Wi, S., Frydman, V. & Frydman, L. Residual dipolar couplings between quadrupolar nuclei in solid state nuclear magnetic resonance at arbitrary fields. *J. Chem. Phys.*, 2001, **114**(19), 8511–8519.

32. Wi, S. & Frydman, L. Residual dipolar couplings between quadrupolar nuclei in high resolution solid state NMR: Description and observations in the high-field limit. *J. Chem. Phys.*, 2000, **112**(7), 3248–3261.

33. Radaev, S.F., *et al* Atomic-Structure and Electron-Density of Lithium Tetraborate Li$_2$B$_4$O$_7$. *Kristallografiya*, 1989, **34**(6), 1400–1407.

34. Bartl, H. & Schuckmann, W. Zur Struktur des Magnesiumdiborats, MgO(B$_2$O$_3$)$_2$. *Neues Jahrb. Mineral.*, 1966, 142–148.

35. Martinez-Ripoll, M., Martinez-Carrera, S. & Garcia Blanco, S. Crystal Structure of Zinc Diborate, ZnB$_4$O$_7$. *Acta Cryst. B*, 1971, **B27**(Mar), 672–677.

36. Ihara, M. & Krogh-Moe, J. The crystal structure of cadmium diborate, CdO.2B$_2$O$_3$. *Acta Cryst.*, 1966, **20**, 132–134.

37. Brinkmann, A., Kentgens, A. P. M., Anupõld, T. & Samoson, A. Symmetry-based recoupling in double-rotation NMR spectroscopy. *J. Chem. Phys.*, 2008, **129**(17), 174507.

38. Barrow, N. S. *Homonuclear Correlation in Solid-State NMR: Developing Experiments for Half-Integer Quadrupolar Nuclei*. PhD Thesis, Department of Physics, University of Warwick, Coventry, 2009.

39. Medek, A., Harwood, J. S. & Frydman, L. Multiple-quantum magic-angle spinning NMR: A new method for the study of quadrupolar nuclei in solids. *J. Am. Chem. Soc.*, 1995, **117**(51), 12779–12787.

40. Frydman, L. & Harwood, J. S. Isotropic Spectra of Half-Integer Quadrupolar Spins from Bidimensional Magic-Angle-Spinning NMR. *J. Am. Chem. Soc.*, 1995, **117**(19), 5367–5368.

41. Kanellopoulos, J., Freude, D. & Kentgens, A. A practical comparison of MQMAS techniques. *Solid State Nucl. Magn.*, 2007, **32**(4), 99–108.

42. Hung, I., Wong, A., Howes, A. P., Anupõld, T., Samoson, A., Smith, M. E., Holland, D., Brown, S. P. & Dupree, R. Separation of isotropic chemical and second-order quadrupolar shifts by multiple-quantum double rotation NMR. *J. Magn. Reson.*, 2009, **197**(2), 229–236.

43. *Borate Glasses, Crystals and Melts*, Eds N. Umesaki & A. C. Hannon, Himeji, Japan, 2008, Society of Glass Technology, Sheffield, 2010.

Phys. Chem. Glasses: Eur. J. Glass Sci. Technol. B, October 2012, **53** (5), 191–204

A neutron diffraction study of $2M_2O.5B_2O_3$ (M=Li, Na, K, Rb, Cs & Ag) and $2MO.5B_2O_3$ (M=Ca & Ba) glasses

Adrian C. Wright,[1] *Roger N. Sinclair, Cora E. Stone, Joanna L. Shaw*

J.J. Thomson Physical Laboratory, University of Reading, Whiteknights, Reading, RG6 6AF, UK

Steven A. Feller, T. J. Kiczenski, Richard B. Williams, Heidi A. Berger

Physics Department, Coe College, Cedar Rapids, IA 52402, USA

Henry E. Fischer

Institut Laue-Langevin, 6 Rue Jules Horowitz, B.P. 156, 38042 Grenoble Cedex 9, France

Natalia M. Vedishcheva

Institute of Silicate Chemistry of the Russian Academy of Sciences, Nab. Makarova 2, St. Petersburg, 199034, Russia

Manuscript received 24 February 2012
Manuscript accepted 15 March 2012

A neutron diffraction investigation has been performed to investigate the effect of the network modifying cation size and charge on the structure of a series of binary borate glasses of composition $2M_2O.5B_2O_3$ (M=Li, Na, K, Rb, Cs & Ag) and $2MO.5B_2O_3$ (M=Ca & Ba). The diffraction patterns were recorded using a twin-axis diffractometer, and Fourier transformed to yield the real-space correlation function, $T(r)$. The first peak in the corrected, normalised diffraction pattern, $I(Q)$, is employed to estimate the spacing between the network-modifying cations occupying adjacent network cages, and the fraction of four-fold co-ordinated boron atoms, x_4, is obtained from the area under the first (B–O) peak in $T(r)$. With the exception of $2CaO.5B_2O_3$, and within the experimental uncertainty, no evidence is found to indicate the presence of significant numbers of nonbridging oxygen atoms. The way in which the distribution of superstructural unit species present varies with the network modifying cation is investigated by comparison with theoretical calculations based on the model of associated solutions. Finally, conclusions are drawn in respect of the first M(O) co-ordination shell and the important role of both the network modifying cation and superstructural units in determining the structure of binary alkali and alkaline earth borate glasses.

1. Introduction

To date, structural studies of alkali and alkaline earth borate glasses have mainly concentrated on the borate network, and especially on the fraction, x_4, of the boron atoms in four-fold co-ordination and the role played by superstructural units.[1,2] The detailed environment of the network modifying cations is frequently neglected, even though the differences between the structures and properties of the various M_2O–B_2O_3 and MO–B_2O_3 systems, and in the way they vary with the network modifier mole fraction, x_M, are *entirely* due to the network modifying cations and their influence on the borate network! The present paper, therefore, will examine the effect of the network modifying cation on the structure of a series of alkali, silver and alkaline earth borate glasses of the di-pentaborate composition $2M_2O.5B_2O_3$ (M=Li, Na, K, Rb, Cs & Ag) and

$2MO.5B_2O_3$ (M=Ca & Ba). As may be seen from Refs 1 and 2, this composition ($x_M=^2/_7=0.286$) corresponds to the point beyond which NMR data indicate that x_4 starts to deviate from the value

$$x_4=R=x_M/(1-x_M) \qquad (1)$$

where R is the network modifier:B_2O_3 ratio ($R=0.4$ for the present glasses); i.e. the maximum network-modifier content without significant numbers of nonbridging oxygen atoms (NBO).

It is important to realise that the network modifying cations do not passively occupy cages formed by the surrounding vitreous network, but that they modify their local environment to suit their own particular requirements, as determined by their charge, ionic radius, r_M, and polarisability. Thus the minimum size of the cage is determined by their ionic radius, the cage size governs the local network packing efficiency, and the requirement for efficient packing influences the distribution of (super)structural unit species present.

---

[1] Corresponding author. Email a.c.wright@reading.ac.uk
Original version presented at VII Int. Conf. on Borate Glasses, Crystals and Melts, Dalhousie University, Halifax, Nova Scotia, Canada, 21–25 August 2011

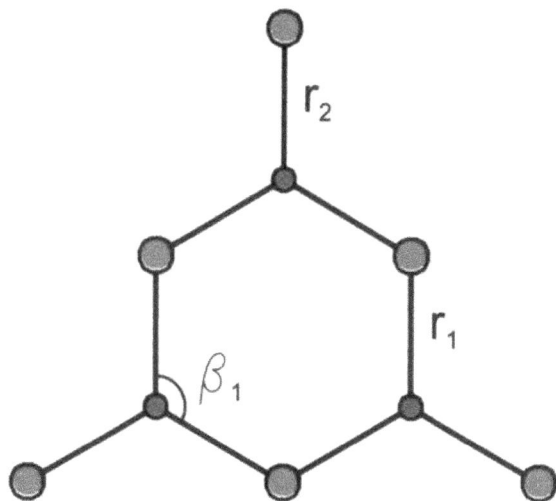

Figure 1. Definition of the bond lengths and angles in a boroxol group. The boron atoms are shown in cyan and the oxygen atoms in red [in colour online]

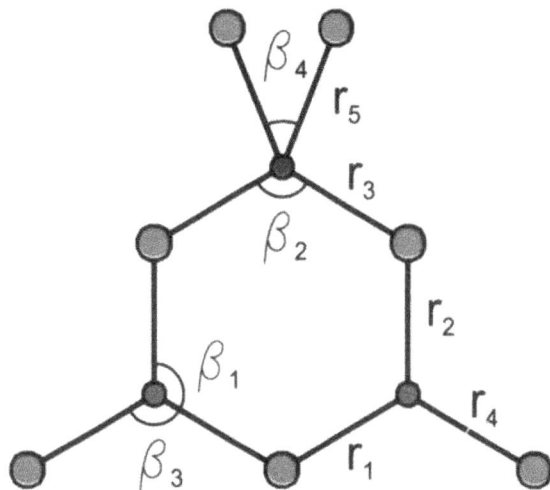

Figure 2. Definition of the bond lengths and angles in a triborate group. Cyan, three-fold co-ordinated boron; blue, four-fold co-ordinated boron and red, oxygen atoms [in colour online]

2. Superstructural units

The superstructural units considered in this paper are the boroxol $\{B_3O_3\varnothing_3^{(1\,Fig.\,7(A))}\}$, pentaborate $\{B_5O_6\varnothing_4^{-(1\,Fig.\,8(A))}\}$, triborate $\{B_3O_3\varnothing_4^{-(1\,Fig.\,8(B))}\}$ and diborate $\{B_4O_5\varnothing_4^{2-(1\,Fig.\,9(B))}\}$ groups. The so-called tetraborate group is not a superstructural unit, since its component pentaborate and triborate groups are linked via a single shared oxygen atom with variable bond and torsion angles. Hence, when modelling the real space neutron total correlation function, $T(r)$,[2] it will be considered as a pentaborate plus a triborate group. In addition to superstructural units, the modelling of $T(r)$ will involve three independent (i.e. not part of a superstructural unit) basic borate structural units, viz. $B\varnothing_3$, $B\varnothing_4^-$ and $B O\varnothing_2^-$. Note that the convention used here for the (super)structural unit formulæ follows that of Wright and co-workers,[1,2] in that the symbol $\varnothing$ is reserved for those bridging oxygen atoms that are shared between adjacent (super)structural units. The fractions of the boron atoms involved in each superstructural unit are denoted x_B $(B_3O_3\varnothing_3)$, x_P $(B_5O_6\varnothing_4^-)$, x_T $(B_3O_3\varnothing_4^-)$ and x_D $(B_4O_5\varnothing_4^{2-})$, whilst the fraction of the boron atoms involved in the independent basic structural units are given by x_{4I} $(B\varnothing_4^-)$, x_{3I} $(B\varnothing_3)$ and x_{2I} $(BO\varnothing_2^-)$.

Whereas it is well known that the average B–$\varnothing$ bond length for a $B\varnothing_4^-$ tetrahedron (~1·47 Å) is longer than that for a $B\varnothing_3$ triangle (~1·37 Å), it is not generally realised that the situation for superstructural units is much more complicated, in that there are systematic variations in both bond lengths and angles,[4] which it is necessary to take into account when analysing diffraction data. Figures 1 to 3 specify the independent parameters (bond lengths r_i and bond angles, β_i) for symmetrised versions[5] of the boroxol, triborate and diborate groups, whilst the pentaborate

group comprises two triborate groups sharing a common $B\varnothing_4^-$ tetrahedron, and so the number of independent parameters is two less than that for a triborate group, since r_3 and r_5, and β_2 and β_4, are symmetrically equivalent. The values for the independent (super)structural unit parameters employed for the modelling in Section 6.2 have been estimated from those for (super)structural units in crystalline alkali borates, as summarised in Wright et al,[5] and are given in Table 1. The interatomic distances within each of the superstructural units are summarised in Figure 4, the height at each distance being proportional to the contribution to the relevant component correlation function, $t_{ij}(r)$.

3. Experimental procedure and results

The powdered samples (Table 2) were prepared in a dry box at Coe College, using boric acid enriched in

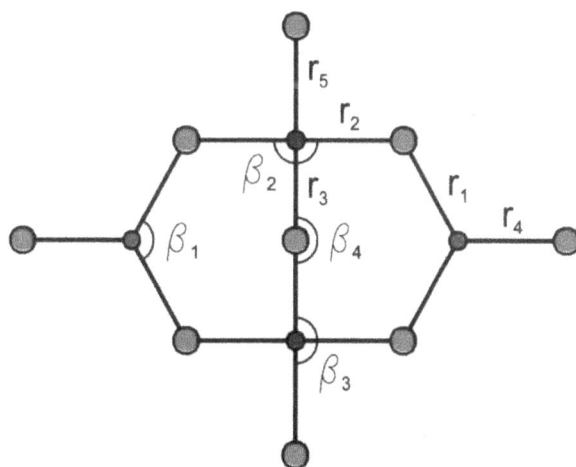

Figure 3. Definition of the bond lengths and angles in a diborate group. (Key as Figure 2) [in colour online]

---

[2] Definitions of the various real- and reciprocal-space functions used in this paper can be found in Wright.[3]

Table 1. Structural parameters for the superstructural units, as defined in Figures 1–3, and for the $B\emptyset_3$, $B\emptyset_4^-$ and $BO\emptyset_2^-$ basic structural units. For the $BO\emptyset_2^-$ structural unit, r_1 and r_2 are respectively the B–$\emptyset$ and B–O^- bond lengths, and β_1 is the $\emptyset$–$\hat{B}$–$\emptyset$ bond angle

Unit	r_1 (Å)	r_2 (Å)	r_3 (Å)	r_4 (Å)	r_5 (Å)	β_1 (°)	β_2 (°)	β_3 (°)	β_4 (°)
$B\emptyset_3$	1·368	-	-	-	-	120·00*	-	-	-
$B_3O_3\emptyset_3$	1·378	1·334	-	-	-	119·0	-	-	-
$B_5O_6\emptyset_4^-$	1·387	1·351	1·474	1·361	-	120·5	111·85	116·5	-
$B_3O_3\emptyset_4^-$	1·389	1·347	1·486	1·367	1·463	121·5	110·3	115·5	111·0
$B_4O_5\emptyset_4^{2-}$	1·352	1·495	1·450	1·3835	1·471	122·6	108·75	115·9	108·5
$B\emptyset_4^-$	1·467	-	-	-	-	109·47*	-	-	-
$BO\emptyset_2^-$	1·400	1·324	-	-	-	115·88	-	-	-

* Fixed by symmetry

[11]B (Li, Na, K, Cs, Ca & Ba, 99·98%; Rb & Ag 99·27%) and, in the case of the $2Li_2O.5B_2O_3$ sample, lithium carbonate enriched in [7]Li (99·94%). Their mass densities were obtained by helium pycnometry, and used to calculate the average number densities, ρ°, given in Table 2. All of the samples except vitreous $2Ag_2O.5B_2O_3$, which was much less water-sensitive, were sealed in 5·2 mm internal diameter thin walled (0·1 mm) vanadium sample cans at Coe College before shipment. Weight loss measurements indicated that the uncertainty on R is ±0·02.

The neutron diffraction patterns were recorded using the D4b diffractometer on the high flux reactor at the Institut Laue-Langevin at an incident neutron wavelength, λ, of ~0·5 Å (0·5027±0·0001 Å for Li, Na, K, Rb, Cs & Ba; 0·5031±0·0001 Å for Ag and 0·5017 ±0·0001 Å for Ca), the composite raw diffraction pattern, $I^E(Q)$ for vitreous $2Na_2O.5B_2O_3$ being shown in Figure 5 (upper pattern) together with a cubic spline fit, which was extrapolated to zero scattering vector magnitude, Q, using a polynomial fit in Q^2. In addition to the normal corrections[6] for the experimental background, scattering from the sample can, neutron absorption, multiple scattering and self-shielding, a small correction was also made for {0·02% (Ca) to 0·07% (Cs)} 'water', using the data of Beyster[7] for liquid water. This latter correction was necessary due

Table 2. Sample data

Sample	$r_M^{(14,15)}$ (Å)	r_{M-O}* (Å)	$\bar{b}_M$ (10^{-12} cm)	[11]B enrichment	ρ° (atom Å^{-3})
$2Li_2O.5B_2O_3$	0·59	1·94	−0·222	99·98%	0·10150
$2Na_2O.5B_2O_3$	1·02	2·37	0·363	99·98%	0·09176
$2K_2O.5B_2O_3$	1·46	2·81	0·367	99·98%	0·07939
$2Rb_2O.5B_2O_3$	1·60	2·95	0·709	99·27%	0·07221
$2Cs_2O.5B_2O_3$	1·78	3·13	0·542	99·98%	0·06559
$2Ag_2O.5B_2O_3$	1·15	2·50	0·5922	99·27%	0·09026
$2CaO.5B_2O_3$	1·10	2·45	0·470	99·98%	0·09219
$2BaO.5B_2O_3$	1·50	2·85	0·507	99·98%	0·08749

* Based on an anionic radius for O^{2-} of 1·35 Å[14]

to the large incoherent scattering cross-section for hydrogen, which contributes to the self-scattering and therefore affects the magnitude of the interference function, $Q_i(Q)$, and hence of the resulting differential correlation function, $D(r)$. Since the samples were in the form of powders, the corrected diffraction patterns were normalised to the calculated self-scattering using the Krogh-Moe[8]–Norman[9] method. The final corrected and normalised diffraction pattern, $I(Q)$, for the $2Na_2O.5B_2O_3$ sample is shown as the lower curve in Figure 5, together with the self-scattering and the contribution from the 'water', the data be-

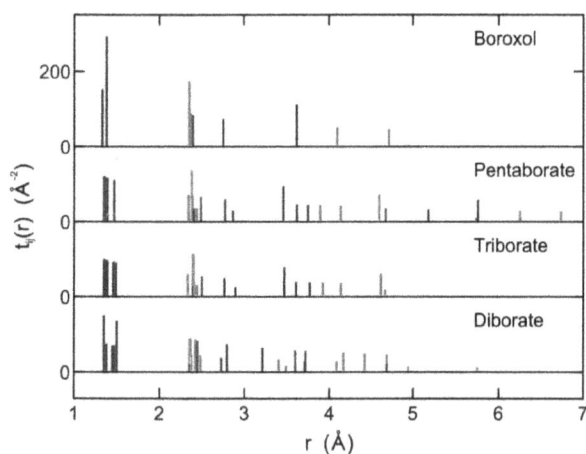

Figure 4. Contributions, $t_{ij}(r)$, to the real space correlation function from the four superstructural units, normalised to one structural unit. B–O distances are in blue, O–O in red and B–B in green [in colour online]

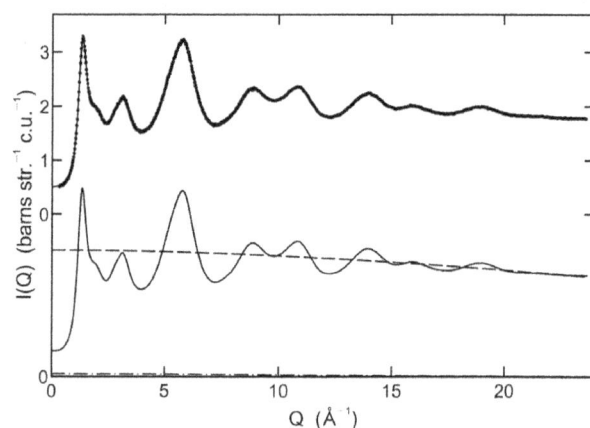

Figure 5. Upper points/curve: composite raw diffraction pattern (points) and cubic spline fit (solid line) for vitreous $2Na_2O.5B_2O_3$. Lower curves: Corrected, normalised diffraction pattern for the same sample, together with the self-scattering (dashed line) and 'water' correction (dash-dotted line) (The patterns are displaced vertically, and the successive ordinate zeros denote the origin for each pattern. The ordinate scale, which is consistent throughout, is indicated for the top pattern.)

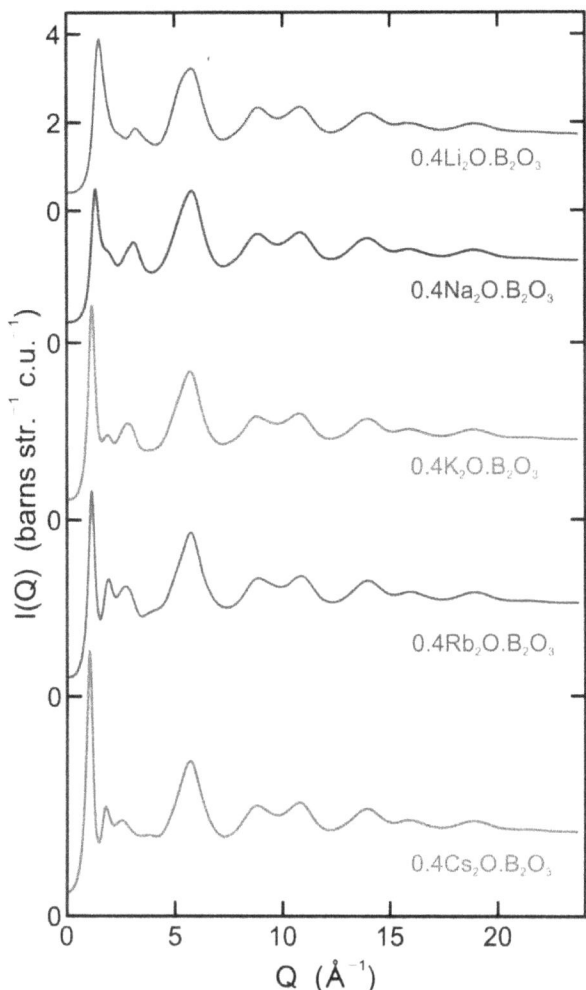

Figure 6. Corrected normalised diffraction patterns for the alkali borate glasses. Green, $2Li_2O.5B_2O_3$; blue, $2Na_2O.5B_2O_3$; red, $2K_2O.5B_2O_3$; cyan, $2Rb_2O.5B_2O_3$ and magenta, $2Cs_2O.5B_2O_3$ (See note for Figure 5.) [in colour online]

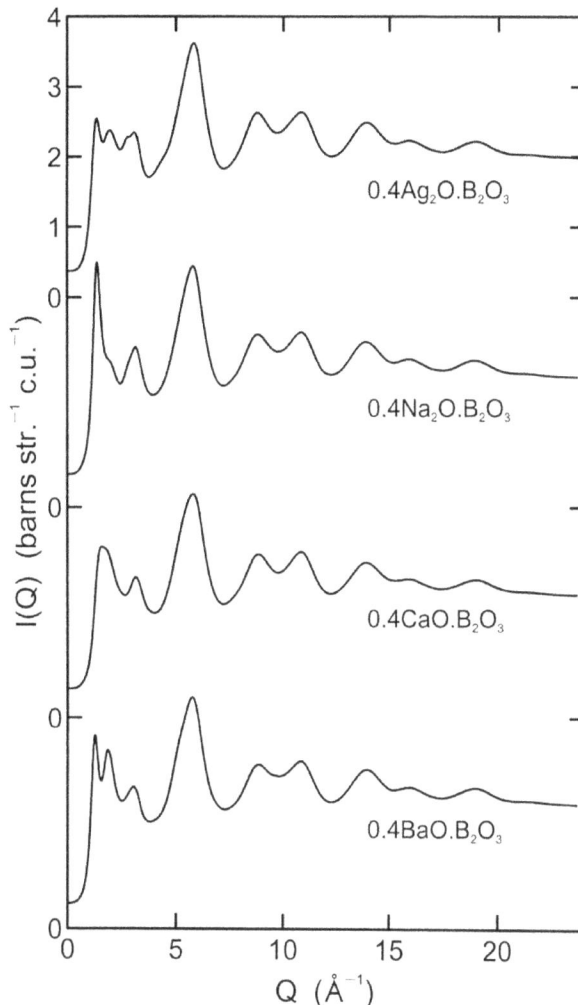

Figure 7. Corrected normalised diffraction patterns for the silver, sodium and alkaline earth borate glasses (See note for Figure 5.)

ing normalised to a $0.4Na_2O.B_2O_3$ composition unit. Neutron scattering lengths for the network modifying cations, $\bar{b}_M$, are given in Table 2, and are taken from Sears[10], the value for lithium being for a ^7Li enrichment of 99·94%. The boron scattering lengths were calculated to be 0.6648×10^{-12} cm (99·98% Enrichment) and 0.6594×10^{-12} cm (99·27% Enrichment), and that for oxygen is 0.5803×10^{-12} cm.

The interference functions, $Q_i(Q)$, calculated from the corrected, normalised diffraction patterns, were Fourier transformed to yield the real-space correlation functions $D(r)$ and $T(r)$, using Filon's quadrature[11] (an improved version of Simpson's quadrature) and the Lorch[12] modification function, the maximum scattering vector magnitude, Q_{max}, being 23·68 Å^{-1}. The full width at half maximum (fwhm) of the corresponding peak function, $P(r)$, which defines the experimental resolution in real space, is 0·230 Å. The corrected, normalised diffraction patterns for all eight samples are shown in Figures 6 (alkali borates) and 7 (silver, sodium and alkaline earth borates), and the

corresponding real-space total correlation functions, $T(r)$, in Figures 8 and 9.

4. Diffraction patterns

The most obvious differences between the diffraction patterns occur below the peak at ~5·8 Å^{-1}, and especially in the region of the first diffraction peak. The latter has been associated with the periodicity associated with adjacent network cages[13] and hence gives information concerning the average (M–M) distance between the network modifying cations that occupy adjacent cages. Note, however, that this may not be the shortest M–M distance, since more than one network modifying cation may be present in the same (highly non-spherical) cage. Above the peak at ~5·8 Å^{-1}, the diffraction patterns reflect the well-defined distances within the borate network {i.e. within the (super)structural units} and are very similar.

A Lorentzian fit[13] has been performed to the first diffraction peak for all eight glasses, that for vitreous $2Rb_2O.5B_2O_3$ being shown in Figure 10. The

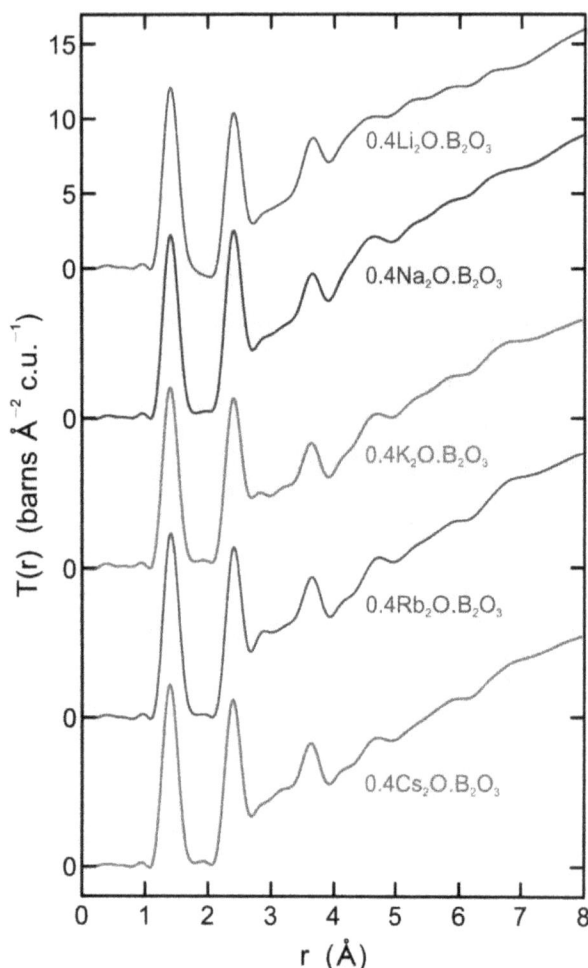

Figure 8. Real space total correlation functions for the alkali borate glasses (Key as Figure 6) (See note for Figure 5.) [in colour online]

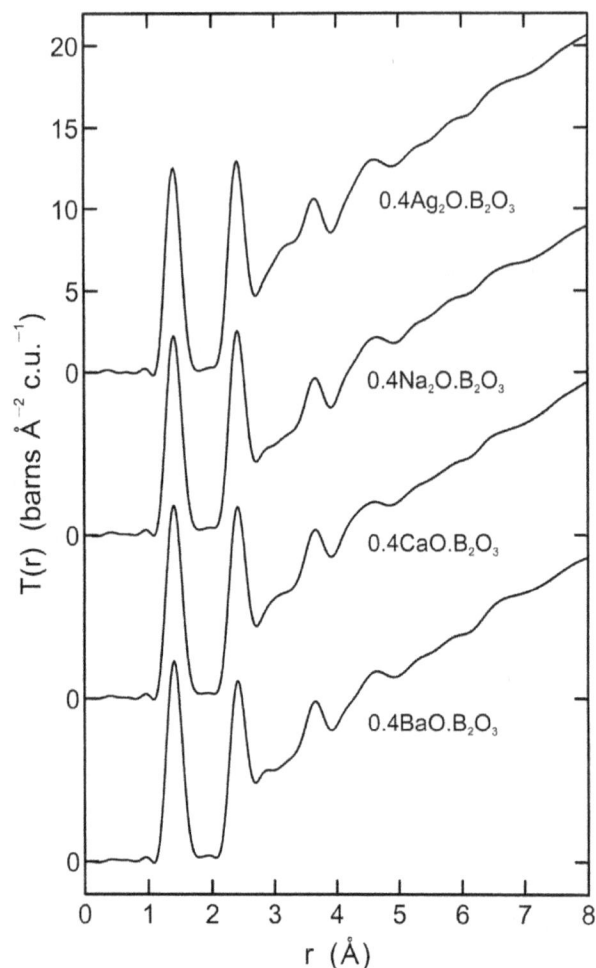

Figure 9. Real space total correlation functions for the silver, sodium and alkaline earth borate glasses (See note for Figure 5.)

resulting peak positions, Q_1, and the corresponding M–M distances, r_{M-M}. are summarised in Table 3, the uncertainties reflecting the quality of the fit. The best fits were obtained for the larger monovalent cations, K^+, Rb^+ and Cs^+, since these force the cages to expand and therefore increase the cage size beyond that occurring naturally in a borate network. The fit for Li^+ was not very good, since the Li^+ cation is well within the natural distribution of cage sizes, as may be seen from the discussion of network number density in Wright and coworkers.[1-2] The Na^+ and Ag^+ cations, too, though larger than Li^+, still have a limited effect on the cage-size distribution. In the case of the alkaline earth glasses, there is a factor of two less cations, and so the first diffraction peak is less prominent. In addition, the Ca^{2+} cationic radius is very similar to that for Na^+ (Table 2), which means that the influence of the Ca^{2+} cation on the cage-size distribution will also be limited.

If the thickness of the walls of the cages containing the network modifying cations is independent of the cation species, a plot of r_{M-M} versus the cation radius, r_M, should yield a straight line of gradient 2·0, with an intercept equal to twice the wall thickness. Cationic

radii have been tabulated by Shannon & Prewitt[14,15] as a function of co-ordination number, $n_{M(O)}$, and so the optimum values for r_M in Table 2 were estimated from the co-ordination numbers found in the corre-

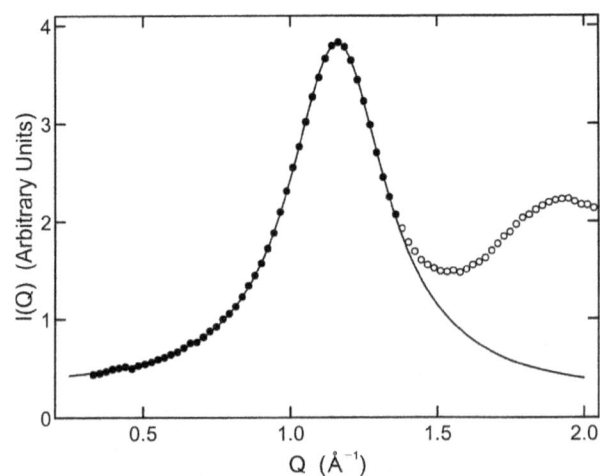

Figure 10. Lorentzian fit to the first diffraction peak for vitreous $2Rb_2O.5B_2O_3$. Circles, experimental data and solid line, fit. The fitting range is indicated by the closed circles

Table 3. Fits to the first diffraction peak

Sample	Q_1 (Å^{-1})	r_{M-M} (Å)
$2Li_2O.5B_2O_3$	1·513±0·020	4·15±0·06
$2Na_2O.5B_2O_3$	1·358±0·010	4·63±0·04
$2K_2O.5B_2O_3$	1·215±0·005	5·17±0·02
$2Rb_2O.5B_2O_3$	1·159±0·005	5·42±0·02
$2Cs_2O.5B_2O_3$	1·094±0·005	5·75±0·03
$2Ag_2O.5B_2O_3$	1·301±0·030	4·83±0·11
$2CaO.5B_2O_3$	1·479±0·030	4·25±0·09
$2BaO.5B_2O_3$	1·253±0·010	5·01±0·04
B_2O_3	1·583±0·020	3·97±0·05

sponding crystalline structures[1] in the composition range between triborate and diborate ($\frac{1}{4} \leq x_M \leq \frac{1}{3}$). The resulting plot can be seen in Figure 11. The straight line least squares fit through the points for K$^+$, Rb$^+$ and Cs$^+$ yields a gradient of 1·797±0·008 and the intercept indicates a 'cage-wall' thickness of 2·546±0·012 Å, which is slightly larger than the O–O distances within the basic borate structural units. (2·37 Å for BØ$_3$ and 2·40 Å for BØ$_4^-$). The fact that the gradient is rather less than 2·0 indicates that the 'cage-wall' thickness decreases slightly with increasing r_M. The data points for Ca^{2+} and Ba^{2+} lie below the straight line for the fit, and a straight line through these points indicates a reduced 'cage-wall' thickness of 2·15 Å, although the gradient (1·91) is closer to 2·0. Note that the M–M spacings (inter-cationic distances) derived from the position of the first diffraction peak are not at all obvious from the correlation functions of Figures 8 and 9.

The width of the first diffraction peak (fwhm), ΔQ_1, is inversely proportional to the associated correlation length L_{10}, which denotes the point at which the correlation has fallen to 10% of that at the origin cage wall. Similarly, the quantity L_{10}/r_{M-M}, gives information concerning the average number of adjacent cages containing a network modifying cation. For K$^+$, Rb$^+$ and Cs$^+$, ΔQ_1 is respectively 0·414, 0·402 and 0·394 (±0·005) Å^{-1}, leading to values for L_{10} of 11·1, 11·4 and 11·7 (±0·1) Å, corresponding to just over two adjacent M$^+$-containing cages.

5. Real space correlation functions

The real space total correlation functions, $T(r)$, for the eight glasses are also very similar and, to examine the relatively small differences, the differential correlation functions, $D(r)$, for the five alkali borate glasses, together with that for vitreous B_2O_3, are over-plotted in Figure 12. {The B_2O_3 sample is the same as that in Ref. 16, but the diffraction pattern was remeasured as part of the present series of experiments (λ=0·5027±0·0001 Å).} The first peak at ~1·4 Å corresponds to the B–O bond length and, for the alkali borate glasses, is shifted to higher r, relative to that for vitreous B_2O_3, due to the presence of BØ$_4^-$ tetrahedra. The second peak at ~2·4 Å includes contributions from the O–O distances within the BØ$_3$ and BØ$_4^-$ basic structural units, the B–B distances between adjacent structural units and, for the smaller

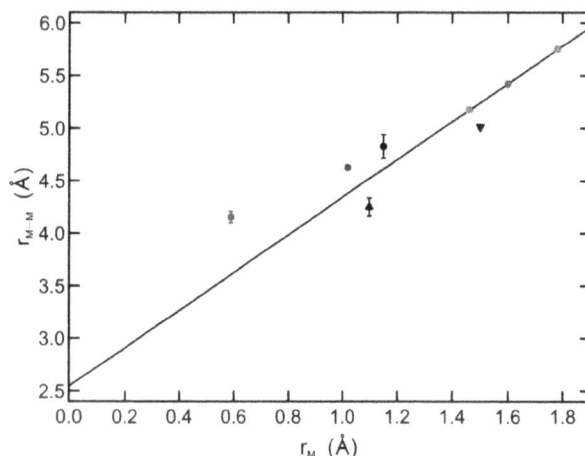

Figure 11. Plot of the cation–cation periodicity, r_{M-M}, versus cation radius, r_M [in colour online]

network modifying cations, M–O distances associated with first M(O) co-ordination shell. For the larger network-modifying cations, these M–O distances mainly occur in the region between the peak at ~2·4 Å and that at ~3·6 Å, which is often associated with well defined B–O distances within superstructural units. Further well defined interatomic distances within superstructural units will contribute to $D(r)$ up to the maximum (O–O) spacing within the pentaborate group at 6·73 Å.

5.1. Real space peak fits

The normal method of extracting the details of the basic structural units for a network glass is by performing a fit to the first two peaks in $T(r)$. However, as indicated above, the second peak in $T(r)$ for the present glasses is extremely complex in that it involves contributions from four different interactions, and so the fits were limited to the first peak. A single distance fit for vitreous $2Na_2O.5B_2O_3$ is shown as the upper curves in Figure 13 and, as expected, the fit

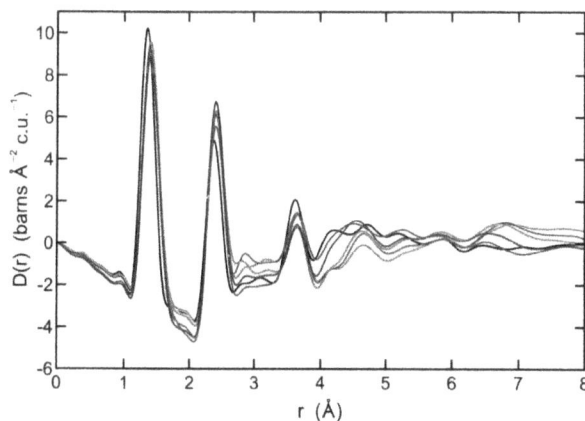

Figure 12. Over-plot of the differential correlation functions, $D(r)$, for the five alkali borate glasses (colours as Figure 6) and vitreous B_2O_3 (black) [in colour online]

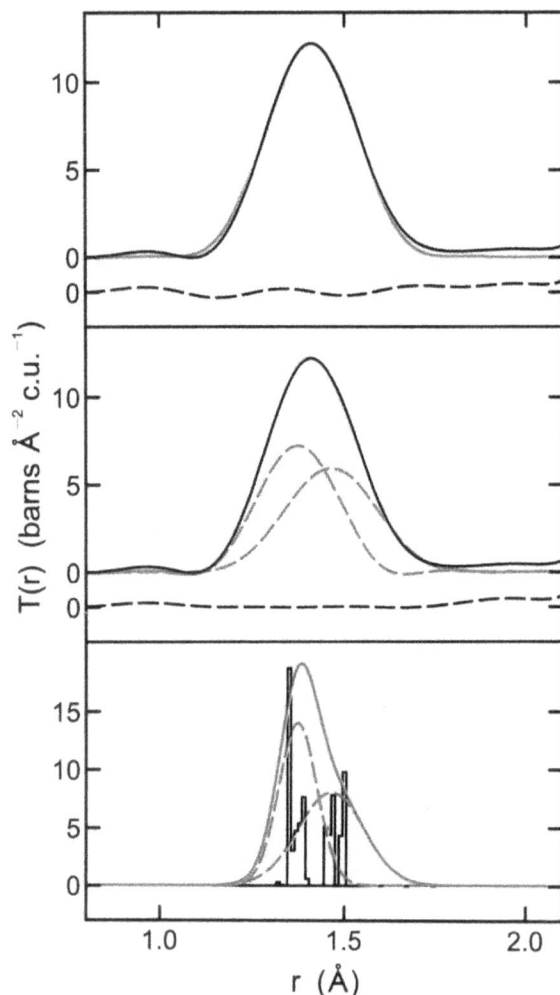

Figure 13. Peak fit to the experimental real space correlation functions, T(r), for vitreous 2Na₂O.5B₂O₃. Top, single peak fit; centre, two-peak fit and bottom, a comparison of the unbroadened fit with the B–O bond lengths predicted by the thermodynamic model in Section 6.2 (histogram). Black solid lines, experiment; red solid lines, fit; dashed red lines, individual peaks and dashed black lines, residual [in colour online]

Table 4. Fit to the first peak in T(r)*. The estimated uncertainty on* $n_{B(O)}$ *is* ±0·05*, whilst those on* r_1 *and* r_2 *are discussed in Section 5.1*

Sample	r_1 (Å)	r_2 (Å)	$n_{B(O)}$	R_χ
2Li₂O.5B₂O₃	1·366	1·474	3·30	0·051
2Na₂O.5B₂O₃	1·377	1·468	3·36	0·011
2K₂O.5B₂O₃	1·383	1·460	3·33	0·010
2Rb₂O.5B₂O₃	1·368	1·475	3·37	0·007
2Cs₂O.5B₂O₃	1·369	1·473	3·31	0·007
2Ag₂O.5B₂O₃	1·364	1·461	3·38	0·004
2CaO.5B₂O₃	1·373	1·464	3·24	0·009
2BaO.5B₂O₃	1·373	1·465	3·34	0·008
B₂O₃	1·366	-	2·90	0·028

$n_{B(O)}$, are given in Table 4, together with the results from similar fits for the other seven glasses. {Note that the higher R_χ factor for the lithium borate sample is almost certainly due to interference between the overlapping wings of the B–O and (negative) Li–O peaks.} However, whilst the parameters listed in Table 4 seem reasonable, a careful examination of the associated rms bond length variations, $\overline{u_1^2}^{1/2}$ and $\overline{u_2^2}^{1/2}$, and co-ordination numbers, n_1 and n_2, for the individual fitted peaks reveals a problem in that some of these values are clearly incorrect.

Assuming the absence of nonbridging oxygen atoms, the total B(O) co-ordination number for the present glasses (x_4=R=0·4) should be 3·4 (3+x_4), with n_1 and n_2 equal to 1·8 and 1·6, respectively, and the rms bond length variations should be of the same order as that for pure vitreous B₂O₃ (0·054 Å). In practice, however, n_2 was always **greater** than n_1, and $\overline{u_1^2}^{1/2}$ and $\overline{u_2^2}^{1/2}$ were significantly larger than 0·054 Å; e.g. for 2Na₂O.5B₂O₃, n_1=1·64, n_2=1·71, $\overline{u_1^2}^{1/2}$=0·052 Å and $\overline{u_2^2}^{1/2}$=0·091 Å. Similarly, for 2Li₂O.5B₂O₃, n_1=1·59 and n_2=1·71. The reason for these discrepancies can be seen from the bottom section of Figure 13, which compares the unbroadened peaks from the fit with the distribution of B–O bond lengths (before thermal and experimental broadening) predicted by the thermodynamic modelling of Section 6.2. Clearly, in the presence of superstructural units, the assumption of just two B–O bond lengths, one for BØ₃ triangles and the other for BØ₄⁻ tetrahedra, is much too simplistic, even though the distances do split into two well-defined groups. Fitting the experimental peak profile with two symmetric Gaussian peaks in the B–O component correlation function, $t_{B–O}(r)$, results in an interchange of area between the two contributions and a consequent effect on their average bond lengths and rms variations.

As may be seen from Figure 13 (centre), the overall two-distance fit to the profile of the first peak in $T(r)$ is very good and therefore the total B(O) co-ordination number should be accurate to within ±0·05. However, as is normally the case, the total B(O) co-ordination numbers are slightly too low and $T(r)$ does not quite oscillate about the abscissa between the first and second peaks. That this is an experimental artefact is demonstrated by the fact that the same effects were

was poor, the real peak being asymmetric, indicating more than one B–O bond length. The average B–O bond length, $r_{B–O}$, root mean square (rms) bond length variation, $\overline{u_{B–O}^2}^{1/2}$ and total B(O) co-ordination number, $n_{B(O)}$, were 1·415 Å, 0·085 Å and 3·34, respectively, and the reliability factor, $R_\chi^{(3)}$, for the fit (1·0≤r≤1·8 Å) was 0·034. The average bond length (1·415 Å) is slightly longer than that calculated for R=0·4 from the BØ₃ and BØ₄⁻ bond lengths in Table 1 (1·408 Å), which is consistent with the fact that the position of the maximum for the fit is displaced to slightly higher r than that for the experimental peak.

A much improved fit to the peak profile (R_χ=0·011) was obtained by assuming different bond lengths for BØ₃ triangles and BØ₄⁻ tetrahedra, as demonstrated by the centre curves of Figure 13, and the two distances, r_1 and r_2, and the total B(O) co-ordination number,

observed for pure vitreous B_2O_3, which yielded a co-ordination number of 2·90 (Table 4), rather than the correct value of 3·0. The problem seems to have its origins in reciprocal space resolution, but that of the D4b diffractometer at the first diffraction peak (fwhm ~0·1 $Å^{-1}$) would only reduce the magnitude of the differential correlation function, $D(r)$, at the first (B–O) peak by ~0·2%, which is much too small an effect to explain the decrease in the co-ordination number. Another possibility was that the quadrature interval used in performing the Fourier transformation ($\Delta Q = 0·02$ $Å^{-1}$) was too large, but reducing this by a factor of 4 had a negligible effect on the magnitude of $D(r)$.

With the exception of $2CaO.5B_2O_3$, $n_{B(O)}$ falls within the range 3·30 to 3·38, with an average value of 3·34. Allowing for the experimental artefact discussed in the previous paragraph, it can therefore be concluded that the experimental data for the remaining seven glasses are consistent with a value of 0·4 for x_4, and hence the absence of significant numbers of non-bridging oxygen atoms. Vitreous $2CaO.5B_2O_3$, on the other hand, yields a significantly lower co-ordination number, which suggests that $BO\varnothing_2^-$ structural units may play a more important role in its structure.

5.2. Borate network and superstructural units

Superstructural units play and important role in the structures of both crystalline and vitreous alkali and alkaline-earth borates up to the diborate composition ($x_M=\frac{1}{3}$).[1] In the case of pure vitreous B_2O_3,[16] it was possible to extract the fraction of the boron atoms involved in boroxol groups, x_B, by fitting the interference function, $Q_i(Q)$. However, the situation for the present glasses is much more complex, due to the fact that there is more than one superstructural unit species present, and also that the other superstructural unit species have many more independent structural parameters (cf. Table 1) and hence well defined interatomic distances. For this reason, it is not possible to extract the fractions of the boron atoms involved in each (super)structural unit directly from the experimental data and so, to investigate the distribution of superstructural units within each glass, it is necessary to resort to modelling.

Relative to the peak at 3·6 Å {B–(3)O distance within the boroxol group}, that for the modified glasses is broadened, and the maximum is shifted

to slightly higher r (~3·7 Å), due to a distribution of superstructural unit species. An inspection of Figure 4 shows that both the pentaborate and triborate groups give rise to B–O distances on either side of the boroxol distance of 3·6 Å whereas, for the diborate group, the extra contributions are biased to higher r. Hence the peak shift to slightly higher r may reflect the presence of significant numbers of diborate groups.

6. Structural modelling

Various models have been proposed to predict the distribution of superstructural unit species in the vitreous state. The Krogh-Moe[17]–Griscom[18] (KMG) model[1,2] is based on the former's melting point depression data for the sodium borate system, and so strictly only applies to this system. Thermodynamic modelling using the model of associated solutions,[19] on the other hand, considers the chemical equilibria in the supercooled melt at the fictive temperature and discusses the structure in terms of chemical groupings, which are assumed to consist of (super)structural unit species similar to those found in related crystalline phases. Thus, from the distribution of chemical groupings in a given glass, it is possible to derive the corresponding distribution of superstructural unit species. Full details of both approaches have been discussed elsewhere,[1,2,19] and will not be repeated here.

6.1. Krogh-Moe–Griscom model

At the $2Na_2O.5B_2O_3$ composition ($x_M=2/7$), the KMG model predicts a structure based solely on tetraborate (i.e. equal numbers of pentaborate and triborate superstructural units) and diborate groups. The fractions of the boron atoms involved in each superstructural unit species are therefore defined by the composition, and are given in Table 5.

6.2. Model of associated solutions

Unfortunately, the modelling was limited to the five alkali borate glasses, because a full set of the necessary thermodynamic input parameters {Gibbs free energies[20,21] (Li, Na & K) or enthalpies[22] (Rb & Cs) of formation of the crystalline phases} was only available for these glasses. Krogh-Moe[23–25] has suggested that the structure of vitreous $Na_2O.2B_2O_3$ is based on

Table 5. Fractions of the boron atoms in (super)structural units, as predicted by the model of associated solutions and the KMG model

Glass	$B\varnothing_3$	$B_3O_3\varnothing_3$	$B_5O_6\varnothing_4^-$	$B_3O_3\varnothing_4^-$	$B_4O_5\varnothing_4^{2-}$	$B\varnothing_4^-$	$BO\varnothing_2^-$
$2Li_2O.5B_2O_3$	0·0081	0·0242	–	0·5180	0·4448	–	0·0049
$2Na_2O.5B_2O_3$	0·0006	0·0018	0·1727	0·3226	0·4887	–	0·0136
$2K_2O.5B_2O_3$	0·0080	–	0·0403	0·5162	0·4313	0·0040	0·0002
$2Rb_2O.5B_2O_3$	0·0210	–	0·1059	0·3780	0·4845	0·0105	0·0001
$2Cs_2O.5B_2O_3$	–	–	0·0006	0·6087	0·3874	–	0·0033
KMG Model	–	–	0·2500	0·1500	0·6000	–	–

interpenetrating networks of diborate groups, rather than incorporating the superstructural units found in crystalline α-Na$_2$O.2B$_2$O$_3$ {di-triborate (NBO) and di-pentaborate groups}. This, combined with the fact that pentaborate and triborate groups are the most commonly found superstructural units in crystalline alkali and alkaline earth borates, suggests that the more unusual superstructural units found in the crystalline state are the result of the imposition of periodicity on the crystalline structure. The present samples correspond to the di-pentaborate composition, and so it might be expected that di-pentaborate groups would play a significant role in their vitreous network, especially since they are reasonably common in the crystalline state.[1,Table 9] However, as discussed in Wright,[1] the di-pentaborate group is 5-connected and thus is not consistent with Zachariasen's criteria[26] as modified for networks containing superstructural units.[1] Similarly, di-pentaborate (NBO) groups would result in too many nonbridging oxygen atoms. Hence in the thermodynamic modelling,[27] it was assumed that the pentaborate, triborate and diborate chemical groupings in the super-cooled liquid state close to the fictive temperature, T_f, are based on the corresponding superstructural units, and not on those appropriate to the crystal.[2,27]

The fractions of the boron atoms involved in the various (super)structural units were calculated from the chemical structure for each glass, and are given in Table 5. Although the uncertainty in the boron atom fractions varies, depending on the system, between ±0·03 and ±0·05, the calculations were performed to four decimal places to ensure an accurate composition and value of x_4. In every case, the number of (super)structural unit species present is greater than for the KMG model, although the major contributions are still from pentaborate, triborate and diborate groups. The absence of pentaborate groups for vitreous 2Li$_2$O.5B$_2$O$_3$ can be explained by the difficulty of packing such large groups around the small Li$^+$ cations. The distance distributions for the superstructural units (Figure 4) and independent basic structural units were added in the proportions indicated by the appropriate boron atom fractions, x_i (Table 5), thermally broadened and then folded with the peak function $P(r)$ (Δr_{fwhm}=0·230 Å). The thermal/disorder broadening for the B–O bond lengths (0·054 Å) and O–O distances (0·082 Å) within the basic structural units were set equal to those for vitreous B$_2$O$_3$ and those at higher r were calculated assuming isotropic rms displacements for the B and O atoms of 0·055 and 0·075 Å, respectively.

7. Discussion

7.1. Borate network

As discussed in Section 5.1, with the exception of vitreous 2CaO.5B$_2$O$_3$, the B(O) co-ordination numbers

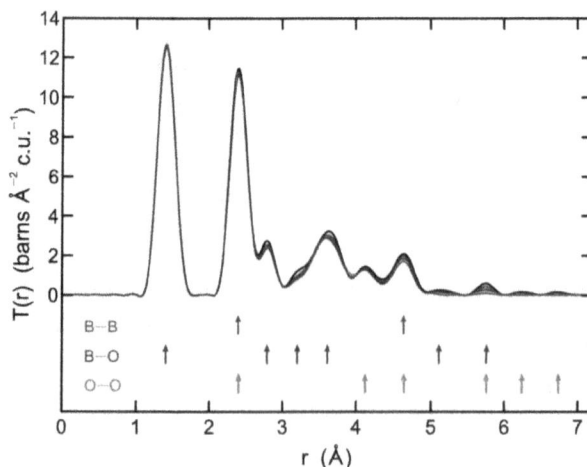

Figure 14. Over-plot of the contributions to T(r) *calculated from the (super)structural unit species distributions in Table 5. The KMG model is shown in black and the colour key for the five alkali borate glasses is the same as in Figure 6. The arrows indicate the components making a significant contribution to each peak [in colour online]*

(Table 4) are consistent with the expected value of 3·4, which indicates that the concentration of nonbridging oxygen atoms must be low, and this is supported by the thermodynamic modelling (Table 5). The highest boron atom fraction predicted for the BØ$_2^-$ structural unit (0·0136) is for vitreous 2Na$_2$O.5B$_2$O$_3$, whilst those for 2K$_2$O.5B$_2$O$_3$ and 2Rb$_2$O.5B$_2$O$_3$ are less than 0·001. All of these values are less than the uncertainty of ±0·03 to ±0·05 estimated from those on the thermodynamic input parameters.

In order to ascertain the sensitivity of $T(r)$ to the distribution of (super)structural unit species, their contributions to the correlation functions for the five alkali borate glasses, calculated from the predicted fractions in Table 5, are over-plotted in Figure 14. The differences between the six curves are small (comparable to the experimental noise), and only involve the magnitudes of the various peaks. This accounts for the similarity between the five experimental correlation functions (Figure 8) in the region of the peak at 3·7 Å, and at higher r, and that between the diffraction patterns (Figure 6) above the peak at ~5·8 Å^{-1}. Hence it must be concluded that the present diffraction data are relatively insensitive to the detailed distribution of (super)structural unit species for these glasses.

7.2. Network modifying cation environment

Whereas the distance between the network modifying cations in adjacent network cages can be inferred from the periodicity associated with the first diffraction peak, it is much more difficult to extract details of the oxygen atom co-ordination shell surrounding the M$^+$ cations, due to the number of different interactions that contribute in the region of, and just above, the second peak in $T(r)$. In the crystalline state, the difficulty of efficiently packing the large (cf. the basic structural units) superstructural units around particularly the small M^+ and M^{2+} cations means that the M(O) co-ordination polyhedra tend to be much less regular than would be the case in the absence of superstructural units. Even in the crystalline state, it is frequently difficult to define the cut-off distance for the first M(O) co-ordination shell, and a given network modifying cation can have different co-ordination numbers in the same crystalline structure. Thus, for example, in β-$Na_2O.3B_2O_3$,[28] the three Na^+ cations in the asymmetric unit have co-ordination numbers of 6, 7 and 8 ($2\cdot237 \leq r_{Na-O} \leq 3\cdot130$ Å) whereas, in $K_2O.2B_2O_3$,[29] two K^+ cations have six neighbours ($2\cdot631 \leq r_{K-O} \leq 2\cdot838$ and $2\cdot668 \leq r_{K-O} \leq 3\cdot056$ Å) plus a further neighbour at $3\cdot245$ or $3\cdot388$ Å and two have eight neighbours ($2\cdot643 \leq r_{K-O} \leq 2\cdot934$ and $2\cdot709 \leq r_{K-O} \leq 3\cdot106$ Å) plus one at $3\cdot325$ Å. In general these effects increase with the size of the network modifying cation, but even the Li^+ cation can have an extra neighbour at a longer distance. {In $Li_2O.2B_2O_3$,[30] Li^+ has four neighbours (distorted tetrahedron) at a mean distance of $2\cdot061$ Å ($1\cdot967 \leq r_{Li-O} \leq 2\cdot170$ Å) plus one at $2\cdot611$ Å}. The fact that the packing of the borate network around the network modifying cations is much poorer than would otherwise be expected in the absence of superstructural units provides further evidence in favour of a superstructural unit stabilisation energy (to offset the energy penalty associated with the inefficient packing).

As has been emphasised elsewhere,[1] the chemistry/bonding in the vitreous and crystalline states is the same and so much can be learned about the glass by considering the corresponding crystalline phases. The main difference is that the periodic symmetry of the crystalline state requires that every unit cell has the same stoichiometry, and hence the same number of each (super)structural unit species, whereas the glass exhibits spatial fluctuations in composition, which allows both a wider range of (super)structural unit species and also means that a given species can be present in small concentrations. (The requirement that every unit cell is the same means that, in the perfect crystalline state, a given species must be present in every unit cell or not at all.) Thus the contribution of a wider range of (super)structural unit species, combined with the disorder of the vitreous state, may allow slightly better packing around the network modifying cations than can be achieved in the crystalline state with only one, or a limited number of, (super)structural unit species. However, it is to be expected that any given network modifying cation will exhibit a range of co-ordination numbers/polyhedra in the same glass and that it is again much too simplistic to fit the M(O) first neighbour shell with a single symmetric Gaussian peak in $t_{M-O}(r)$.

Two different approaches were employed in an attempt to extract information concerning the first

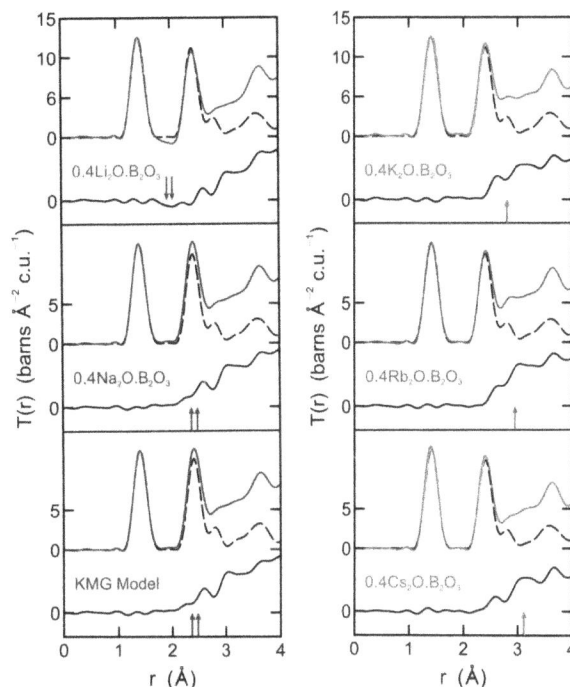

Figure 15. Left: Differences between the thermodynamic model (black dashed lines) and experiment (colours as Figure 6) for vitreous $2Li_2O.5B_2O_3$ (top) and $2Na_2O.5B_2O_3$ (centre), and between the KMG model and experiment for vitreous $2Na_2O.5B_2O_3$ (bottom). Right: Differences between the thermodynamic model and experiment for vitreous $2K_2O.5B_2O_3$, $2Rb_2O.5B_2O_3$ and $2Cs_2O.5B_2O_3$. The arrows indicate the M–O distances calculated from the ionic radii (Table 2), and the black solid lines are the differences between the models and experiment [in colour online]

M(O) co-ordination shell. The first considers the difference, $\Delta T(r)$, between the experimental total correlation functions and the superstructural unit contribution calculated from the thermodynamic and KMG models, whereas the second involves the differences between the experimental total correlation functions for pairs of glasses predicted to have similar diborate group boron atom fractions, x_D.

7.3. Thermodynamic and KMG models versus experiment

The thermodynamic and KMG models are compared with experiment in Figure 15, the arrows indicating the expected M–O distances calculated from the cationic radii assuming an anionic radius for O^{2-} of $1\cdot35$ Å[13] (Table 2). In the case of $2Li_2O.5B_2O_3$ and $2Na_2O.5B_2O_3$, the arrow at higher r denotes the Li–O (distorted tetrahedron, $r_{Li-O}=2\cdot04$ Å) and Na–O (six-fold, $r_{Na-O}=2\cdot48$ Å) distances estimated from the corresponding crystalline phases in the composition range $\frac{1}{4} \leq x_M \leq \frac{1}{3}$ (i.e. from triborate to diborate). Note that the neutron scattering length for Li is negative and so features in $\Delta T(r)$ due to Li–X (X=B or O) interactions are also negative. The problem with this approach is that the intra-superstructural unit contribution to $T(r)$ rapidly decreases with increasing r (cf. Figure 14) and so, at higher r, the correlation function is dominated by the interactions between atoms in different superstructural units and/or independent basic structural units.

The peak at 2·6 Å in the difference correlation functions, $\Delta T(r)$ (solid black lines), is due to the B–B distance associated with the B–O–B bridges between adjacent superstructural and/or independent basic structural units. Krogh-Moe[29] has noted that a broad range of inter-superstructural unit B–Ô–B bond angles is observed in the crystalline state, presumably because this is the easiest distortion to improve local network packing around the network-modifying cations.

Similarly, in the crystalline state $B^{[4]}–O–B^{[4]}$ bridges outside superstructural units, where $B^{[4]}$ indicates a tetrahedrally bonded boron atom, have not been found in alkali and alkaline earth borates below $x_M=0·3$,[1] and hence are also unlikely to occur in significant numbers in the vitreous state at the di-pentaborate composition; i.e. such bridges will mainly be of the $B^{[3]}–O–B^{[3]}$ and $B^{[3]}–O–B^{[4]}$ type. The KMG model, for example, predicts that 20% of the inter-structural unit bridges will be $B^{[3]}–O–B^{[3]}$ and 80% $B^{[3]}–O–B^{[4]}$.

In some cases, there is a feature in $\Delta T(r)$ that corresponds at least approximately to the expected M–O distance, although it is only partially due to the

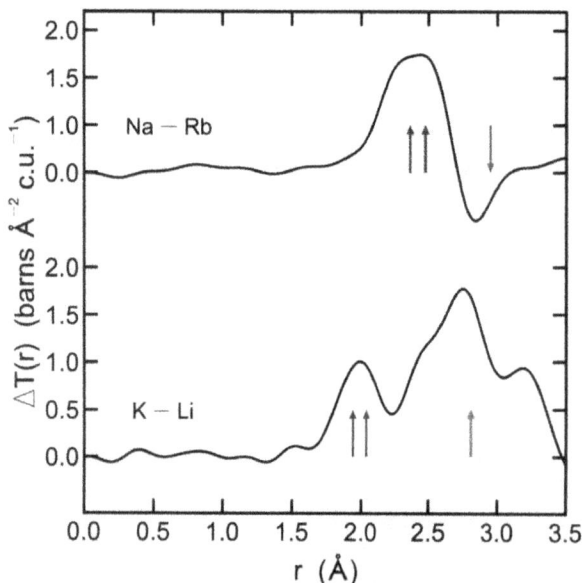

Figure 17. Difference (black dashed line) between the experimental total correlation functions for $2Ag_2O.5B_2O_3$ (black) and $2Na_2O.5B_2O_3$ (blue). The arrows indicate M–O distances calculated from the ionic radii (Table 2) and estimated from crystalline phases (see text)

contribution from the $t_{M-O}(r)$ component correlation function. Hence, given the number of contributions to $\Delta T(r)$ and the expected broad distribution of M–O distances, it is not possible to extract their detailed distribution via a peak fit.

7.4. Difference correlation functions

Figure 16 shows the experimental Na–Rb and K–Li difference correlation functions and, in each case, there are clear features associated with the M–O distance. (The direction of the arrows indicates the expected sign of the peak.) The K–O and Rb–O peaks occur at a slightly lower r than that predicted by the cationic radius appropriate to the co-ordination number estimated for the network modifying cations from the corresponding crystalline phases ($\frac{1}{4} \leq x_M \leq \frac{1}{3}$). This may indicate a lower average co-ordination number in the glass as a result of improved packing around the network modifying cations.

The final figure (Figure 17) compares the experimental total correlation functions for $2Na_2O.5B_2O_3$ and $2Ag_2O.5B_2O_3$. Given the similar size of the Na^+ and Ag^+ cations, it might be expected that they would occupy similar sites (network cages) and that the vitreous network for the two glasses would be similar. However, even in the crystalline state, the Ag^+ cation behaves differently to Na^+ in that in $Ag_2O.4B_2O_3$[31] two of the Ag^+ cations in the asymmetric unit have three close oxygen neighbours plus others at longer

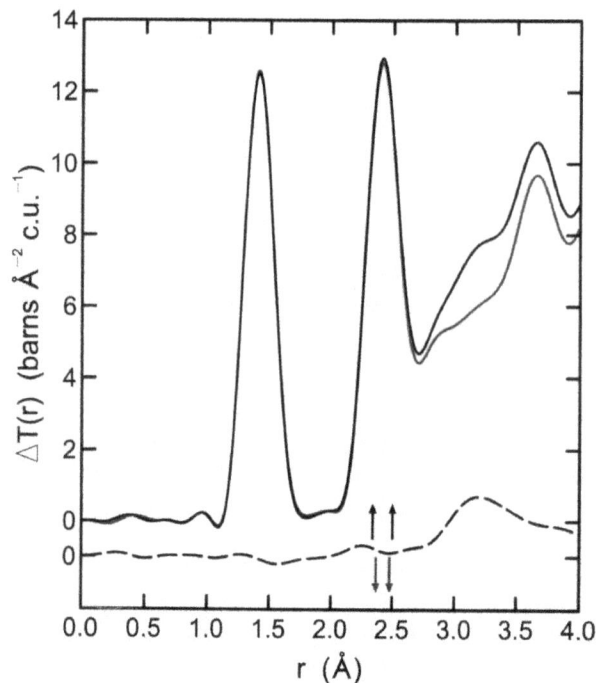

Figure 16. Differences between the experimental total correlation functions for $2Na_2O.5B_2O_3$ and $2Rb_2O.5B_2O_3$, and for $2K_2O.5B_2O_3$ and $2Li_2O.5B_2O_3$. The arrows indicate M–O distances calculated from the ionic radii (Table 2) and estimated from crystalline phases (see text) [in colour online]

distances, the average M–O distance for the three close neighbours (2·34 Å) being indicated by the black arrow at lower r in Figure 17. As a result, the two arrows for each cation occur at similar values of r and hence the feature in the difference correlation function, $\Delta T(r)$, in this region is positive due to the larger neutron scattering length for Ag (Table 2). The main difference between the two correlation functions, however, is represented by the large peak in $\Delta T(r)$ (dashed line) at 3·2 Å, which clearly demonstrates that the two glass networks are *not* the same.

7.5. Element specific techniques

In order to unambiguously extract more detailed information concerning the first M–O distances, it would be necessary to resort to an element specific technique, such as isotopic substitution (neutrons), anomalous dispersion (x-rays) or EXAFS spectroscopy, although it is not clear as to the sensitivity of the last of these to the expected broad distribution. There is also the problem that the longer M–O distances will almost certainly overlap the shorter M–B distances.

Cormier and co-workers[32,33] have employed the single-difference isotopic substitution technique to investigate the Li^+ cation environment in a series of Li_2O–B_2O_3 glasses. Their Li–Li+Li–X (X=B or O) first order difference correlation function exhibits a clear Li–O peak at ~2·0 Å, but any larger Li–O contribution at ~2·6–2·7 Å, as found in the crystalline state, would be 'lost' under the leading edge of the second (mainly Li–B) peak at ~3·0 Å. Cormier et al[34] have also extended this study, without isotopic substitution, to a series of Rb_2O–B_2O_3 glasses, together with a Li_2O–Rb_2O–B_2O_3 mixed alkali glass. In particular, by invoking the heavy element hypothesis, they identify a contribution to $T(r)$ at 4·7 Å as being due to Rb–Rb correlations, which is significantly shorter than the Rb–Rb distance in Table 3 (5·42±0·02 Å). However, as pointed out by Prins,[35] it is not always correct to associate a (prominent) peak in $T(r)$ with M–M, rather than M–O, interactions.[3] In addition, it should also be pointed out that the heavy element technique was initially proposed for x-rays, where the difference in scattering amplitude between the network modifying (heavy metal) cation and those of the boron and oxygen atoms forming the vitreous network is very much greater than for neutrons. It is also interesting to note that the total correlation functions, $T(r)$ obtained by Cormier and co-workers[32,34] will similarly yield B(O) co-ordination numbers that are slightly too low, like those of the present study, as indicated by the fact that the average value of $T(r)$ lifts slightly above the r axis approaching the first (B–O) peak.

The negative neutron scattering length for ^7Li allows the use of the null technique to directly measure

$t_{X-X}(r)$ (X=B or O), without interference from the Li–Li component correlation function, and this approach has been adopted by Swenson et al[36] in a study of Li_2O–B_2O_3, Na_2O–B_2O_3 and Ag_2O–B_2O_3 glasses. In their data analysis, Swenson et al[36] assume that the short range order of the borate network is the same for all three systems and use the ^0Li–^NLi, Na–^0Li and Ag–^0Li first order differences to investigate the M–M+M–X (M=Li, Na or Ag; X=B or O) correlations, where ^NLi indicates natural lithium and ^0Li lithium with a zero neutron scattering length. Their Li–Li+Li–X difference is similar to that of Majérus et al,[32] but of poorer quality, and they do not show a Ag–Na difference curve. However, inspection of their Ag–^0Li and Na–^0Li curves does indicate an extra contribution at ~3·2 Å, as found in the present study (Figure 17). It should also be noted that the thermodynamic modelling of Section 6.2 (Table 5) suggests that the assumption that the borate network is the same for the Li_2O–B_2O_3 and Na_2O–B_2O_3 systems is highly suspect, given the lack of pentaborate groups and the correspondingly higher triborate group contribution for vitreous $2Li_2O.5B_2O_3$.

It is, perhaps, also worthwhile to add a few comments to explain why reverse Monte Carlo (RMC) modelling has not been employed here to interpret the diffraction data, as in some other studies. The problem with this type of modelling is that the number of degrees of freedom for the fit ($3N$, where N is the total number of atoms in the unit cell) is vastly more than is necessary to fit the data, and so the result is that the final arrangement of atoms is merely one of an effectively infinite number that yield an equally good fit. In general, this atomic arrangement will not be stable to realistic interatomic potentials, and hence will change on relaxation. However, even if an inter-atomic potential were to be added as a constraint, this would not include the (mostly unknown) superstructural unit stabilisation energies, as is the case for most molecular dynamics simulations that yield superstructural unit fractions that are far too low. Hence, in its present state, the RMC technique is not capable of providing a meaningful interpretation, neither for the network modifying cation environment nor for the surrounding borate network, since both accurate (many body) interatomic potentials and the correct superstructural unit stabilisation energies *must* be incorporated to ensure that the correct subset is isolated from the essentially infinite number of possible sets of atomic co-ordinates.

7.6. Comparison with far infrared data

In a series of papers, Kamitsos et al[37–40] have investigated the environment of the network modifying cations in sodium magnesium borate and binary alkali borate glasses, as a function of composition, by far infrared spectroscopy. The fitting of Gaussian

---

[3] See also Section 10.2 of Wright.[1]

peaks to the measured spectra indicates the existence of two distinct distributions of network-modifying cation sites for the Li_2O–B_2O_3, Na_2O–B_2O_3 and K_2O–B_2O_3 glasses, and also for the Rb_2O–B_2O_3 and Cs_2O–B_2O_3 glasses with modifier contents in excess of the triborate composition. At lower modifier contents, the Rb_2O–B_2O_3 and Cs_2O–B_2O_3 glasses exhibit a single distribution of cation sites. However, Kamitsos *et al*[38] do not specify in detail the differences expected between the widths and mean values for the associated M–O bond length distributions and, in any case, these would not be detectable from the present neutron data.

Whilst Kamitsos *et al*[38] eliminate phase separation as an explanation for the two distributions, they do not consider the nanoheterogeneous structure arising from the chemical equilibria that characterise the melt of binary or multicomponent systems, as embodied in the thermodynamic modelling (Section 6.2). Indeed, from the latter, it is to be *expected* that the distribution of network modifying cation sites will vary between the different species of chemical grouping, as a result of the necessity to pack the relevant superstructural units (pentaborate, triborate and/or diborate groups) of different sizes around a given network modifying cation (*cf.* Section 7.2). In fact, the chemical structure, as determined from the model of associated solutions, would suggest more than two distributions, although some of these may make only a minor contribution to the total, and thus not be detectable by far infrared spectroscopy. That this effect is more pronounced for the smaller network modifying cations (Li^+, Na^+ and K^+) is also to be expected since, the smaller the cation, the greater is the difficulty of efficiently packing large superstructural units around it. The extreme case is afforded by the lack of pentaborate groups in the chemical structure of $2Li_2O.5B_2O_3$, which can be explained by the inability to efficiently pack the largest superstructural unit (pentaborate group) around the smallest network modifying cation. The larger sizes of the Rb^+ and Cs^+ cations makes it easier to pack superstructural units around them, and so any variation in packing efficiency for the different superstructural units, and hence in the cation sites themselves, will be less significant than for the small network modifying cations, especially at lower modifier contents.

8. Conclusions

A neutron diffraction and modelling study has been performed of a series of alkali, silver and alkaline-earth di-pentaborate glasses to investigate the structural role of the network modifying cations, M^+ or M^{2+}. The spacing between the M^+ or M^{2+} cations in adjacent network cages has been inferred from the position of the first peak in the reciprocal space diffraction pattern, $I(Q)$, although this may not be the shortest M–M distance. Modelling the distribution of superstructural unit species for the five alkali borate glasses indicates that the real space total correlation function, $T(r)$, is relatively insensitive to this distribution, due to the complex distribution of interatomic distances associated with each superstructural unit.

Although it has not been possible to extract the detailed distribution of M–O distances associated with the first oxygen atom co-ordination shell around the network-modifying cations, the use of difference correlation functions demonstrates that the M–O distances are consistent with those inferred from the M^+ and M^{2+} cationic radii, or from relevant crystalline phases. However, a consideration of such crystalline phases indicates that the presence of superstructural units leads to M(O) first co-ordination shells that are both distorted and ill defined with respect to the cut off M–O distance. Thus the extraction of further information concerning the detailed geometry of the first M(O) co-ordination shell will require the use of element-specific techniques such as isotopic substitution, anomalous dispersion or EXAFS spectroscopy.

Acknowledgements

This work was supported financially by the UK EP-SRC, and CES and JLS would like to thank the EPSRC and BNFL, respectively, for PhD studentships. The sample preparation at Coe College was funded under US National Science Foundation grant DMR 0904615, and the authors are also grateful to the Institut Laue-Langevin for neutron beam time and experimental assistance, and to 'Reviewer #1' for his/her helpful comments.

References

1. Wright, A. C. *Phys. Chem. Glasses: Eur. J. Glass Sci. Technol. B*, 2010, **51** (1), 1–39.
2. Wright, A. C., Dalba, G., Rocca, F. & Vedishcheva, N. M. *Phys. Chem Glasses: Eur. J. Glass Sci. Technol. B*, 2010, **51** (5), 233–65.
3. Wright, A. C. In *Experimental Techniques of Glass Science*. Edited by C. J. Simmons and O. H. El Bayoumi, American Ceramics Society, Westerville, 1993, Chap. 8, p. 205.
4. Krogh-Moe, J. *Acta Crystallogr.*, 1972, **B28**, 168.
5. Wright, A. C., Shaw, J. L., Sinclair, R. N., Vedishcheva, N. M., Shakhmatkin, B. A. & Scales, C. R. *J. Non-Cryst. Solids*, 2004, **345–346**, 24.
6. Johnson, P. A. V., Wright, A. C. & Sinclair, R. N. *J. Non-Cryst. Solids*, 1983, **58**, 109.
7. Beyster, J. R. *Nucl, Sci. Eng.*, 1968, **31**, 254.
8. Krogh-Moe, J. *Acta Crystallogr.*, 1956, **9**, 951.
9. Norman, N. *Acta Crystallogr.*, 1957, **10**, 370.
10. Sears, V. F. *Neutron News*, 1992, **3**, 26.
11. Filon, L. N. G., *Proc. Roy. Soc. (Edinburgh)*, 1929, **49**, 38.
12. Lorch, E. A. *J. Phys. C*, 1969, **2**, 229.
13. Wright, A. C. *Phys. Chem. Glasses: Eur. J. Glass Sci. Technol. B*, 2008, **49** (3), 103–17.
14. Shannon, R. D. & Prewitt, C. T. *Acta Crystallogr.*, 1969, **B25**, 925.
15. Shannon, R. D. & Prewitt, C. T. *Acta Crystallogr.* 1970, **B26**, 1046.
16. Hannon, A. C., Grimley, D. I., Hulme, R. A., Wright, A. C. & Sinclair, R. N. *J. Non-Cryst. Solids*, 1994, **177**, 299.
17. Krogh-Moe, J. *Phys. Chem. Glasses*, 1962, **3** (4), 101–10.
18. Griscom, D. L. In *Borate Glasses: Structure, Properties, Applications*. Edited by L. D. Pye, V. D. Fréchette & N. J. Kreidl, Plenum, New York, 1978, p. 11.

19. Shakhmatkin, B. A., Vedishcheva, N. M. & Wright, A. C., In *Proc. XIX Int. Congr. on Glass, Invited Papers*. Society of Glass Technology, Edinburgh, 2001, p. 52.
20. Chase, M. W., Davis, C. A., Downey, J. R., Frurip, D. J., McDonald, R. A. & Sywerud, A. N. *JANAF Thermochemical Tables*. Third Edition, American Chemical Society, American Institute of Physics and National Bureau of Standards, 1985.
21. Barin, I. *Thermochemical Data of Pure Substances*. Second Edition, VCH, Weinheim, 1993.
22. Shultz, M. M., Vedishcheva, N. M. & Shakhmatkin, B. A. In *Fizika I Khimiya Silikatov*. Edited by M. M. Shultz & R. G. Grebenshchikov, Nauka, Leningrad, 1987, p. 5.
23. Krogh-Moe, J. *Phys. Chem. Glasses*, 1962, **3** (1), 1–6.
24. Krogh-Moe, J. *Phys. Chem. Glasses*, 1965, **6** (1), 46-54.
25. Krogh-Moe, J. *Arkiv Kemi*, 1958, **12**, 475.
26. W.H. Zachariasen, W. H. *J. Amer. Chem. Soc.* **5** (1932), 3841.
27. Vedishcheva, N. M. & Wright, A. C., to be published.
28. Krogh-Moe, J. *Acta Crystallogr.*, 1972, **B28**, 1571.
29. Krogh-Moe, J. *Acta Crystallogr.*, 1972, **B28**, 3089.
30. Radaev, S. F., Muradyan, L. A., Malakhova, L. F., Burak, Ya. V. & Simonovo, V. I. *Sov. Phys. Crystallogr.*, 1989, **34**, 842.
31. Penin, N., Touboul, M. & Nowogrocki, G. *Solid State Sci.*, 2003, **5**, 559.
32. Majérus, O., Cormier, L., Calas, G. & Beuneu, B. *J. Phys. Chem.*, 2003, **107**, 13044.
33. Cormier, L., Calas, G. & Beuneu, B. *Phys. Chem. Glasses: Eur. J. Glass Sci. Technol. B,* 2009, **50** (3), 195–200.
34. Cormier, L., Calas, G. & Beuneu, B. *J. Non-Cryst. Solids*, 2007, **353**, 1779.
35. Prins, J. A. In *Physics of Non-Crystalline Solids*, Edited by J. A. Prins, North-Holland, Amsterdam, 1965, 39.
36. Swenson, J., Börjesson, L. & Howells, W. S. *Phys. Rev. B*, 1995, **B52**, 9310.
37. Kamitsos, E. I., Chryssikos, G. D. & Karakassides, M. A. *J. Phys. Chem.*, 1987, **91**, 1067.
38. Kamitsos, E. I., Karakassides, M. A. & Chryssikos, G. D. *J. Phys. Chem.*, 1987, **91**, 5807.
39. Kamitsos, E. I. *J. Phys. Chem.*, 1989, **93**, 1604.
40. Kamitsos, E. I. & Chryssikos, G. D. *Solid State Ionics*, 1998, **105**, 75.

Phys. Chem. Glasses: Eur. J. Glass Sci. Technol. B, October 2012, **53** (5), 210–218

Structure and properties of barium and calcium borosilicate glasses

Steve Feller,* Tyler Mullenbach, Maranda Franke, Suarav Bista, Anthony O'Donovan-Zavada, Kris Hopkins, Daken Starkenberg, Jedidiah McCoy, Desirae Leipply, Jessica Stansberry, Evan Troendle, Mario Affatigato

Department of Physics, Coe College, Cedar Rapids, Iowa, USA 52402

Diane Holland, Mark E. Smith

Department of Physics, University of Warwick, Coventry, UK CV4 7AL

Scott Kroeker, Vladimir K. Michaelis§ & John E. C. Wren

Department of Chemistry, University of Manitoba, Winnipeg, Manitoba, Canada R3T 2N2

Manuscript received 16 November 2011
Revised version received 7 February 2012
Accepted 14 March 2012

Structure–property relationships of alkaline earth borosilicate glasses were determined using solid state NMR, the glass transition temperature and density. A sharing of alkaline earth oxide amongst the borate and silicate networks was observed and modelled. Density and T_g data were compared with the constructed models providing support to structural ideas. Comparisons to structure–property relations of alkali borosilicate glasses were also made.

A. Introduction

The early ^{11}B nuclear magnetic resonance (NMR) work of Dell & Bray,[1] and Yun & Bray[2] produced semi-empirical structural models for sodium borosilicate glasses. These models used the structural information provided by NMR to describe the manner in which the alkali oxide modifier is shared between borate and silicate local coordination units. In our previous work on the alkali borosilicate systems, we analyzed these glasses using these structural models and others to predict density.[3–7] Barium and calcium borosilicate glasses are a natural extension of this research to the alkaline-earth borates. An effort has been made to determine the effects of alkaline earth oxide modifiers on the structure of the borosilicate network and glass properties. We report NMR derived values of N_4, the fraction of four-coordinate boron with oxygens. Measurements of the glass density and glass transition temperature (T_g) are also reported. Data are reported in terms of R (molar ratio of alkaline earth oxide to boron oxide, MO/B_2O_3) for fixed K (molar ratio of silica to boron oxide, SiO_2/B_2O_3) families where the glass composition is $RMO.B_2O_3.KSiO_2$. We go on to consider similarities and differences between structures and properties.

B. Experimental procedures

B1. Sample preparation

Barium and calcium borosilicate glasses were made from reagent grade chemicals purchased from Sigma Aldrich, each with purities above 99%. The modifier oxide, MO, was introduced to the mix via carbonates, $BaCO_3$ and $CaCO_3$. Boric acid (H_3BO_3) and silica (SiO_2) were used to complete the chemical formulation, with dehydration of boric acid to boron trioxide.

The reagents were weighed in a platinum crucible using electronic balances to the nearest ten thousandths of a gram to create between 6 and 8 g batches. Powdered samples were mixed for at least five minutes to ensure homogeneity. The batches were then placed in either a Thermoline 46100 or Carbolite 1400 high temperature furnace and heated between 15 and 20 min. Glasses were heated between 1200 and 1300°C depending on the composition. The melts were then removed from the furnace and allowed to air cool to room temperature. Weight loss was recorded for each sample. Appropriate weight loss was accepted to be within a tenth of a gram of the expected, and indicated that the glass sample had the correct composition.

Correct glass compositions were placed in the furnace for a second heat treatment, enabling the sample to be quenched rapidly, faster than traditional air cooling. The mixture was heated for 10 min at the same temperature as the initial melt. Upon completion of the second heating, the melts were removed

---

* Corresponding author. Email sfeller@coe.edu
Original version presented at VII Int. Conf. on Borate Glasses, Crystals and Melts, Dalhousie University, Halifax, Nova Scotia, Canada, 21–25 August 2011
§ Current Address: Department of Chemistry and Francis Bitter Magnet Laboratory, Massachusetts Institute of Technology, Cambridge, MA 02139, USA

from the furnace and the bottom of the crucible was placed in a beaker containing a water and ice slurry. If this method of quenching failed to produce glass, the samples were roller quenched; such samples typically had high modifier content (high R). High silica content required higher melting temperatures. The samples were found to be non-hygroscopic.

Calcium and barium binary borate glasses were made similarly but cooled using a plate or roller quenching method. This allowed glass formation for compositions with R as high as 1·9 (66 mol% modifier oxide).

B2. NMR spectroscopy

Boron-11 MAS NMR data were acquired using a Bloch sequence and a short tip angle (~13°) at fields between 11·74 T (Univeristy of Warwick) and 14·1 T (University of Manitoba). The fraction of four-coordinate boron, N_4 was determined using these data for both barium and calcium borates, and borosilicates. Typical spinning frequencies varied between 10 and 16 kHz using a 3·2 mm rotor which typically contained 30 to 40 mg of sample. Recycle delays were checked for each sample with averages being 5 s and 32 to 256 co-added transients were collected. To determine N_4 the relative area method was used corrected for spinning side band effects. N_4 data found for samples with the same composition showed the same values within experimental error (±0·02). All boron data were referenced with respect to 0·1 M boric acid to 19·6 ppm (relative to BF_3–Et_2O at 0·0 ppm).

In addition, some samples were run on a Brucker Avance II+ 600 MHz (14·1 T) at 192·4 MHz, with Varian T3 4 mm probe and MAS frequency from 12 to 15 kHz. A 1·5 µs pulse length and 8 s pulse delay were used. BPO_4 was used as a secondary reference at −3·3 ppm with respect to the primary reference boron trifluoride ether.

B3. Density

The densities were measured using 0·5 to 1·5 cm^3 of sample. Either a Quantachrome Micropycnometer or a Quantachrome Ultrapycnometer 1000 was used with helium gas. The density of each sample was determined 10 to 15 times and the average of the last five runs were used as the actual datum. To ensure accuracy and to minimize effects due to temperature each run was calibrated against ultrapure aluminium. Glass densities are accurate to within ±1%.

B4. T_g

Glass transition temperatures, T_g were measured for each glass by the onset method using a Perkin-Elmer DSC-7 or a Perkin-Elmer DTA-7 using 40°C/s as the

Table 1. Short range borate units

F_i unit	Structure	R value
F_1	trigonal boron with three bridging oxygens	0·0
F_2	tetrahedral boron with four bridging oxygens	1·0
F_3	trigonal boron with two bridging oxygens (one NBO)	1·0
F_4	trigonal boron with one bridging oxygen (two NBOs)	2·0
F_5	trigonal boron with no bridging oxygens (three NBOs)	3·0

heating rate. The samples were run in duplicate multiple times. References were frequently run. The error associated with this method and heating rate is ±5°C.

C. Structural models

C1. Binary models

Structural ideas for alkaline earth borosilicate glasses were based on borate and silicate binary glass models. Alkali borates are comprised of five short range structural units. These five units will be labelled as F_1, F_2, F_3, F_4 and F_5, and are defined in Table 1. The units appear and disappear as a function of R (our numerical model ignores the F_5 (orthoborate) unit because glass formation with roller quenching does not reach high enough R values to need it). Pure B_2O_3 glass is completely composed of F_1 units and as R is increased each unit appears as numbered. At the initial stage of introducing alkaline earth modifier, F_1 units undergo a change in coordination from three-coordinate to four-coordinate boron. This produces the F_2 unit with an alkaline earth (i.e. Ca^{2+} or Ba^{2+}) ion to maintain charge balance from the negative ion now present on the BO_4^- unit. Increasing the concentration of alkaline earth modifier will eventually cause a coordination change back to three-coordinated boron, forming amounts of nonbridging oxygens (NBOs). We assign these structural units as F_3, F_4, and F_5 groups, forming sequentially. This model has been examined by Bray and others.[8–10]

The silicate system works on a similar principle. The two main differences are notation and the lack of a coordination change. Binary alkali silicate glasses contain five tetrahedral units (Q^4, Q^3, Q^2, Q^1 and Q^0) with decreasing numbers of bridging oxygen atoms from 4 to 0, and each NBO is paired with an alkali ion for charge balance,[11–13] see Table 2. Similar to the borate units, the silicate units appear and disappear depending on J, the ratio of modifier oxide to silicate (MO/SiO_2).

Table 2. Short range silicate units

Q^i unit	Structure	J value
Q^4	tetrahedral silica with four bridging oxygens	0·0
Q^3	tetrahedral silica with three bridging oxygens (one NBO)	0·5
Q^2	tetrahedral silica with two bridging oxygens (two NBOs)	1·0
Q^1	tetrahedral silica with one bridging oxygen (three NBOs)	1·5
Q^0	tetrahedral silica with no bridging oxygens (four NBOs)	2·0

C2. Proportional sharing ternary model

A simple model of alkali borosilicate glasses depicts the modifier oxide being shared proportionally between the borate and silicate networks. In this model the modifier changes the borate and silicate networks according to binary models with

$$R = R_B + R_{Si} \tag{1a}$$

$$R_{Si} = R(K/(1+K)) \tag{1b}$$

$$R_B = R(1/(1+K)) \tag{1c}$$

^{29}Si MAS NMR data by MacKenzie et al support this simple idea with some evidence that the borate network has a preference for the alkali in alkali borosilicate glasses.[14]

This, however, is not the case according to Dell.[1] Dell stated that in these glass systems, the borate network initially uses the incoming alkali oxide to form F_2 units. This happens until $R=0.5$ when the alkali oxide begins being shared among the borate and silicate networks forming reedmergnerite groups (BSi_4O_{10}, a tetrahedral boron where each oxygen is bonded to a silica tetrahedron, Q^4). At $R_{max}=\frac{1}{2}+\frac{1}{16}K$, the alkali oxide begins to form NBOs on the silica tetrahedra of the reedmergnerite groups until $R_{d1}=\frac{1}{2}+\frac{1}{4}K$. After R_{d1} the alkali oxide starts being shared between the borate and silicate networks.[1] The sharing for these processes are summarized in the following equations:

$$R_{max} = \frac{1}{2} + \frac{1}{16}K \tag{2a}$$

$$R_{d1} = \frac{1}{2} + \frac{1}{4}K \tag{2b}$$

$$R_{d2} = \frac{3}{2} + \frac{3}{4}K \tag{2c}$$

$$R_{d3} = 2 + K \tag{2d}$$

$0 \leq R \leq R_{max}$

$$F_1 = 1 - R \tag{3a}$$

$$F_2 = R \tag{3b}$$

$$F_3 = 0 \tag{3c}$$

$$F_4 = 0 \tag{3d}$$

$R_{max} \leq R \leq R_{d1}$

$$F_1 = 1 - R_{max} \tag{4a}$$

$$F_2 = R_{max} \tag{4b}$$

$$F_3 = 0 \tag{4c}$$

$$F_4 = 0 \tag{4d}$$

$R_{d1} \leq R \leq R_{d2}$

$$F_1 = \left(1 - \frac{1}{8}K\right)\left(\frac{3}{4} - \frac{R}{2+K}\right) \tag{5a}$$

$$F_2 = (8+K)\left(\frac{1}{12} - \frac{R}{24+12K}\right) \tag{5b}$$

$$F_3 = \frac{1}{3}(R - R_{d1})\left(\frac{2 - \frac{1}{4}K}{2+K}\right) \tag{5c}$$

$$F_4 = \frac{1}{2}(R - R_{d1})\left(\frac{2 - \frac{1}{4}K}{2+K}\right) + \frac{2}{3}\left(\frac{KR}{8+4K} - \frac{K}{16}\right) \tag{5d}$$

$R_{d2} \leq R \leq R_{d3}$

$$F_1 = 0 \tag{6a}$$

$$F_2 = (8+K)\left(\frac{1}{12} - \frac{R}{24+12K}\right) \tag{6b}$$

$$F_3 = \left(\frac{4}{3} - \frac{1}{6}K\right)\left(1 - \frac{R}{2+K}\right) \tag{6c}$$

$$F_4 = \frac{1}{2}\left(1 - \frac{1}{8}K\right) + (R - R_{d2})\left(\frac{2 - \frac{1}{4}K}{2+K}\right) \tag{6d}$$

$$+ \frac{2}{15}\left(\frac{5KR}{8+4K} - \frac{5K}{16}\right)$$

Once knowledge of the borate modification from NMR is determined the silicate structures were determined using a binary level rule model. Evidence for this is discussed in Ref. 14.

Table 3. Fraction of four-coordinate boron in calcium and barium borates (± 0.02)

R	$N_4(Ca)$	$N_4(Ba)$
0	0	0
0.2	0.39	0.25
0.25	0.40	
0.33	0.40	
0.4	0.40	0.40
0.4		0.42
0.43	0.45	
0.5	0.46	
0.6	0.45	0.48
0.67	0.49	
0.7		
0.8		0.45
0.8	0.46	0.40
0.9		0.38
1	0.36	
1.2	0.21	
1.3		0.17
1.4	0.22	0.17
1.6	0.13	0.13
1.8		0.05
1.9		0.05

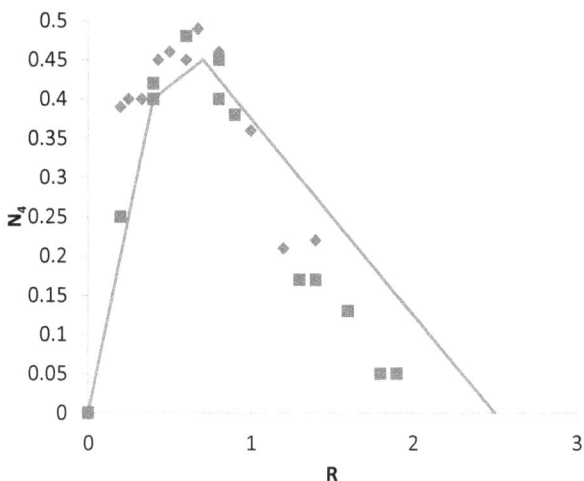

Figure 1. N_4 as a function of R for calcium (diamonds) and barium (square) borate glasses. Lithium borate system[15] is defined by the line [colour available online]

Table 4. Fraction of four-coordinate boron in calcium borosilicate glasses (±0·02)

R	K=0	K=0·5	K=1	K=2
0	0			
0·2	0·39			
0·25	0·40			
0·33	0·40			
0·4	0·40			
0·43	0·43			
0·5	0·46	0·31	0·31	
0·6	0·45			0·31
0·67	0·49			
0·7		0·40		
0·8	0·46		0·36	
0·9		0·44		0·36
1	0·36	0·44		
1·1			0·40	
1·2	0·21	0·42		0·38
1·4	0·22			
1·5				0·38
1·6	0·13	0·33		
1·8			0·37	0·39
2		0·24	0·33	
2·1				0·40
2·4				0·36
2·6			0·19	
3·2				0·24
4				0·12

D. Results and discussion

D1. ^{11}B MAS NMR

Table 3 lists while Figure 1 compares the tetrahedral boron fraction of barium and calcium borates. The three line fit depicted in Figure 1 is from lithium borate ^{11}B MAS NMR data from Jellison, Feller & Bray.[15] The Ba and Ca data agree reasonably well but with some small deviations, especially at high R values. This indicates that the conversion of F_1 to F_2 groups and subsequently to F_3, F_4, and F_5 proceeds along similar lines in the three binary borate systems. The deviation from the lithium trend at higher R is consistent with the end of glass formation by roller quenching near R=2 rather than R=3 in the lithium borate case. Somewhat similar trends were found in results reported from neutron scattering, molecular dynamics simulations and FTIR.[16,17]

Figure 2 depicts the N_4 fraction for the calcium borosilicates as a function of R for various K families and Table 4 lists all experimental data. Overall the data in Figure 2 indicate that the borate network is being diluted by the silicate network. To test this hypothesis we examine Figures 3(a)–(d) which show N_4

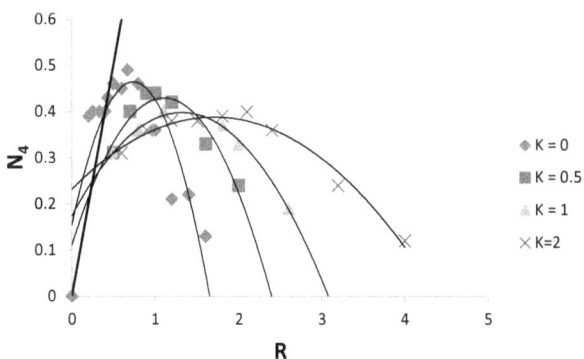

Figure 2. N_4 for calcium borosilicate glasses as a function of R. The curves are guides to the eye, the straight line segment is the curve $N_4=R$ [colour available online]

predicted by the proportional sharing of the calcium with the borate and silicate networks (red lines) as well as comparisons with the Dell model[1] (red line for K=0, blue lines otherwise). In Figure 3(a) we plot the calcium borate N_4 data (CaBSi, K=0) superimposed with the following straight lines taken from both the proportional sharing and Dell models for N_4:

$$N_4 = R \qquad R \leq 0·5 \qquad (7a)$$

$$N_4 = (2-R)/3 \qquad 0·5 \leq R \leq 2 \qquad (7b)$$

The fit is reasonable for both models. Low R samples have anomalously high N_4 values. A possible explanation is that this anomaly may be due to the presence of macroscopic phase separated samples. It is also possible that the anomalous N_4 data for R<0·4 is indicative of three-fold coordinate oxygen. These samples were tested on samples prepared independently at Coe College and at the University of Warwick and were found to have very close N_4 values. This adds credence to these data.

To test proportional sharing we modify Equations 7(a) and 7(b) by

$$N_4(R,K) = R_B \qquad R \leq 0·5 \qquad (8a)$$

$$N_4(R,K) = (2-R_B)/3 \qquad 0·5 \leq R \leq 2 \qquad (8b)$$

where R_B is defined by Equation (1c). Figures 3(b)–(c) show that proportional sharing is a reasonably good description of the calcium borosilicates (see red straight lines derived using proportional sharing superimposed on the data) with some evidence for disproportionation occurring at high K. Note that the Dell model (Equations (2a–d) through (6a–d) predicts ever increasing N_4 values as K increases, which

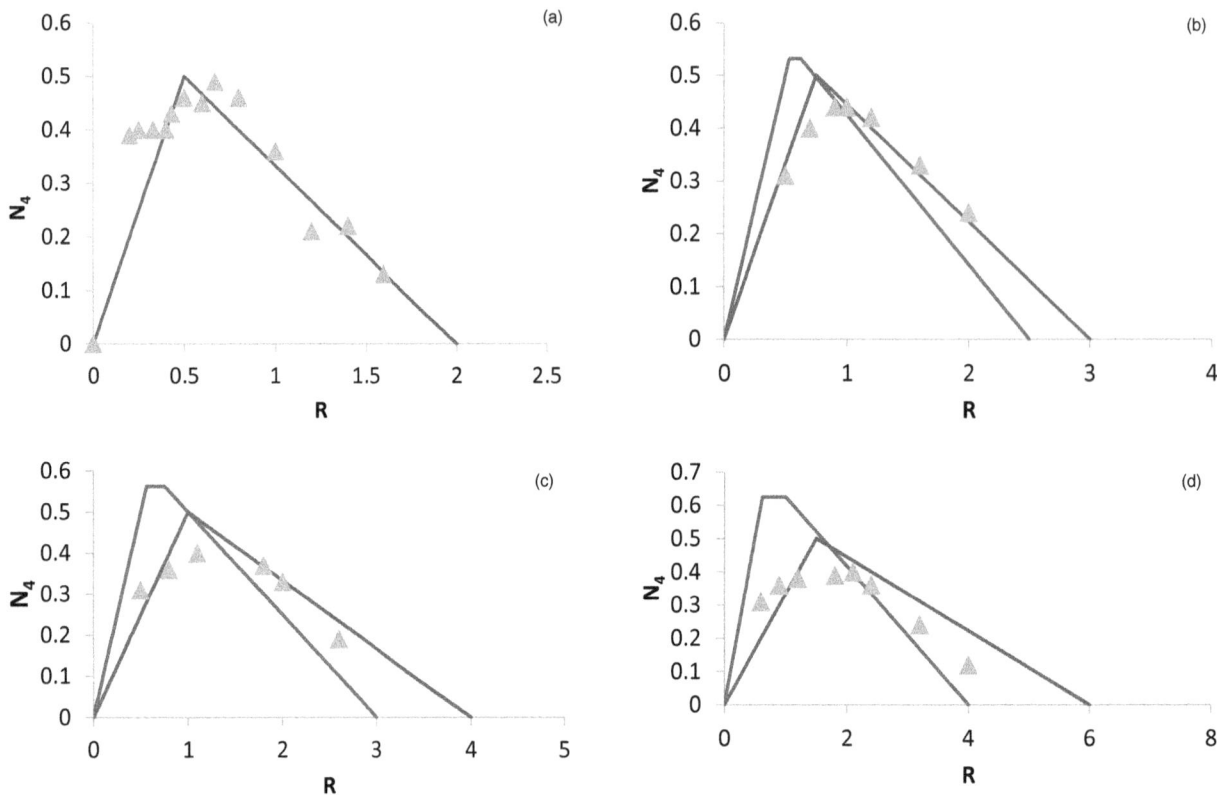

Figure 3. N_4 for calcium borate glasses as a function of R. (a) The lines represent Equations (7a) and (7b), a simplified proportional sharing model for N_4 discussed in the text, and the Dell model.[1] (b) K=0·5, (c) K=1·0, (d) K=2·0. The red lines represent Equations (8a) and (8b), a simplified proportional sharing model for N_4 discussed in the text. The blue lines represent the Dell model[1] [colour available online]

is a rather poor description for N_4 in the calcium borosilicate system (blue lines in Figures 3(b)–(d)). This implies that the silicate and borate networks have little interaction with each other and the borate network evolution moves to higher R as K increases.

In contrast in the barium borosilicate system N_4 is not simply the dilution of a non-interacting silicate network with the borate network but rather we see a strong interaction between the networks resulting in N_4 increasing with K, see the data of Stallworth[18] in Figure 4 and Figures 5(a)–(d), which compares the Dell model and simple proportional sharing (note the

data in Figure 5(a) are not the same data as shown in Figure 1, this was done to facilitate a more meaningful comparison to Figures 5(b)–(d)). Neither the Dell model nor the simple proportional model accurately describe these data. To explain these data Stallworth proposed a complicated barium oxide sharing model with the borate part taking a disproportionate share of the moderator; see Ref. 18 for details. It is similar in complexity but yields different values for N_4 when compared to the Dell model for sodium borosilicates.

D.2 Density

The density data are given in Tables 5 and 6 for calcium and barium borosilicate glasses, respectively. The density model that was found to fit best[19] was a proportional model that assumed uniform sharing of modifier oxide between the borate and silicate networks. This model, as described above, effectively treats the borosilicate glass as a nanoscale phase separated network. The following proportional sharing equation was used for this model:

$$\rho(R,K) = \frac{1}{1+K}\left[\rho\left(\frac{R}{1+K}MO.B_2O_3\right)\right]$$

$$+ \frac{K}{1+K}\left[\rho\left(\frac{R}{1+K}MO.SiO_2\right)\right]$$

(9)

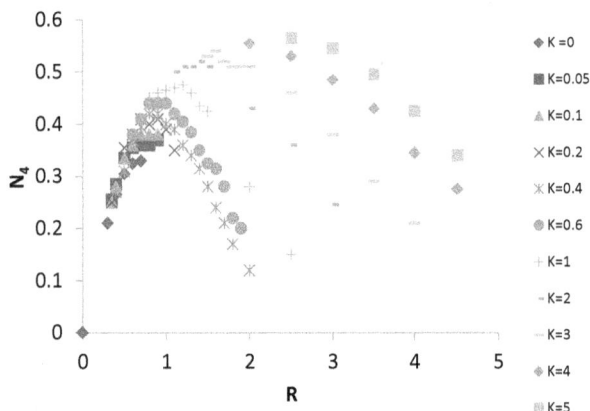

Figure 4. The fraction of four-coordinated borons in barium borosilicate glasses[18] [colour available online]

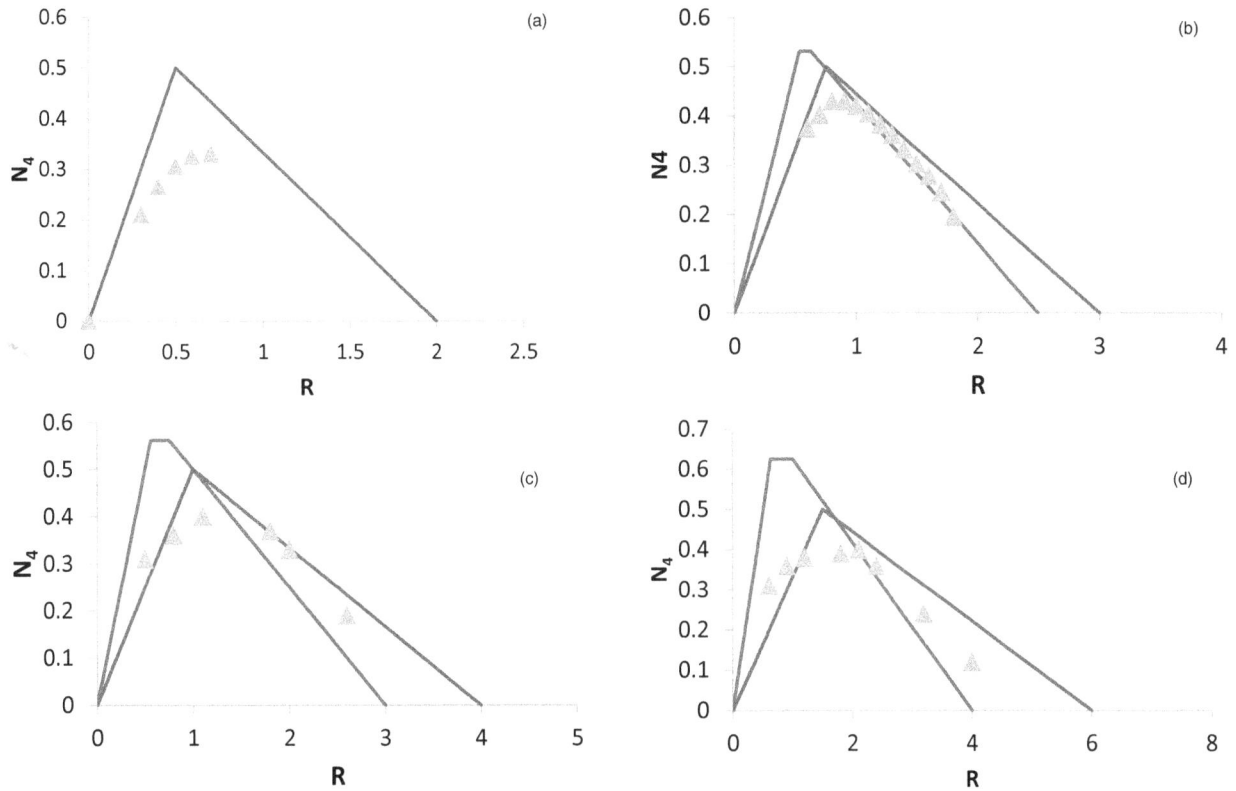

Figure 5. N_4 for barium borate glasses[18] as a function of R. (a) The lines represent Equations (7a) and (7b), a simplified proportional sharing model for N_4 discussed in the text, and the Dell model.[1] (b) K=0·5, (c) K=1·0, (d) K=2·0. The red lines represent Equations (8a) and (8b), a simplified proportional sharing model for N_4 discussed in the text. The blue lines represent the Dell model[1] [colour available online]

The densities of the borate glasses were taken from our previous work[20] while the silicate densities were taken from literature.[21] Least squares fits to the binary density data were used to extrapolate the needed densities at compositions in between where actual data were available.[6,7] The results of this model are

Table 5. Densities for calcium borosilicate glass (in g/cm³, ±1%)

R	K=0	K=0·5	K=1	K=2
0·0	1·81	1·90	1·91	1·87
0·2		2·16		
0·35	2·45			
0·4	2·48			
0·5	2·5	2·59	2·25	2·17
0·6	2·68	2·61	2·43	
0·7		2·69	2·44	
0·8		2·68	2·54	
0·81	2·72			
1·0	2·79	2·74	2·63	2·47
1·2		2·77	2·68	
1·22	2·8			
1·4		2·80	2·80	2·60
1·6		2·81	2·79	
1·8		2·83	2·80	2·68
2·0		2·83	2·83	
2·4			2·86	2·80
2·6			2·87	
2·8			2·86	2·84
3·0			2·87	
3·2				2·86
3·6				2·85
4·0				2·88

Table 6. Densities for barium borosilicate glass (in g/cm³, ±1%)

R	K=0	K=0·5	K=1	K=2
0·0	1·81	1·90	1·91	1·87
0·2	2·68	2·49		
0·2	2·66			
0·4	3·35			
0·4	3·29			
0·5		3·30	3·06	
0·5		3·38		
0·6	3·71	3·46		
0·6	3·68	3·54		
0·7		3·60	3·40	
0·8	3·95	3·71	3·54	
0·8	3·90	3·73		
0·9	4·09		3·61	
1·0		3·89	3·73	3·40
1·0		3·89		3·42
1·2	4·22	4·02	3·86	3·57
1·2				3·56
1·3	4·31			
1·4		4·13	3·97	3·71
1·4			3·99	
1·5	4·4	4·12		3·75
1·6		4·23	4·08	
1·6		4·21		
1·7	4·5			
1·8		4·31	4·15	3·91
2·0	4·53	4·37	4·23	3·97
2·0		4·38	4·23	
2·4		4·51	4·36	4·12
2·5			4·38	4·15
2·8			4·47	4·22
3·0			4·45	4·28
3·2			4·57	4·31
3·6				4·40
4·0				4·45

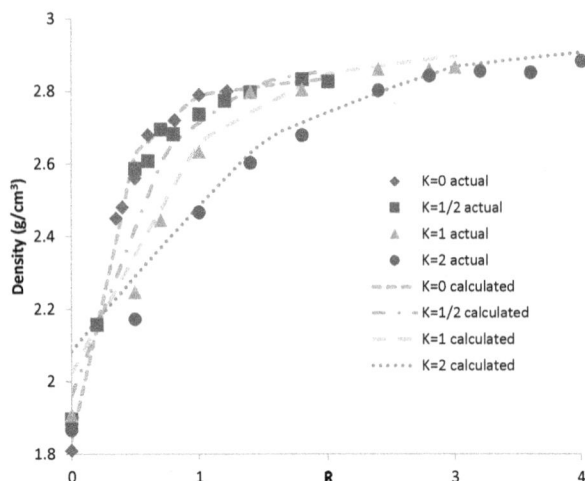

Figure 6. Calcium borosilicate proportional sharing density model [colour available online]

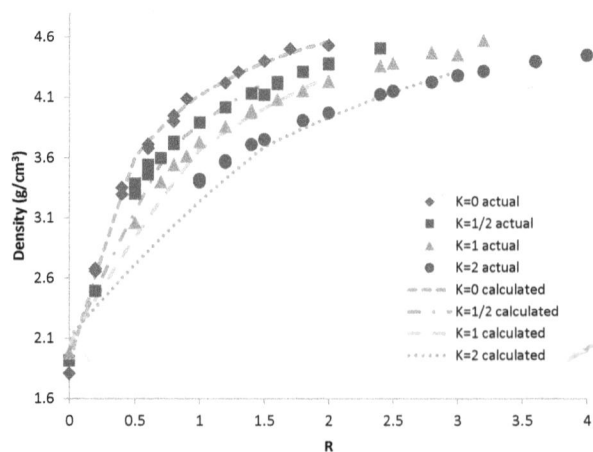

Figure 7. Barium borosilicate proportional sharing density model [colour available online]

shown in Figures 6 and 7 for the calcium and barium borosilicates systems. The fit is quite good. Attempts to model the density using the Dell model failed.[20]

D.3 Glass transition temperature

The T_g values are listed in Tables 7 and 8, and depicted in Figures 8 and 9 for the calcium and barium borosilicate systems, respectively. The calcium trends are consonant with the NMR data presented earlier in the clear observation of a dilution taking place in the T_g trends as K increases. It appears to be the binary calcium borate trend extended to higher R values as K increases. Similar results were found in

the lithium borosilicate system, see Figure 10(a) for comparison. However, in the sodium borosilicate case the behaviour is similar to the barium borosilicates, see Figure 10(b). In both systems T_g is enhanced as K increases indicating the presence of a strong interaction between the borate and silicate glass networks, presumably in the form of Si–O–B bonding.

E. Conclusions

The fraction of four-coordinated boron was determined for binary calcium and barium borate glasses as well as the calcium borosilicate system using ^{11}B magic angle spinning NMR. Our results indicate that calcium and barium borates have similar N_4 trends and each is similar to their alkali counterpart lithium borate system. The calcium borosilicate N_4 values indicate that the borate subnetwork receives an amount of modifier that is proportional to the relative amount of borate compared with silicate present. Based on our modelling of the densities of alkaline earth bo-

Table 7. T_gs for calcium borosilicate glasses (in °C, ±5°C)

R	K=0	K=0·5	K=1	K=2	K=3
0	260	317	373	478	426
0·1	631				
0·2	638				
0·3	645				
0·3	642				
0·4	643				
0·4	649				
0·5	656	658	664	666	
0·6		661	662		
0·7	658				
0·7		669	668		
0·8		663	666		
0·8	652				
1	635	660	667	674	674
1·2	613	651	669		
1·2	595				
1·4		643	668	668	
1·6		632	660		
1·8		624	657	681	
2		627	646		679
2·4			634	671	
2·6			629		
2·8			624	665	
3			625		682
3·2				657	
3·6				656	
4				648	678
5					671

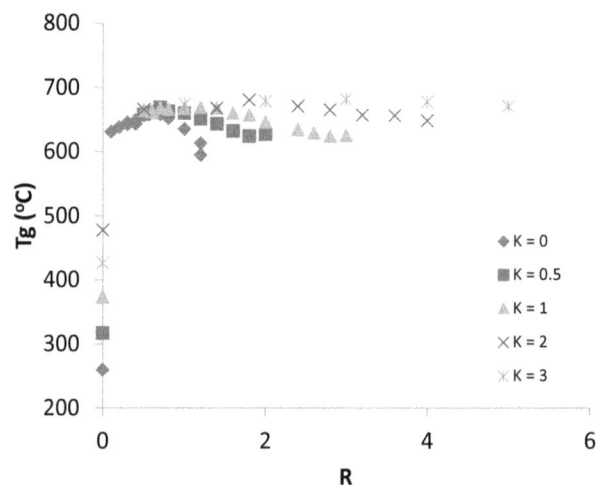

Figure 8. Calcium borosilicate T_gs as a function of R [colour available online]

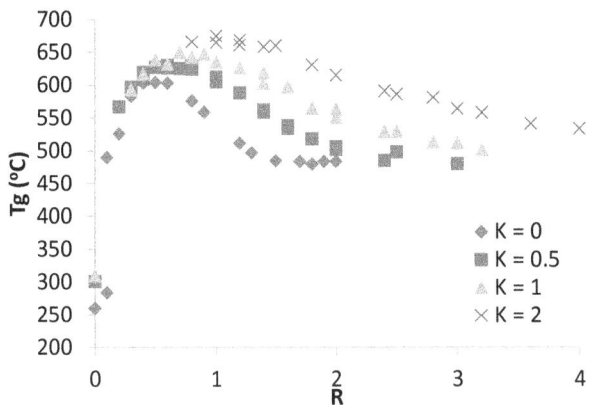

Figure 9. Barium borosilicate T_gs as a function of R [colour available online]

rosilicates, we have found the proportional sharing model to be both simple and accurate. The success of this model suggests that the borosilicate glass consists of a nanoscale phase separated network. The Dell model, originally designed for sodium borosilicates, failed to reproduce the witnessed density trends in the barium and calcium borosilicates, suggesting that the method of sharing modifier oxide amongst the networks in alkaline earth borosilicate glasses is different than that witnessed in the alkali borosilicate systems.

Table 8. T_gs for barium borosilicate glasses (in °C, ±5°C)

R	K=0	K=0·5	K=1	K=2
0	260	301	309	
0·1	284			
0·1	490			
0·2	526	567		
0·3	583	597	593	
0·4		609		
0·4	603	619	617	
0·5	604	627	638	
0·6	603	630	632	
0·6		626		
0·7		625	650	
0·8	576	629	642	666
0·8		624		
0·9	559		647	
1		605	635	665
1		612		675
1·2	512	588	626	669
1·2				661
1·3	497			
1·4		562	618	658
1·4		559	602	
1·5	485			660
1·6		538	597	
1·6		534		
1·7	484			
1·8	480	518	564	631
1·9	484			
2	484	506	563	615
2		502	551	
2·4		485	529	591
2·5		498	529	586
2·8			513	581
3		480	512	564
3·2			501	558
3·6				541
4				533

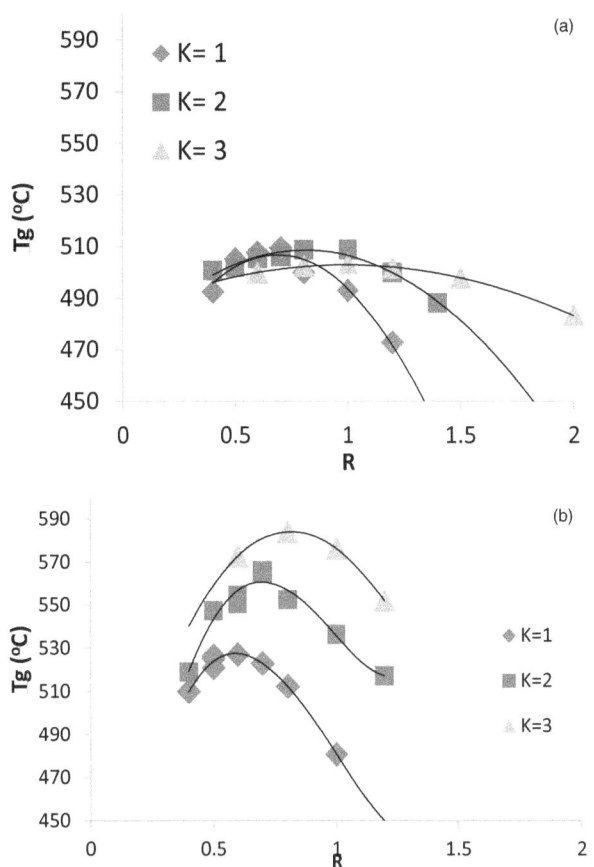

Figure 10. (a) T_gs of lithium borosilicate glasses as a function of R. (b) T_gs of sodium borosilicate glasses as a function of R [colour available online]

F. Acknowledgements

We would like to thank The National Science Foundation, for their support with Grant DMR 0502051 and 0904615; Coe College, for housing and student support; and the Carver Foundation of Eastern Iowa, for student support. VKM acknowledges NSERC for a PGSD3. SK is grateful to NSERC of Canada and the Canada Foundation for Innovation for operating and infrastructure support, respectively.

References

1. Dell, W. J., Bray, P. J. & Xiao, S. Z. *J. Non-Cryst. Solids*, 1983, **58**, 1.
2. Yun, Y. H. & Bray, P. J. *J. Non-Cryst. Solids*, 1978, **27**, 363.
3. Budhwani, K. Physical Properties of and Structures in Lithium, Sodium, and Potassium Borosilicate Glass System, *Coe College Honors Thesis*, 1993.
4. Boekenhauer, R. The Density of Lithium Borosilicate Glasses Related to Atomic Arrangement, *Coe College Honors Thesis*, 1991.
5. Kottke, J. An Examination of the Physical Properties of Rubidium and Cesium Borosilicate Glasses and the Development of a Quantitative Density Model for the Alkali Borosilicate Glass System, *Coe College Honors Thesis*, 1995.
6. Feil, D. & Feller, S. A. *J. Non-Cryst. Solids*, 1990, **119**, 103.
7. Kapoor, S., George, H., Betzen, A., Affatigato, M. & Feller, S. A. *J. Non-Cryst. Solids*, 2000, **270**, 215.
8. Bray, P. J., Feller, S. A., Jellison Jr., G. E. & Yun, Y. H. *J. Non-Cryst. Solids*, 1980, **38/39**, 93.
9. Yun, Y. H. & Bray, P. J. *J. Non-Cryst. Solids*, 1981, **44**, 227.

10. Feller, S. A., Dell, W. J. & Bray, P. J. *J. Non-Cryst. Solids*, 1982, **51**, 21.

11. Emerson, J. F., Stallworth, P. E. & Bray, P. J. *J. Non-Cryst. Solids*, 1989, **111**, 253.

12. Dupree, R., Holland, D. & Martoza, M. G. *J. Non-Cryst. Solids*, 1990, **116**, 148.

13. Larson, C., Doerr, J., Affatigato, M., Feller, S. A., Holland, D. & Smith, M. W.(E?) *J. Phys. Condens. Matter*, 2006, **18**, 11323.

14. MacKenzie, J. W., Bhatnagar, A., Bain, D., Bhowmik, S., Parameswar, C., Budhwani, K., Feller, S. A., Royle, M. L. & Martin. S. W. *J. Non-Cryst. Solids*, 1994 **177**, 269.

15. Jellison, Jr, G. E., Feller, S. A. & Bray P. J. *Phys. Chem. Glasses*, 1978 **19**, 52.

16 Yiannopoulos, Y. D., Chryssikos, G. D. & Kamitsos, E. I. *Phys. Chem. Glasses*, 2001, **42** (3), 164–172.

17. Ohtori, N., Takese, K., Akiyama, I., Suzuki, Y., Handa, K., Sakai, I., Iwadate, Y., Fukunada, T. & Umesaki, N. *J. Non-Cryst. Solids*, 2001, **293–295**, 136–145.

18. Stallworth, P. E. PhD. Thesis, Brown University, 1989.

19 Mullenbach, T., Franke, M., Ramm, A., Betzen, A., Kapoor, S., Lower, N., Munhollon, T., Berman, M., Affatigato, M. & Feller, S. A. *Phys. Chem. Glasses: Eur. J. Glass Sci. Technol. B*, 2009, **50** (2), 89.

20. Feller, S. A., Kottke, J., Welter, J., Nijhawan, S., Boekenhauer, R., Zhang, H., Feil, D., Paramswar, C., Budhwani, K., Affatigato, M., Bhasin, G., Bhowmik, S., Mackenzie, J., Royle, M., Kambeyanda, S., Pandikuthira, P. & Sharma, M. In: *Borate Glasses Crystals and Melts*, Eds A. C. Wright, S. A. Feller, A. C. Hannon, SGT, Sheffield, 1997, p.246.

21. Bansal, N. & Doremus, R. *Handbook of Glass Properties*, Academic Press, New York, 1986.

Phys. Chem. Glasses: Eur. J. Glass Sci. Technol. B, December 2012, **53** (6), 245–253

Structure and properties of lead borophosphate glasses doped with molybdenum oxide

Ladislav Koudelka,* Ivana Rösslerová, Zdeněk Černošek, Petr Mošner

Department of General and Inorganic Chemistry, Faculty of Chemical Technology, University of Pardubice, 532 10 Pardubice, Czech Republic

Lionel Montagne, Bertrand Revel & Gregory Tricot

Université Lille Nord de France – UCCS - UMR CNRS 8181, USTL, 59655, Villeneuve d'Ascq, France

Manuscript received 31 October 2011
Revised version received 3 February 2012
Accepted 5 April 2012

Lead borophosphate glasses doped with MoO_3 were studied in two compositional series $(1-x)[50PbO–10B_2O_3–40P_2O_5]–xMoO_3$ with a constant B_2O_3/P_2O_5 ratio and $(1-x)[50PbO–yB_2O_3–(50-y)P_2O_5]–xMoO_3$ with 0, 20, 40 and 60 mol% MoO_3. Some physical properties were measured and their structure was studied by ^{31}P, ^{11}B MAS NMR and Raman spectroscopies. Thermal behaviour was studied by DTA and dilatometry. The glass transition temperature in the series $(1-x)[50PbO–10B_2O_3–40P_2O_5]–xMoO_3$ has a maximum at $x=0.3$, whereas the thermal expansion coefficient has a minimum at $x=0.4$. The highest thermal stability was observed in glasses with $x=0.1–0.2$. Structural studies were devoted to an investigation of boron coordination versus MoO_3 content and the B_2O_3/P_2O_5 ratio. According to the NMR spectra, molybdenum atoms in borophosphate glasses preferentially form P–O–Mo bonds. Tetrahedral BO_4 units predominate in the studied borophosphate glasses at a low MoO_3 content. ^{11}B MAS NMR revealed the formation of mixed tetrahedral units $B(OP)_{4-n}(OMo)_n$ in the studied glasses. With an increase in the B_2O_3/P_2O_5 ratio and in the MoO_3 content, the relative intensities of resonances ascribed to mixed tetrahedral units and trigonal BO_3 units increase. ^{11}B DQ NMR spectra revealed a complex connectivity scheme between the tetrahedral and trigonal units manifested in the formation of BO_4 units with B–O–P, B–O–Mo, B–O–B^{IV} and B–O–B^{III} bonds.

1. Introduction

Borophosphate glasses rank among important classes of glassy materials as they provide better thermal stability and chemical durability than phosphate glasses.[1] Zinc and lead metaphosphate glasses have greater chemical durabilities than alkali metaphosphate glasses.[2,3] Improvements in the properties of borophosphate glasses are ascribed to the transformation of the linear chain structure of metaphosphate glasses into the three-dimensional structure of borophosphate glasses due to the addition of B_2O_3. When the boron oxide content is low, tetrahedral BO_4 units prevail in the structural network.[1,4,5] At a higher B_2O_3 content, trigonal BO_3 groups are also formed. There has been evidence that tetrahedral BO_4 units mix with phosphate units, thus inducing a reticulation of the global network and an increase in the glass transition temperature and the chemical durability. The connectivity between trigonal boron and phosphate has not been unambiguously investigated thus far, even if the complete separation between the two networks has been derived through a space correlation NMR experiment.[6]

Changes in the network structure of borophos-phate glasses can be deduced from changes in their Raman spectra. Raman scattering by phosphate structural units possesses a much stronger efficiency than that by borate units and thus the Raman spectra of borophosphate glasses reveal only characteristic vibrational bands of phosphate structural units.[7] The most active in the Raman spectra of metaphosphate glasses are symmetric stretching vibrations of P=O and P–O bonds in PO_4 units and symmetric stretching vibrations of bridging oxygen atoms in P–O–P bridges.[8,9] In phosphate glasses, the coordination of phosphorus can also be studied by ^{31}P MAS NMR spectroscopy.[4,10] Q^n notation is usually applied for the description of the coordination of phosphorus atoms in phosphate groups, where n denotes the number of bridging oxygen atoms on the given PO_4 tetrahedra.[11] The application of these principles to describe the bonding relations in phosphate glasses has been applied to binary phosphate glasses[4] as well as some ternary phosphate glasses.[12,13] Nevertheless, in borophosphate glasses ^{31}P MAS NMR spectra are usually broad and poorly resolved, particularly in glasses with a higher B_2O_3 content.[14,15]

In contrast, however, information on the coordination of boron in borophosphate glasses can be obtained from ^{11}B MAS NMR spectroscopy because these spectra possess an ability to discriminate

Corresponding author. Email Ladislav.Koudelka@upce.cz
Original version presented at VII Int. Conf. on Borate Glasses, Crystals and Melts, Dalhousie University, Halifax, Nova Scotia, Canada, 21–25 August 2011

between tetrahedral BO_4 coordination and trigonal BO_3 coordinations[16] due to the different ranges of chemical shift values for BO_4 and BO_3 units. Application of this technique has been reported for several borophosphate glasses[1,4,5,17] ^{11}B MQ-MAS NMR spectra were also applied for the investigation of medium range order in lead borophosphate glasses[5] and for the discrimination of two types of BO_4 units: $B(OP)_4$ and $B(OP)_3(OB)$.[6] In the latter units, boron atoms are linked to three PO_4 groups and one BO_3 group.

The addition of heavy metal oxides such as Nb_2O_5, MoO_3, WO_3 or TeO_2 to borophosphate glasses is of interest in order to modify their properties. Their semiconducting properties, for example, can be enhanced by the presence of transition metal ions in multivalent states. In MoO_3-containing glasses molybdenum may be present as Mo^{5+} ions as well as Mo^{6+} ions.[18] Studies of dc conductivity in copper borophosphate glasses have been aimed at an explanation of electronic transport properties using various models for electron hopping between the multivalent states in transition metals.[18] Structural studies of zinc borophosphate glasses have been reported by Šubčík et al.[19,20] Heavy metal oxide additions also strongly affect the chemical durability of borophosphate glasses since the glass network structure is modified. The incorporation of MoO_3 into zinc metaphosphate glass results in weakening of the bond strength in the glass network and a decrease in the chemical durability of these glasses.[19] The formation of mixed phosphate–molybdate units $B(OP)_{4-x}(OMo)_x$ was revealed by ^{11}B NMR spectroscopy, whereas from Raman spectra the presence of MoO_4 and MoO_6 units was proposed in these glasses.[19,20]

This paper is devoted to $PbO–B_2O_3–P_2O_5–MoO_3$ glasses with B_2O_3/P_2O_5 ratios of 5/45, 10/40 and 15/45. We have measured ^{11}B MAS NMR spectra with a high resolution in order to be able to sensitively investigate changes in boron coordination. Raman spectroscopy has given complementary information about phosphate bonding in these glasses. The relationships between structural changes and changes in physical properties are discussed.

2. Experimental

$PbO–B_2O_3–P_2O_5–MoO_3$ glasses were prepared by conventional melt quenching from analytical grade PbO, MoO_3, H_3BO_3 and H_3PO_4 using a total batch weight of 30 g. The homogenized starting mixtures were slowly calcined up to 600°C for 2 h to remove water. The reaction mixtures were consequently heated up to 1000–1350°C in a platinum crucible with a lid. The melt was held at the maximum temperature for 20 min and then poured into a preheated graphite mould. The resultant glasses were then transferred to an annealing furnace for 2 h at a temperature 5°C below their glass transition temperature, T_g, and then

cooled to room temperature. The volatilization losses checked by weighing were not significant, hence the batch compositions can be considered as reflecting actual compositions. The amorphous character of the glasses was checked by x-ray diffraction. A structureless spectrum was obtained for all the glass compositions.

The glass density, ρ, was determined on bulk samples using Archimedes' method with toluene as the immersion liquid. The molar volume, V_M, was calculated as $V_M = M/r$, where M is the average molar weight of the glass composition $aPbO–bB_2O_3–cP_2O_5–dMoO_3$ calculated for $a+b+c+d = 1$. The chemical durability of the glasses was evaluated from the measurement of the dissolution rate, DR, at 25°C on glass cubes with a dimension of ~5×5×5 mm. The glass cubes were shaken in 100 cm^3 of distilled water (pH=6) for 48 h. The dissolution rate, DR, was calculated from the expression $DR = \Delta\omega/St$, where $\Delta\omega$ is the weight loss (g), S is the area (cm^2) before the dissolution test and t is the dissolution time (min). The thermal behaviour of the glasses was studied using a DTA 404 PC operating in DSC mode at a heating rate of 10°C min^{-1}. The measurements were carried out with 100 mg powder samples under an inert atmosphere of N_2. The thermal expansion coefficient, α, the glass transition temperature, T_g, and the dilatometric softening temperature, T_d, were measured on bulk samples with dimensions of 25×5×5 mm using a dilatometer DIL 402 PC (Netzsch). Based on the obtained dilatation curves, the coefficient of the thermal expansion, α, was determined as a mean value over the temperature range 100–200°C; the glass transition temperature, T_g^b, was determined from the change in the slope of the elongation versus temperature plot and the dilatatometric softening temperature, T_d, was determined as the maximum of the expansion trace corresponding to the onset of viscous deformation. The dilatometric measurements were carried out in air at a heating rate of 3°C min^{-1}.

The Raman spectra were measured on bulk samples at room temperature using a Horiba-Jobin Yvon LaBRam HR spectrometer. The spectra were recorded in back scattering geometry under excitation with He–Ne laser radiation (632·8 nm) at a power of 12 mW on the sample. The spectral slit width was 1·5 cm^{-1} and the total integration time was 50 s. ^{31}P MAS NMR spectra were measured at 9·4 T on a 400 MHz BRUKER Avance spectrometer with a 4 mm probe. The spinning speed was 12·5 kHz and relaxation (recycling) delay was 180 s. The chemical shifts of ^{31}P nuclei are given relative to H_3PO_4 at 0 ppm. ^{11}B MAS NMR spectra were measured at 18·8 T on a BRUKER Avance 800 spectrometer with a 2·5 mm probe. The spectra were acquired with a single short (0·5 μs) radiofrequency pulse ($\pi/12$) in order to compensate for the different nutation behaviour of the two boron sites and use the obtained spectra for BO_3/BO_4

Table 1: Density, ρ, molar volume, V_M, glass transition temperature T_g^a (onset on DTA curve), T_g^b (measured by dilatometry), thermal expansion coefficient, α, and dissolution rate, DR, of $(1-x)[50PbO-yB_2O_3-(50-y)P_2O_5]-xMoO_3$ glasses

$xMoO_3$ [mol%]	yB_2O_3	$\rho\pm0.02$ [$g\,cm^{-3}$]	$V_m\pm0.5$ [cm^3mol^{-1}]	$T_g^a\pm2$ [°C]	$T_g^b\pm2$ [°C]	$T_c\pm2$ [°C]	$\alpha\pm0.3$ (100–200°C) [ppm°C]	$DR\pm0.2$ [$g\,cm^{-2}min^{-1}$]
0	5	4·75	37·6	349	350	570	14·1	2·2×10⁻⁸
0	10	4·97	35·5	396	393	588	13·7	7·9×10⁻⁸
0	15	5·07	33·9	445	447	590	12·6	3·6×10⁻⁸
10	10	4·86	35·4	428	428	620	12·5	4·4×10⁻⁸
20	5	4·69	36·7	419	421	-	12·6	2·5×10⁻⁸
20	10	4·78	35·4	444	442	663	12·0	1·1×10⁻⁸
20	15	4·99	33·3	457	459	606	11·6	3·1×10⁻⁸
30	10	4·70	35·3	449	448	574	11·6	3·6×10⁻⁸
40	5	4·60	35·9	435	437	590	11·0	2·4×10⁻⁸
40	10	4·63	35·2	431	429	529	11·2	5·1×10⁻⁸
40	15	4·70	34·1	417	419	538	11·7	7·1×10⁻⁸
50	10	4·57	34·9	402	400	489	11·8	8·6×10⁻⁸
60	5	4·44	35·6	386	388	496	12·3	8·0×10⁻⁸
60	10	4·48	34·9	376	375	466	12·6	1·4×10⁻⁸
60	15	4·66	33·2	369	371	450	11·2	8·4×10⁻⁸
70	10	4·36	35·1	356	349	429	12·6	-

quantification. The recycling delay was 10 s, and the spinning rate was 20 kHz. The chemical shifts of ^{11}B nuclei are given relative to BPO_4 at −3·6 ppm. The NMR spectra deconvolution was carried out using the Dmfit NMR software.[21] BO_4 lines are known to be subject to a negligible second order quadrupolar effect; hence the decomposition was performed with a Gaussian-type function assuming that the line shape is dominated by the chemical shift distribution.

The interactions between the different boron species have been investigated with the through space correlation DQ-SQ technique.[22,23] The 2D spectrum has been recorded on a 18·8 T spectrometer with a 3·2 mm probe head operating at a spinning frequency of 20 kHz. The 2048×90 acquisition points have been recorded under rotor-synchronized conditions with a selective spin echo sequence using 10 μs pulse length, 512 transients and a recycle delay of 1 s. The 2D coherences have been created with an excitation/reconversion pulse scheme of 750 μs. The correlation signals indicate spatial proximity which can be reasonably interpreted in our case (short excitation/reconversion times) as chemical connectivity.

3. Results

In this study we prepared 16 glass samples from the system $PbO-B_2O_3-P_2O_5-MoO_3$. Eight samples were obtained in the compositional series $(1-x)[50PbO-10B_2O_3-40P_2O_5]-xMoO_3$ and another eight samples were prepared in the glass series $(1-x)[50PbO-yB_2O_3-(50-y)P_2O_5]-xMoO_3$ for glasses with 0, 20, 40 and 60 mol% MoO_3 and B_2O_3/P_2O_5 ratios equal to 5/45 and 15/35 to also study the effect of the replacement of P_2O_5 by B_2O_3 in these glasses. The composition of all the studied glasses is given in Table 1. We have determined the density and molar volume of the prepared glasses and their values are also given in Table 1. The density of glasses in the compositional series $(1-x)[50PbO-10B_2O_3-40P_2O_5]-xMoO_3$ decreases with increasing MoO_3 content, mainly due to the decrease

in the content of heavy oxide PbO. The dissolution rate in the glass series $(1-x)[50PbO-10B_2O_3-40P_2O_5]-xMoO_3$ increases from 2·2×10⁻⁸ $g\,cm^{-2}min^{-1}$ for the glass with $x=0$ up to 8·6×10⁻⁶ $g\,cm^{-2}min^{-1}$ for the glass with 50 mol% MoO_3 (see Table 1). The dissolution rate of the glass with 70 mol% MoO_3 was not determined due to the disintegration of the sample in water.

Thermal properties were studied by DTA and dilatometry. The glass transition temperatures, T_g, were determined from the DTA curves as the onset of a change in the thermal capacity of the samples in the glass transition region and crystallization temperatures, T_c, were determined as the onset of the exothermic crystallization peak. Both values are given in Table 1. All the glasses crystallize on heating and the crystallization temperature reaches its maximum at a sample with 20 mol% MoO_3. No crystallization was observed with the glass of the composition $0·8[50PbO-5B_2O_3-45P_2O_5]-0·2MoO_3$. The thermal stability of the glasses, estimated from the difference T_c-T_g[24] reveals a maximum value of 221°C for the glass containing 20 mol% MoO_3 and slowly decreases with further increases in MoO_3 content.

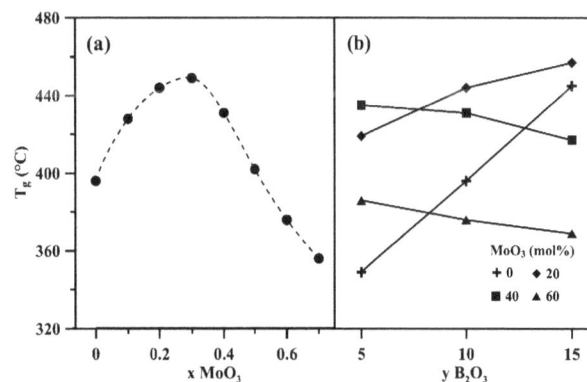

Figure 1. Compositional dependence of the glass transition temperature T_g (obtained from DTA curves) in the glass series $(1-x)[50PbO-10B_2O_3-40P_2O_5]-xMoO_3$ (a) and in the glass series $(1-x)[50PbO-yB_2O_3-(50-y)P_2O_5]-xMoO_3$ (b). The lines are only a guide to the eye

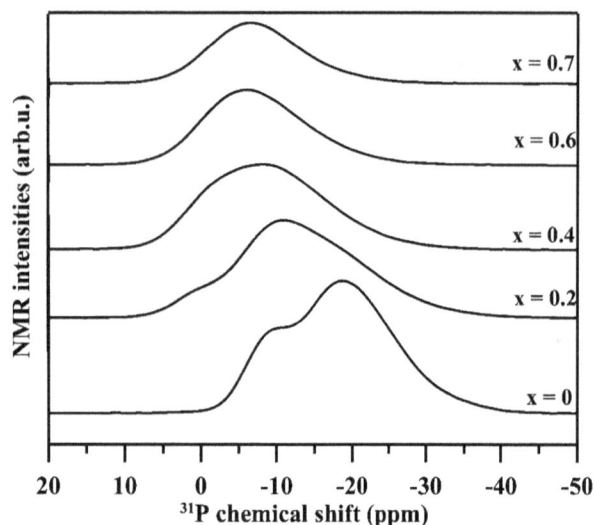

Figure 2. ^{31}P MAS NMR spectra of (1–x)[50PbO–10B$_2$O$_3$–40P$_2$O$_5$]–xMoO$_3$ glasses

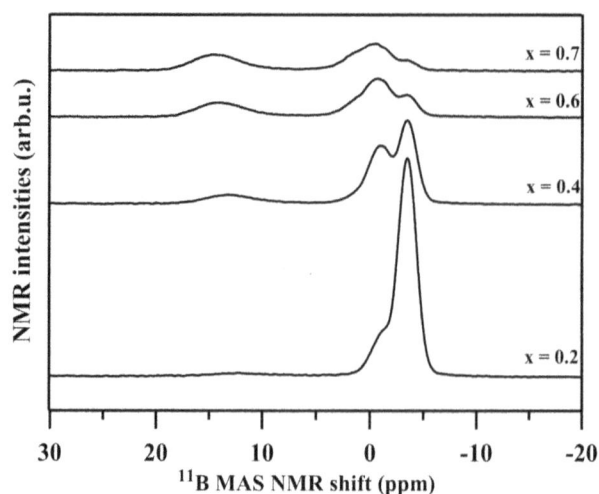

Figure 3. ^{11}B MAS NMR spectra of (1–x)[50PbO–10B$_2$O$_3$–40P$_2$O$_5$]–xMoO$_3$ glasses

The compositional dependence of the glass transition temperature in the series (1–x)[50PbO–10B$_2$O$_3$–40P$_2$O$_5$]–xMoO$_3$ is shown in Figure 1(a) and in the series (1–x)[50PbO–yB$_2$O$_3$–(50–y)P$_2$O$_5$]–xMoO$_3$ is shown in Figure 1(b). The glass transition temperature of the studied glasses (Figure 1(a)) increases with increasing MoO$_3$ content up to 30 mol% MoO$_3$ and with further MoO$_3$ additions it decreases. The effect of changing B$_2$O$_3$/P$_2$O$_5$ ratio in the glass series (1–x)[50PbO–yB$_2$O$_3$–(50–y)P$_2$O$_5$]–xMoO$_3$ on the evolution of T_g values is shown in Figure 1(b). In the glasses with 0–20 mol% MoO$_3$ an increase in the B$_2$O$_3$/P$_2$O$_5$ ratio results in an increase in T_g values, whereas in the glass series with 40 and 60 mol% MoO$_3$ a slow decrease in T_g values can be observed with increasing B$_2$O$_3$/P$_2$O$_5$ ratio.

The dilatometric measurements also gave the values of T_g (see Table 1), along with values of the dilatometric softening temperature, T_d, and values of the thermal expansion coefficient, α. The compositional dependence of T_d values in the glass series (1–x)[50PbO–10B$_2$O$_3$–40P$_2$O$_5$]–xMoO$_3$ is similar to the compositional dependence of T_g values with a maximum at 30 mol% MoO$_3$. The thermal expansion coefficient of the glasses in this glass series decreases with increasing MoO$_3$ content and has a minimum at a glass with 40 mol% MoO$_3$ and then slightly increases with a further increases in MoO$_3$ content.

^{31}P MAS NMR spectra were only measured for the glass series (1–x)[50PbO–10B$_2$O$_3$–40P$_2$O$_5$]–xMoO$_3$ and their evolution with changes in MoO$_3$ content is presented in Figure 2. ^{11}B MAS NMR spectra were measured not only for the (1–x)[50PbO–10B$_2$O$_3$–40P$_2$O$_5$]–xMoO$_3$glass series (Figure 3), but also in three glass series (1–x)[50PbO–yB$_2$O$_3$–(50–y)P$_2$O$_5$]–xMoO$_3$ with constant MoO$_3$ contents of 20, 40 and 60 mol% MoO$_3$ and different ratios of B$_2$O$_3$/P$_2$O$_5$ corresponding to values of y=5, 10 and 15 (Figures 4–6).

Similarly, Raman spectra were obtained for the glass series (1–x)[50PbO–10B$_2$O$_3$–40P$_2$O$_5$]–xMoO$_3$ (Figure 7) but also for the glass series (1–x)[50PbO–yB$_2$O$_3$–(50–y)P$_2$O$_5$]–xMoO$_3$ with constant MoO$_3$ contents of 20, 40 and 60 mol% MoO$_3$ and different ratios of B$_2$O$_3$/P$_2$O$_5$ corresponding to values of y=5, 10 and 15 (Figures 8–10).

4. Discussion

The ^{31}P MAS NMR spectra of (1–x)[50PbO–10B$_2$O$_3$–40P$_2$O$_5$]–xMoO$_3$ glass series with increasing MoO$_3$ content are shown in Figure 2. The spectrum of the glass without MoO$_3$ (x=0) contains two resonances at –19 ppm and –8.4 ppm, which could be ascribed to the presence of Q^2 and Q^1 phosphate units respectively, in accordance with previous studies.[10] In the spectrum of the glass with x=0.2, the resonance of Q^1 units becomes stronger and that of Q^2 units weaker, moreover, a new resonance at +1.3 ppm appears in the spectrum, the position of which lies within the region of the chemical shift characteristic of Q^0 units.[10] Such an evolution of the NMR spectra indicates depolymerization of phosphate chains in the glass structure due to the incorporation of molybdate units and the formation of P–O–Mo bonds within the network structure. A further increase in MoO$_3$ content results in the appearance of a broad resonance signal with a maximum at –6 to –9 ppm. This broad resonance is ascribed to phosphorus atoms with an increasing number of P–O–Mo bonds and a decreasing number of P–O–P bonds.

Structural studies were specifically aimed at the boron coordination in these glasses as the ^{11}B MAS NMR spectra measured with the NMR spectrometer at 18.8 T enables a higher resolution to be obtained, since the quadrupolar effect of ^{11}B (nuclear spin=3/2) is scaled down in high fields. The NMR spectra obtained are shown in Figures 3–6. The previous

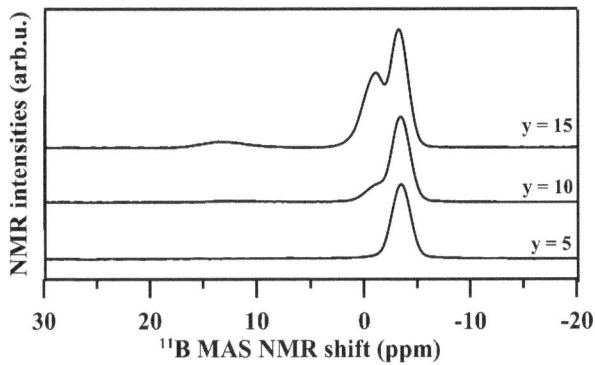

Figure 4. ^{11}B MAS NMR spectra of $0.8[50PbO–yB_2O_3–(50–y)P_2O_5]–0.2MoO_3$ glasses

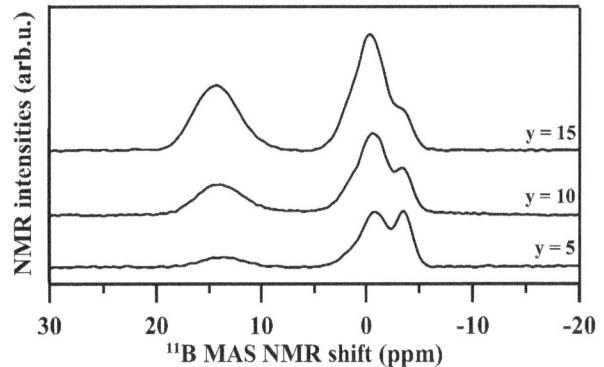

Figure 6. ^{11}B MAS NMR spectra of $0.4[50PbO–yB_2O_3–(50–y)P_2O_5]–0.6MoO_3$ glasses

^{11}B MAS NMR studies of borophosphate glasses[7,14] revealed that BO_3 units in borophosphate glasses have NMR shift values ranging from +10 ppm to +15 ppm, whereas BO_4 units possess a NMR resonance within the region −5 ppm to +1 ppm. Lead borophosphate glasses with a low B_2O_3 content (5–10 mol%) have a single narrow signal with a chemical shift value of −3.6 ppm ascribed to the tetrahedral $B(OP)_4$ units.[7] In the ^{11}B MAS NMR spectra of the $(1–x)[50PbO–10B_2O_3–40P_2O_5]–xMoO_3$ glass series with a constant B_2O_3/P_2O_5 ratio (Figure 3) changes are apparent in the NMR spectra with increasing MoO_3 content both in the BO_3- and BO_4-region of the chemical shift.

In order to discuss these changes the ^{11}B MAS NMR spectra of the glass series $0.8[50PbO–yB_2O_3–(50–y)P_2O_5]–0.2MoO_3$ (Figure 4) can serve as a starting point. The spectra of glasses with the lowest MoO_3 content indicates that there is no BO_3 resonance in the spectra of glasses with y=5 and 10 mol% B_2O_3 and only a weak signal of BO_3 units appearing in the spectrum of the glass with a B_2O_3 content of y=15. More visible changes can be seen in those spectra within the region characteristic of BO_4 units. The spectrum of the glass with y=5 (Figure 4) contains only a single resonance at −3.6 ppm ascribed in previous papers[1,7] to $B(OP)_4$ tetrahedral units. As the B_2O_3 content increases and P_2O_5 content decreases, a new resonance appears at about −1.1 ppm. We assume that this new resonance

can be assigned to boron atoms with B–O–Mo bonds since the ratio of Mo/B increases and the P/B ratio decreases, and considering that the probability of B–O–B bonding is low at such B_2O_3 content (this latter point will be discussed in more detail later). In this manner mixed boro-phosphate/molybdate structural units $B(OP)_{4–n}(OMo)_n$ are formed in the glass structure. We have noticed that only P–O–Mo bonds are formed in the glass with 20 mol% MoO_3 and y=5, but no B–O–Mo bonds are formed, since only the resonance of $B(OP)_4$ units can be observed in this spectrum (Figure 4). This means that molybdenum atoms preferentially form P–O–Mo bonds in these glasses.

In the ^{11}B MAS NMR spectra of the glass series $0.6[50PbO–yB_2O_3–(50–y)P_2O_5]–0.4MoO_3$ (Figure 5) similar trends in the evolution of NMR spectra have

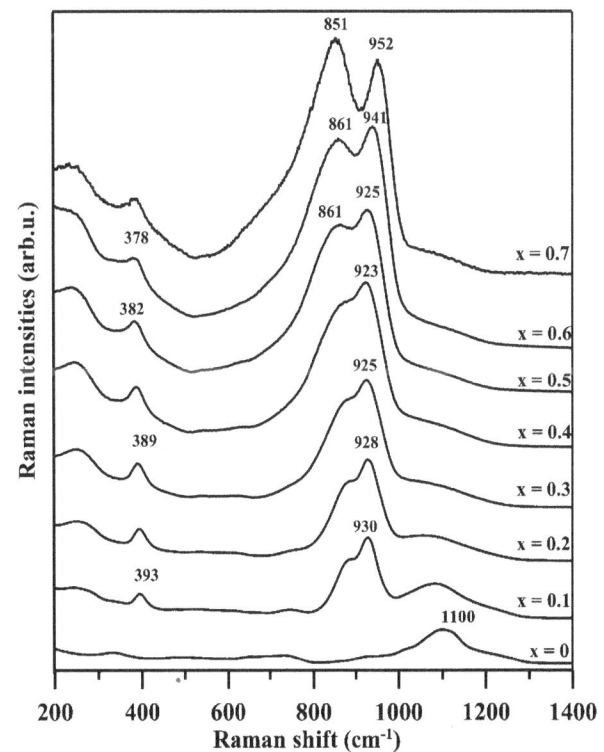

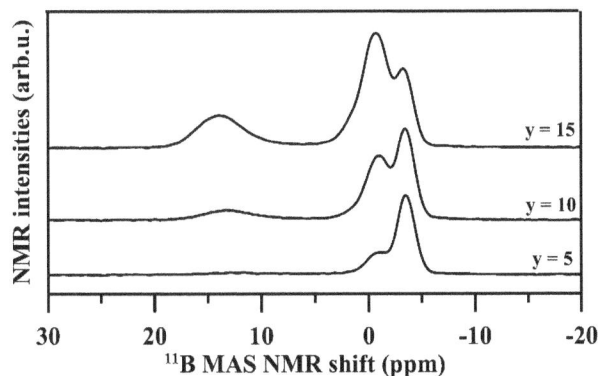

Figure 5. ^{11}B MAS NMR spectra of $0.6[50PbO–yB_2O_3–(50–y)P_2O_5]–0.4MoO_3$ glasses

Figure 7. Raman spectra of $(1–x)[50PbO–10B_2O_3–40P_2O_5]–xMoO_3$ glasses

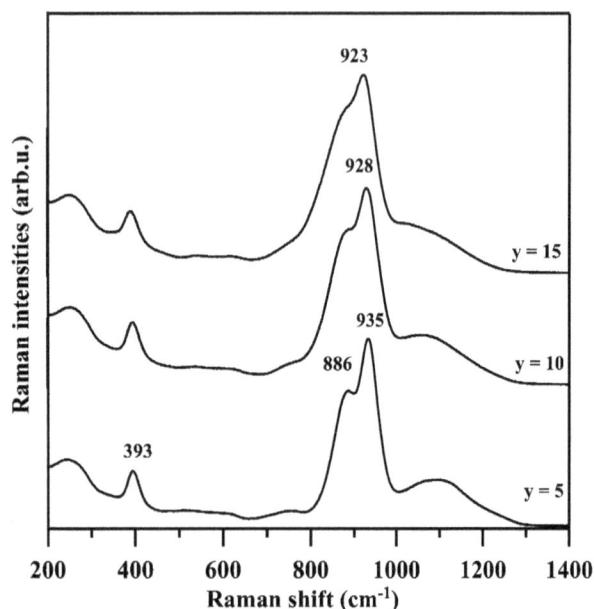

Figure 8. Raman spectra of 0·8[50PbO–yB$_2$O$_3$–(50–y) P$_2$O$_5$]–0·2MoO$_3$ glasses

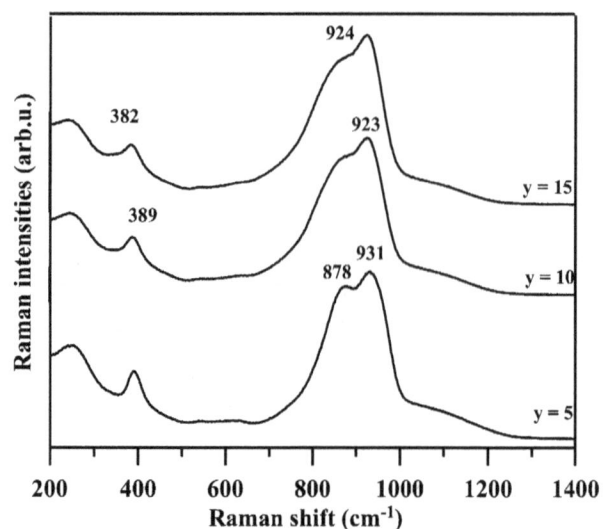

Figure 9. Raman spectra of 0·6[50PbO–yB$_2$O$_3$–(50–y) P$_2$O$_5$]–0·4MoO$_3$ glasses

been observed as in the previous series of glasses with 20 mol% MoO$_3$ (Figure 4), i.e. the number of mixed structural units B(OP)$_{4-n}$(OMo)$_n$ increases with increasing B$_2$O$_3$ content, as expected. Moreover, the relative amount of BO$_3$ units also increases in this glass series with an increasing B$_2$O$_3$/P$_2$O$_5$ ratio.

In the ^{11}B MAS NMR spectra of the glass series 0·4[50PbO–yB$_2$O$_3$–(50–y)P$_2$O$_5$]–0·6MoO$_3$ (Figure 6) the trends in the increasing intensity of the signal of mixed structural units B(OP)$_{4-n}$(OMo)$_n$ and BO$_3$ units are even more pronounced than in the previous two series (Figures 4 and 5). Simultaneously the relative intensity of the signal of B(OP)$_4$ units at –3·6 ppm further decreases.

Previous studies have indicated that the presence of bonds between tetrahedral BO$_4$ units and trigonal BO$_3$ units gives rise to a signal in the same region of the chemical shift[6,24] as the mixed B(OP)$_{4-n}$(OMo)$_n$ units. A decision was consequently made to investigate the mutual connectivity between the boron units, i.e. trigonal BO$_3$ and tetrahedral BO$_4$, by measuring the double quantum ^{11}B MAS NMR spectrum of 0·6[50PbO–15B$_2$O$_3$–35P$_2$O$_5$]–0·4MoO$_3$ glass (Figure 11).

As can be seen in Figure 11, the ^{11}B DQ MAS-NMR

spectrum displays both diagonal and off-diagonal signals. The spectrum provides new insights regarding the chemical environment and the interconnection between the different borate groups. Firstly, the diagonal signal (denoted as 2) suggests some autocorrelation for the borate site centred at –3·8 ppm. This result indicates a linking between the different borate units of that particular site, although this is not expected for the commonly reported assignment to B(OP)$_4$. The off-diagonal signals (denoted as 3/3') suggest the presence of two minor sites at 0 and –2 ppm, revealing a complex borate structure composed of at least four tetra-coordinated species. It is also worthy of note that the site at –2 ppm experiences a connectivity with tri-coordinated boron groups, as shown by the off-diagonal signals (4/4'). This result suggests a particularly complex chemical environment for that particular species which seems to be connected to phosphorus, BO$_4$, BO$_3$ and MoO$_n$ as well, leading to a B(OP)(OB^{IV})(OB^{III})(OMo) grouping (with no assumption on the stoichiometry of this species).

Finally, it is possible to decompose the ^{11}B MAS NMR spectra from DQ-NMR data into four individual components: (i) BO$_3$ at ca. 13 pmm, (ii) B(OP)$_4$ at ca. –3·6 ppm, (iii) mixed units with one B–O–Mo bond (BO$_{4mix}^1$) at ca. –1 ppm, which can also contain

Table 2. Chemical shifts and relative amounts of BO$_n$ structural units, measured with ^{11}B MAS NMR

xMoO$_3$ [mol%]	yB$_2$O$_3$	BO$_3$ δ [ppm]	%	B(OP)$_4$ δ [ppm]	%	BO$_{4mix}^1$ δ [ppm]	%	BO$_{4mix}^2$ δ [ppm]	%
0·2	5	12·6	1	–3·6	96	–1·1	3		
0·2	10	12·7	3	–3·5	73	–1·2	24		
0·2	15	13·1	6	–3·4	51	–1·1	43		
0·4	5	12·6	3	–3·6	71	–0·8	25	1·9	1
0·4	10	13·4	11	–3·5 *	43	–0·7	40	1·8	6
0·4	15	13·9	25	–3·5	23	–0·7	45	1·7	7
0·6	5	13·7	11	–3·5	31	–0·9	51	1·5	7
0·6	10	14·1	30	–3·6	15	–0·7	48	1·7	7
0·6	15	14·3	40	–3·6	7	–0·6	46	1·7	7
0·7	10	14·4	38	–3·7	3	–0·6	52	1·8	7

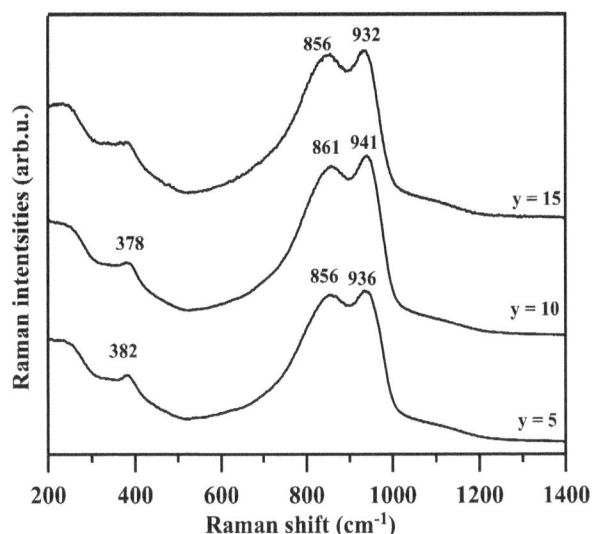

Figure 10. Raman spectra of 0·4[50PbO–yB$_2$O$_3$–(50–y) P$_2$O$_5$]–0·6MoO$_3$ glasses

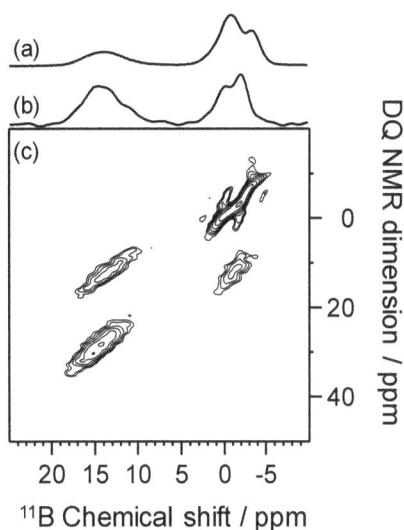

Figure 11. ^{11}B double quantum NMR spectrum of 0·6[50PbO–15B$_2$O$_3$–35P$_2$O$_5$–15B$_2$O$_3$]–0·4MoO$_3$ glass. (a) ^{11}B MAS NMR, (b) projection on MAS-axis, (c) ^{11}B double quantum filtered MAS-NMR spectrum

B^{IV}–O–B^{III} and B^{IV}–O–B^{IV} bonds and (iv) mixed units with two B–O–Mo bonds (BO$_{4mix}^2$) at ~1·8 ppm, which can also contain B^{IV}–O–B^{III} and B^{IV}–O–B^{IV} bonds. The results of the fitting of all ^{11}B MAS NMR spectra are given in Table 2. Since the quadrupolar constant of BO$_4$ is extremely small, we used Gaussian lineshapes without correction for the second order quadrupolar effect. It can be assumed that the Gaussian lineshape reflects the distribution of bond characteristics (length and angle) inherent to the glassy state. As can be seen in Table 2, the relative number of BO$_3$ units increases both with increasing MoO$_3$ content and B$_2$O$_3$/P$_2$O$_5$ ratio, whereas the number of B(OP)$_4$ units decreases, both resulting from the decrease in P$_2$O$_5$ content. The relative number of mixed units of the first type BO$_{mix}^1$ increases with increasing B$_2$O$_3$/P$_2$O$_5$ ratio only in the glasses with 20 and 40 mol% MoO$_3$, whereas in the glass series with 60 mol% MoO$_3$ it slightly decreases.

The evolution of the Raman spectra of (1−x)[50PbO–10B$_2$O$_3$–40P$_2$O$_5$]–xMoO$_3$ glass series is shown in Figure 7. The spectrum of the parent 50PbO–10B$_2$O$_3$–40P$_2$O$_5$ borophosphate glass shows one dominant broad band in the high frequency region with a maximum at 1100 cm^{-1} ascribed to the symmetric stretching vibration of nonbridging phosphorus–oxygen bonds in PO$_4$ units.[7] For the glass with 10 mol% MoO$_3$ the vibration band of phosphate units shifts to 1080 cm^{-1} and a new strong band at 930 cm^{-1} with a shoulder at 880 cm^{-1} appears in its Raman spectrum. These two bands can be ascribed to the vibrations of molybdate groups. These are particularly strong Raman scatterers in comparison with PO$_4$ groups which at a higher MoO$_3$ content suppress the vibrational bands of phosphate units. With increasing MoO$_3$ content the relative strength of both bands changes. At a low MoO$_3$ content the higher-frequency band at 925 cm^{-1} is stronger than the lower frequency band at ~880 cm^{-1}, while with increasing

MoO$_3$ content the latter band becomes stronger than the first band as well as broader and asymmetrical, shifting to shorter wavelengths down to 851 cm^{-1} for the glass with x=0·7. The strong band at 925–952 cm^{-1} is usually ascribed to the vibrations of Mo-O terminal bonds in MoO$_x$ polyhedra.[25] Sekiya et al[26] assume that sharp and intense peaks in the region of 890–1000 cm^{-1} can be assigned to stretching vibration between the molybdenum atom and unshared oxygen atoms. It can be assumed that the strong band 925–952 cm^{-1} can be assigned to the vibrations of Mo–O terminal bonds in MoO$_6$ octahedra.[13,20] The reason for this conclusion follows from the existence of the crystalline compound Pb(MoO$_2$)$_2$(PO$_4$)$_2$,[27,28] the structure of which consists of chains of inter-connected PO$_4$ tetrahedra and MoO$_6$ octahedra linked together via Mo–O–P bridges between their corners.[27]

The position of the Raman band of Mo–O vibrations at 925–930 cm^{-1} does not change significantly with composition in the studied glasses and therefore in accordance with Hardcastle & Wachs[29] changes in the deformation of MoO$_6$ octahedra with composition can be neglected. In the Pb(MoO$_2$)(PO$_4$)$_2$ compound MoO$_6$ octahedron has two shorter bonds (1·710 and 1·713 Å) and four longer bonds (1·977–2·136 Å). In the Raman spectra of the crystals of this compound[28] the symmetric stretching mode v_1(A$_{1g}$) is split into three components (880, 903 and 923 cm^{-1}), which is ascribed to the presence of different bond lengths existing in the MoO$_6$ octahedra in agreement with Hardcastle & Wachs.[29] The shape of dominant bands in the Raman spectra of the glasses with 0–40 mol% MoO$_3$ covers the spectral range of 840–950 cm^{-1} and therefore both main vibrational bands at 882 and 925 cm^{-1} can be ascribed to the vibrations of Mo–O bonds. Within the compositional range of 50–70 mol%

MoO_3 the formation of Mo–O–Mo bonds occurs and their vibrational band at ~840 $cm^{-1(30)}$ increases its strength with increasing MoO_3 content; it's shift to lower wavelengths is ascribed to an increase in the number of Mo–O–Mo bonds connecting MoO_6 octahedra. The structure of crystalline MoO_3 is composed of MoO_6 octahedra and therefore it can be assumed that the studied glasses with a high MoO_3 content will also contain MoO_6 octahedra which tend to form clusters via Mo–O–Mo bonds. A decrease in the number of P–O–Mo bonds with a decrease in the phosphorus content and the formation of Mo–O–Mo bonds between MoO_6 octahedra are accompanied by a decrease in the glass transition temperature in the studied glasses in this compositional range.

In the Raman spectra of glasses containing molybdenum oxide a medium band at 378–393 cm^{-1} can be seen (Figure 7) which can be ascribed to the vibrations of P–O–Mo bonds. A similar band was also observed in the Raman spectra of $ZnO–P_2O_5–MoO_3$ glasses[12] and $ZnO–P_2O_5–MoO_3$ glasses[13] with its intensity the highest with glasses with large contents of both P_2O_5 and MoO_3 where the number of P–O–Mo bonds is the highest. In the $(1-x)[50PbO–10B_2O_3–40P_2O_5]–xMoO_3$ glasses, it is apparent from their Raman spectra that its intensity is also the highest in glasses with large contents of both P_2O_5 and MoO_3 and therefore its assignment to vibrations of P–O–Mo bonds seems to be logical.

The changes in Raman spectra with changes in the B_2O_3/P_2O_5 ratio are not as pronounced, as can be seen from the Raman spectra of glass series $(1-x)[50PbO–yB_2O_3–(50-y)P_2O_5]–xMoO_3$ for glasses with 20, 40 and 60 mol% MoO_3 shown in Figures 8–10, respectively. In Figure 8 in the spectra of the $0·8[50PbO–yB_2O_3–(50-y)P_2O_5]–0·2MoO_3$ glasses with 20 mol% MoO_3 a shift of the maxima of the vibrational bands of phosphate units can be seen in the region of 1000–1200 cm^{-1} towards lower wavelengths, indicating a depolymerization of phosphate chains with increasing B_2O_3/P_2O_5 ratio and also slight changes in the vibrational bands of molybdate units, where the dominant band at 935 cm^{-1} shifts to lower wavelengths. The latter shift is actually due to a shift in the vibrational band.

The described structural changes are also reflected in changes in the physical properties of the studied glasses, particularly in the compositional dependence of the glass transition temperature in the compositional series $(1-x)[50PbO–10B_2O_3–40P_2O_5]–xMoO_3$, where a maximum was observed for the glass with 30 mol% MoO_3. The initial increase in T_g is ascribed to the reticulation of the structural network by the formation of P–O–Mo bonds, which are bonded to nonbridging, bridging and terminal oxygen atoms of the phosphate groups. The formation of molybdate clusters by the formation of Mo–O–Mo bonds in the glasses with a higher MoO_3 content and a decrease in the number of P–O bonds nevertheless results in a decrease in cohesive forces within the glass network. The effect of B_2O_3/P_2O_5 ratio on the T_g values in the glass series $(1-x)[50PbO–yB_2O_3–(50-y)P_2O_5]–xMoO_3$ is most pronounced in the glasses with the lowest MoO_3 content and the highest lead borophosphate content. Generally, with increasing B_2O_3/P_2O_5 ratio T_g increases as the BO_4 tetrahedra also reticulate the network structure. In the glasses with higher MoO_3 contents reticulation of the network involves P–O–Mo bonds and the replacement of P–O–Mo bonds by the formation of weaker B–O–Mo bonds has the reverse effect and thus we can observe a decrease in T_g values in the dependence on MoO_3 content at the glasses with 40 and 60 mol% MoO_3.

5. Conclusion

Homogeneous glasses can be prepared in the series $(1-x)[50PbO–10B_2O_3–40P_2O_5]–xMoO_3$ for $x=0$–$0·7$. The incorporation of molybdate units into the borophosphate structural network results in the depolymerisation of phosphate chains and the gradual transformation of metaphosphate into diphosphate and orthophosphate units. Both ^{31}P and ^{11}B MAS NMR spectra reveal a preferential formation of P–O–Mo bonds in glasses with a low B_2O_3 content. In addition to the formation of P–O–Mo bonds, the formation of B–O–B bonds between tetrahedral BO_4 units and trigonal BO_3 units also takes place as shown by ^{11}B DQ NMR spectroscopy. The formation of mixed tetrahedral structural units with B–O–P, B–O–Mo and B–O–B bonds was deduced from the analysis of ^{11}B MAS NMR and ^{11}B double quantum NMR spectra. DQ NMR spectra revealed the formation of complex BO_4 units with B–O–P, B–O–Mo and B–O–B bonds. The Raman spectra indicated that MoO_6 coordination prevails in the studied glasses and on the formation of Mo–O–Mo bonds connecting MoO_6 octahedra in the glass structure. Decreasing chemical durability and thermal stability of glasses with a higher MoO_3 content, results from the decreasing bonding forces inside the structural network with the replacement of P–O–P and P–O–B bonds by weaker P–O–Mo and B–O–Mo bonds.

Acknowledgements

The Czech authors are grateful for the financial support from the research project No. P106/10/0283 of the Grant Agency of Czech Republic. LM thanks the Feder, the USTL, the Région Nord Pas de Calais and the CNRS for funding of NMR spectrometers.

References

1. Brow, R. K. & Tallant, D. R. *J. Non-Cryst. Solids*, 1997, **222**, 396.
2. Koudelka, L., Jirák, J., Mošner, P., Montagne, L. & Palavit, G. *J. Mater. Sci.*, 2006, **41**, 4636.
3. Mošner, P., Koudelka, L., Jirák, J. & Vlček, M. *Adv. Mater. Res.*, 2008,

39–40, 181.

4. Brow, R. K. *J. Non-Cryst. Solids*, 2000, **263–264**, 1.
5. Koudelka, L., Mošner, P., Jirák, J., Zeyer-Düsterer, M. & Jäger, C. *Phys. Chem. Glasses: Eur. J. Glass Sci. Technol. B*, 2006, **47**, 471.
6. Elbers, S., Strojek, W., Koudelka, L. & Eckert, H. *Solid State Nucl. Magn. Resonance*, 2005, **27**, 65–76.
7. Koudelka, L., Mošner, P., Zeyer, M. & Jäger, C. *Phys. Chem. Glasses*, 2002, **43C**, 102.
8. Bobovich, Ya. S. *Opt. Spektrosk.*, 1962, **13**, 459.
9. Nelson, B. N. & Exarhos, G. J. *J. Chem. Phys.*, 1979, **71**, 2739.
10. Mustarelli, P. *Phosphorus Res. Bull.*, 1999, **10**, 25.
11. Van Wazer, J. *Phosphorus and its Compounds*, Interscience, New York, 1951.
12. Šubčík, J., Koudelka, L., Mošner, P., Montagne, L., Trickot, G., Delevoye, L. & Gregora, I. *J. Non-Cryst. Solids*, 2010, **356**, 2509.
13. Koudelka, L., Rösslerová, I., Holubová, J., Mošner, P., Montagne, L. & Revel, B. *J. Non-Cryst. Solids*, 2011, **357**, 2816.
14. Zeyer-Düsterer, M., Montagne, L., Palavit, G. & Jäger, C. *Solid State Nucl. Magn. Reson.*, 2005, **27**, 50.
15. Koudelka, L., Mošner, P., Zeyer-Düsterer, M. & Jäger, C. *J. Phys. Chem. Solids*, 2007, **68**, 638.
16. Mackenzie, K. J. D. & Smith, M. E. *Multinuclear Solid-State NMR of Inorganic Materials*, Pergamon, Amsterdam 2002.
17. Koudelka, L., Šubčík, J., Mošner, P., Montagne, L. & Delevoye, L. *J. Non-Cryst Solids*, 2007, **353**, 1828.
18. Kumar, B. V., Sankarappa, T., Kumar, M. P. & Kumar, S. *J. Non-Cryst. Solids*, 2009, **355**, 229.
19. Šubčík, J., Koudelka, L., Mošner, P. & Černošek, Z. *Adv. Mater. Res.*, 2008, **39–40**, 97.
20. Šubčík, J., Koudelka, L., Mošner, P., Montagne, L., Revel, B. & Gregora, I. *J. Non-Cryst. Solids*, 2009, **355**, 970.
21. Massiot, D., Fayon, F., Capron, M., King, I., Le Calvé, S., Alonso, B., Durand, J. O., Bujoli, B., Gan, Z. & Hoatson, G. *Magn. Reson. Chem.*, 2002, **40**, 70.
22. Wang, Q., Hu, B., Lafon, O., Trebosc, J., Deng, F. & Amoureux, J. P. *J. Magn. Reson.*, 2009, **200**, 251.
23. Raguenet, B., Tricot, G., Silly, G., Ribes, M. & Pradel, A. *J. Mater. Chem.*, 2011, **21**, 17693.
24. Dietzel, A. *Glasstech. Ber.*, 1968, **22**, 41.
25. Chowdari, B. V. R., Tan, K. L. & Chia, W. T. *Mater. Res. Soc. Symp. Proc.*, 1993, **293**.
26. Sekiya, T., Mochida, N. & Ogawa, S. *J. Non-Cryst. Solids*, 1995, **185**, 135.
27. Masse, R., Averbuch-Pouchot, M. T. & Durif, A. *J. Solid State Chem.*, 1985, **58**, 157.
28. Isaac, M., Jayasree, V., Suresh, G. & Nayar, V. U. *Indian J. Phys. B*, 1992, **66**, 65.
29. Hardcastle, F. D. & Wachs, I. E. *J. Raman Spectrosc.*, 1990, **21**, 683.
30. Santagnelli, S. H., de Araujo, C. C., Strojek, W., Eckert, H., Poirier G., Ribeiro, S. J. L. & Messadeq, Y. *J. Phys. Chem. B*, 2007, **111**, 10109.

*Phys. Chem. Glasses: Eur. J. Glass Sci. Technol. B, February 2013, **54** (1), 20–26*

Clustering in borate-rich alkali borophosphate glasses: a ¹¹B and ³¹P MAS NMR study

Vladimir K. Michaelis,[§] *Palak Kachhadia & Scott Kroeker**

Department of Chemistry, University of Manitoba, Winnipeg, Manitoba, R3T 2N2 Canada

Manuscript received 1 December 2011
Revised version received 27 July 2012
Accepted 6 August 2012

¹¹B and ³¹P magic angle spinning nuclear magnetic resonance (MAS NMR) spectroscopy are used to quantify short and medium range order in a series of borate-rich K, Rb and Cs borophosphate glasses, with a B_2O_3:P_2O_5 ratio of 5. At low alkali loadings (ca. <10 mol%), BPO_4-like regions segregate from the predominantly borate matrix. Above this level of alkali, a homogeneous mixed network glass forms with depolymerization of phosphate and borate species. The ¹¹B chemical shift of four-coordinated boron is sensitive to the number of bonded phosphate units – as determined by ¹¹B{³¹P} double resonance NMR and density functional theory calculations – and confirms the presence of phase separation for low alkali glasses.

Introduction

Borophosphate glasses have been intensively studied for potential applications ranging from solid state electrolytes to biomaterials.[1–4] The presence of two network forming cations confers on the resulting materials superior chemical inertness than either of the parent glasses alone, without sacrificing the inherently low processing temperatures.[5] Much effort has been expended to understand the physical origin of the nonlinearity exhibited by the glass transformation temperatures and ionic conductivities with variable B:P ratios.[6–8]

Nuclear magnetic resonance (NMR) spectroscopy has been used extensively to study borophosphate glasses.[1, 9–15] ¹¹B and ³¹P are abundant and receptive NMR nuclei, and decades of experimental work have established reliable correlations between spectral characteristics and local network structure. ¹¹B MAS NMR at suitably high magnetic fields (i.e. 11·7 T and above) can be simply interpreted in terms of the fraction of three- and four-coordinated boron, with additional insight about connectivity and depolymerization sometimes available under close scrutiny. ³¹P MAS NMR readily yields information about the degree of phosphate depolymerization and – under favourable circumstances – about the nature of neighbouring network formers. Higher magnetic fields often provide enhanced peak resolution and correspondingly finer structural detail. Double resonance experiments can be used to infer connectivity information based on the measurement of direct dipolar interactions between spin active nuclei.[16,17]

The focus of this work differs from previous studies of borophosphate glasses in two ways. First, we explore the borate-rich region of the phase diagram for its superior chemical durability, maintaining a B_2O_3:P_2O_5 ratio of five for all compositions. This permits us to isolate the specific impact of increasing modifier loading on the overall network structure. Second, we use the heavier alkali cations K, Rb and Cs in place of the more common Li and Na. Whereas much of the prior work has been directed at applications in ionic conductivity using Li, Na and Ag, we are interested in determining whether the size dependent differences observed in binary alkali borate[18, 19] and ternary alkali borosilicate[20] glasses are reproduced in borophosphates. Finally, our experimental NMR data are complemented by density functional theory (DFT) calculations of chemical shielding in model glass clusters to aid with peak assignments.

Materials and methods

Synthesis

Glass samples (500–600 mg) were synthesized using reagent grade P_2O_5, H_3BO_3, K_2CO_3, Rb_2CO_3 and Cs_2CO_3, and dried in an oven at 140°C for 12 h. H_3BO_3 was dehydrated at 650°C for 1 h to form the B_2O_3 starting material. Stoichiometric quantities of these reagents were placed into a platinum/gold (5%) crucible and heated between 1100 and 1300°C for 30 min. Mass loss measurements were made after heat treatment to ensure decarbonation was complete and to identify any additional volatilization losses. Melts were quenched to transparent glasses by cooling the crucibles in deionized water. Glasses were inspected by polarized light microscopy to ensure no crystallinity was present, and ground into fine powders using an agate mortar-and-pestle prior to NMR analysis.

* Corresponding author. Email Scott.Kroeker@ad.umanitoba.ca
Original version presented at VII Int. Conf. on Borate Glasses, Crystals and Melts, Dalhousie University, Halifax, Nova Scotia, Canada, 21–25 August 2011
§ Current Address: Department of Chemistry and Francis Bitter Magnet Laboratory, Massachusetts Institute of Technology, Cambridge, MA, 02139, USA

Quantum chemical calculations

Density functional theory (DFT) calculations were performed using Gaussian03[21] on anionic borophosphate model clusters incorporating four-coordinated boron ([4]B) bonded through bridging oxygens to x[3]B and $(4-x)$[4]P (x=0–4), each of which was hydroxyl terminated unless bonded to additional [3]B or [4]P. Clusters were assembled using Gaussview 3.09 and geometry optimized using the B3LYP[22–24] level of theory and 6311++g(d,p) and aug-cc-pVTZ basis sets.[25–28] NMR calculations were done on optimized clusters at the same level of theory. Calculated shieldings for the central [4]B were converted to chemical shifts by comparison with an optimized boric acid (B(OH)$_3$) cluster calculated at the same level of theory.

Nuclear magnetic resonance

Magic angle spinning (MAS) nuclear magnetic resonance experiments were performed for all samples on a Varian ^{UNITY}Inova 600 (14·1 T) using 3·2 mm double and triple resonance Varian-Chemagnetics probes for [11]B (ν_L=192·4 MHz) and [31]P (ν_L=242·3 MHz). Ground samples were packed into 3·2 mm outer diameter ZrO$_2$ rotors (22 μL fill volume) and spun at 16000±4 Hz. ^{11}B MAS NMR spectra used 0·4 μs excitation pulses (9° tip angle, ν_{rf}=63 kHz), 2 s recycle delays and 128 co-added transients. ^{31}P MAS NMR spectra used 1·3 μs excitation pulses (23° tip angle, ν_{rf}=50 kHz), 150 s recycle delays and 64 co-added transients. Recycle delays were optimized for each sample.

[11]B{[31]P} REDOR experiments were performed with a spinning frequency of 6000±1 Hz, 2 s recycle delays and 128 co-added transients. Due to the difference in nutation frequencies for [3]B and [4]B sites, selective $\pi/2$ pulses were optimized separately for [3]B (3·7 μs, ν_{rf}=63 kHz), [4]B (5·1 μs, ν_{rf}=63 kHz) and [31]P (4·5 μs, ν_{rf}=56 kHz), resulting in the acquisition of two independent REDOR datasets. Spectra were referenced to 0·1 M boric acid ([11]B) at +19·6 ppm, a secondary reference relative to BF$_3$–Et$_2$O (0·0 ppm) and 0·1 M phosphoric acid (^{31}P) at 0·0 ppm.

Ultrahigh field ^{11}B MAS NMR (ν_L=288·6 MHz) spectra were acquired on a Bruker Avance II 900 (21·1 T) spectrometer using a 3·2 mm double resonance boron free MAS probe. Powdered samples were packed in 3·2 mm ZrO$_2$ rotors and spun at 15 kHz. Single pulse experiments used a 0·4 μs pulse (~10° tip angle, ν_{rf}=63 kHz), 5 s recycle delay and 512 co-added

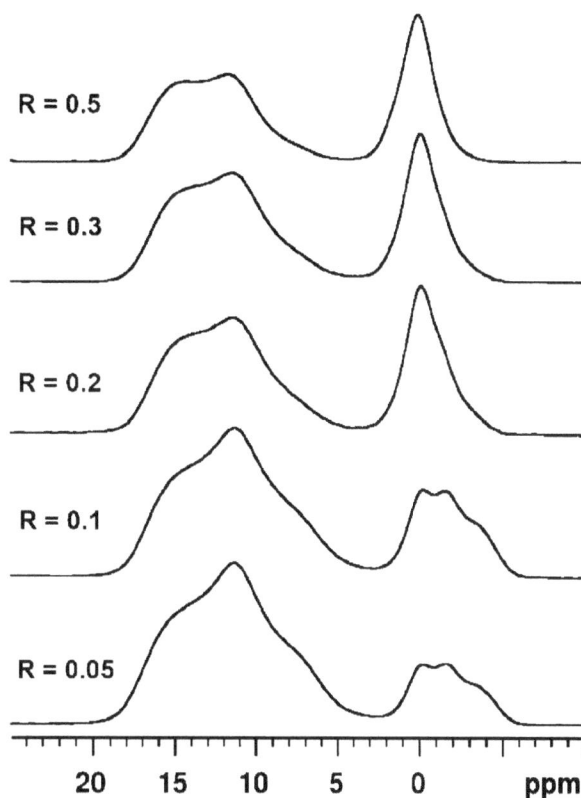

Figure 1. ^{11}B MAS NMR spectra of caesium borophosphate glasses (with a B$_2$O$_3$:P$_2$O$_5$ ratio of ~5) recorded at 14·1 T. Compositions are indicated by the alkali to boron ratio, R

transients. Spectra were referenced to solid NaBH$_4$ (−42·06 ppm) relative to BF$_3$–Et$_2$O (0·0 ppm).

Results

^{11}B MAS NMR spectra were recorded for the alkali borophosphate glasses listed in Table 1. Figure 1 depicts typical spectral changes associated with increasing alkali content. The three-coordinated boron region (10

Table 2. ^{11}B MAS NMR peak fitting (B$_0$=14·1 T) of alkali borophosphate glasses with overall four-coordinated boron fraction (N$_4$, ±0·01) and relative contributions (%, ±5%) from [4]B peaks located approximately at 0, −1·5 and −3·5 ppm

	R	N$_4$	0 ppm	−1·5 ppm	−3·5 ppm
K	0·05	0·15	28	36	36
	0·1	0·23	45	37	18
	0·2	0·29	69	35	5
	0·3	0·32	85	24	1
	0·5	0·36	94	5	1
Rb	0·05	0·14	15	39	46
	0·1	0·23	55	38	7
	0·2	0·29	65	31	4
	0·3	0·32	80	19	1
	0·5	0·37	95	3	2
Cs	0·05	0·17	22	57	19
	0·1	0·25	44	41	27
	0·2	0·3	69	24	4
	0·3	0·34	81	21	1
	0·5	0·41	95	3	1

Table 1. Nominal alkali (M=K, Rb, Cs) borophosphate glass compositions

M$_2$O (mol%)	B$_2$O$_3$ (mol%)	P$_2$O$_5$ (mol%)	K (P$_2$O$_5$/B$_2$O$_3$)	R (M$_2$O/B$_2$O$_3$)
4	80	16	0·2	0·05
8	77	15	0·2	0·1
14·5	71	14·5	0·2	0·2
20	67	13	0·2	0·3
29	59	12	0·2	0·5

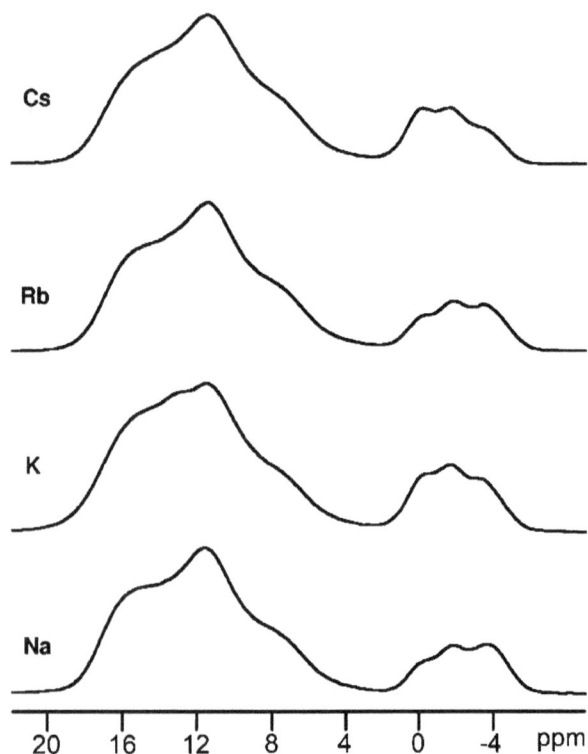

Figure 2. ^{11}B MAS NMR spectra of alkali borophosphate glasses (R=0·05) recorded at 14·1 T

to 20 ppm) exhibits features associated with ring and non-ring boron,[29,30] whereas the four-coordinated region (0 to −4 ppm) consists of a multiplicity of peaks governed by the degree of phosphate connectivity, especially for low alkali compositions. Only subtle spectral differences were observed for different alkali modifiers (Figure 2). Ultrahigh field MAS NMR was used in an attempt to achieve better resolution amongst the different three- and four-coordinated boron species (Figure 3). No improvement was observed in the four-coordinated boron region, and the spectral changes in the three-coordinated boron

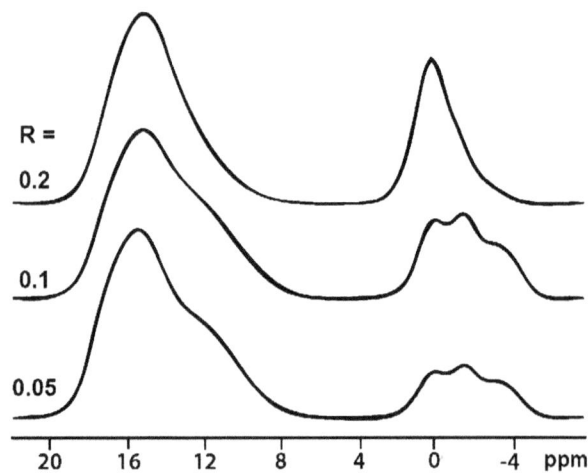

Figure 3. ^{11}B MAS NMR spectra of selected caesium borophosphate glasses recorded at 21·1 T

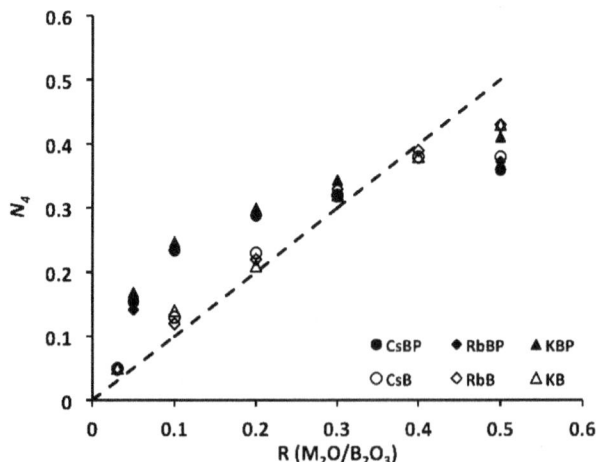

Figure 4. Fraction of four-coordinated boron, N_4, as a function of alkali content for alkali borate (open symbols) and alkali borophosphate glasses (closed symbols). Dotted line indicates N_4=R

region confirmed the presence of ring and non-ring boron without yielding appreciably better quantitative accuracy. The overall fraction of boron in four-coordination (N_4) was evaluated by peak integration of the 14·1 T data, and corrected for spinning effects.[31] These data are plotted as a function of alkali content in Figure 4, along with analogous data for binary alkali borates. Individual $^{[4]}$B sites were also quantified by peak fitting to obtain their evolution as a function of alkali loading (Figure 5).

^{11}B$\{^{31}$P$\}$ rotational echo double resonance (REDOR) experiments[32] were performed to determine which of the $^{[4]}$B peaks are subject to the greatest ^{31}P dipolar field, and hence to aid in assigning these peaks in terms of the number of bonded phosphate groups. Figure 6(a) indicates that in a caesium borophosphate glass (R=0·1), $^{[4]}$B sites experience a greater ^{31}P dipolar field than $^{[3]}$B sites. Figure 6(b) shows REDOR dephasing data for the three $^{[4]}$B peaks, with the inset illustrating the difference between the normal spin echo spectrum and the ^{31}P-dephased spin echo spectrum. The stronger dephasing observed for the lower frequency peaks reveals that they correspond to the most P-rich local environment.

^{31}P MAS NMR spectra are depicted in Figure 7 for caesium borophosphate glasses. At low alkali content two peaks are observed in approximately equal abundance, while higher alkali contents shift the peaks to higher frequency, indicating progressive depolymerization of phosphate groups in the network.

Quantum chemical calculations of clusters modelling $^{[4]}$B with increasing numbers of phosphate neighbours yield monotonically decreasing ^{11}B isotropic chemical shifts (i.e. more negative) with the number of bonded phosphate polyhedra (Figure 8). Combining peak assignments based on REDOR data and DFT calculations, the experimental ^{11}B chemical shift for

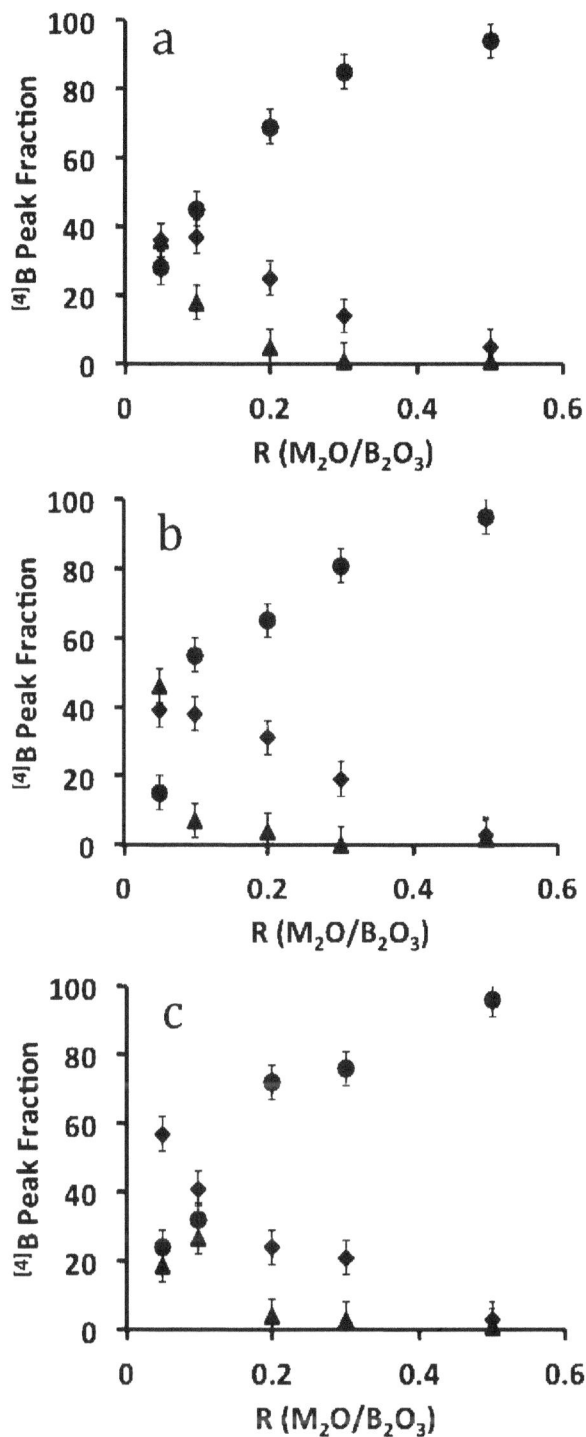

Figure 5. *Quantification of [4]B NMR peaks as a function of alkali content: (a) K, (b) Rb, (c) Cs*

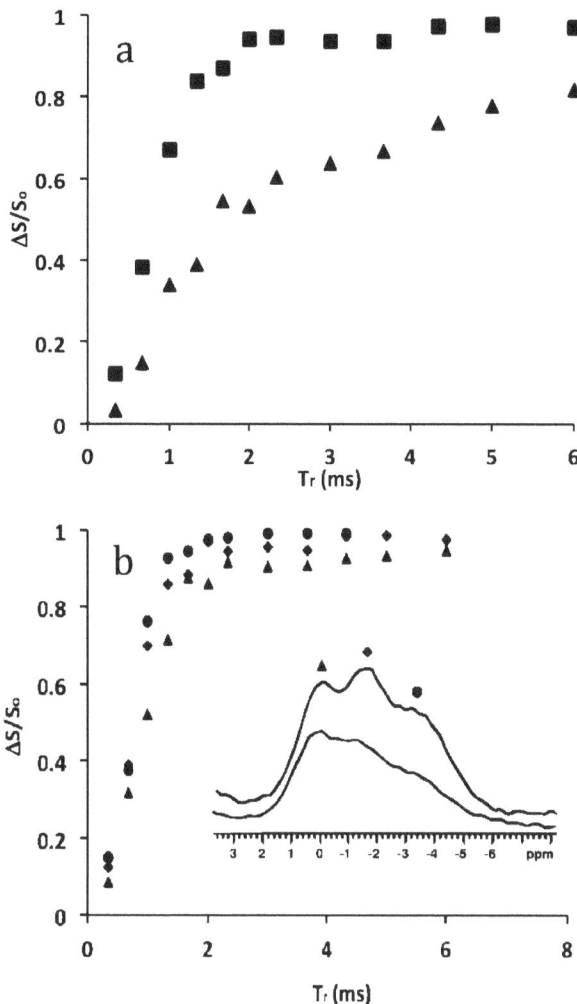

Figure 6. *[11]B{[31]P} REDOR dephasing curve for caesium borophosphate glass (R=0·1), depicting: (a) distinct dephasing behaviour for [3]B (triangles) and [4]B (squares); (b) dephasing curves for three [4]B signals. Inset compares [11]B spin-echo spectra with and without six ^{31}P dephasing pulses*

four-coordinated boron decreases by about 1·8 ppm with each additional bonded phosphate group.

Discussion

The fraction of four-coordinated boron provides the first clue that these mixed network former glasses have structural features that differ from either pure borate or pure phosphate glasses. The positive devia-

tion from the expected N_4=R relationship at low alkali content demands a charge-balancing mechanism in addition to the added alkali cations (Figure 4). The only such possibility in glasses of this composition is a positively charged phosphorus species, Q^4 (where Q^n indicates a PO_4 unit with n bridging oxygens), such as that found in crystalline BPO_4,[33] where it exists in strict alternation with anionic [4]B units. This species is unknown in pure phosphate glasses, where phosphorus is found as a neutral species with three bridging oxygens or as some anionic depolymerized form.

In principle, [4]B-Q^4 dimers (i.e. a [4]B and a Q^4 which are bonded to each other, and are otherwise bonded to [3]B units) could satisfy local charge balance requirements. However, the presence of three distinct [4]B peaks reflecting the number of bound phosphate polyhedra indicates that at least some of the anionic [4]B are surrounded by more than one Q^4. Since the REDOR and computational data provide only the

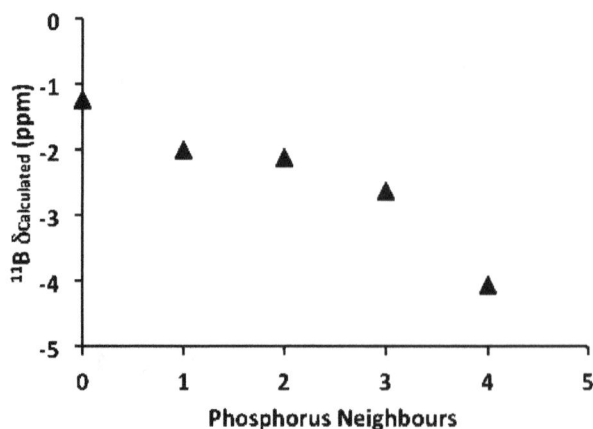

Figure 8. Correlation between calculated ^{11}B chemical shift of $^{[4]}B$ and number of bonded phosphate tetrahedra

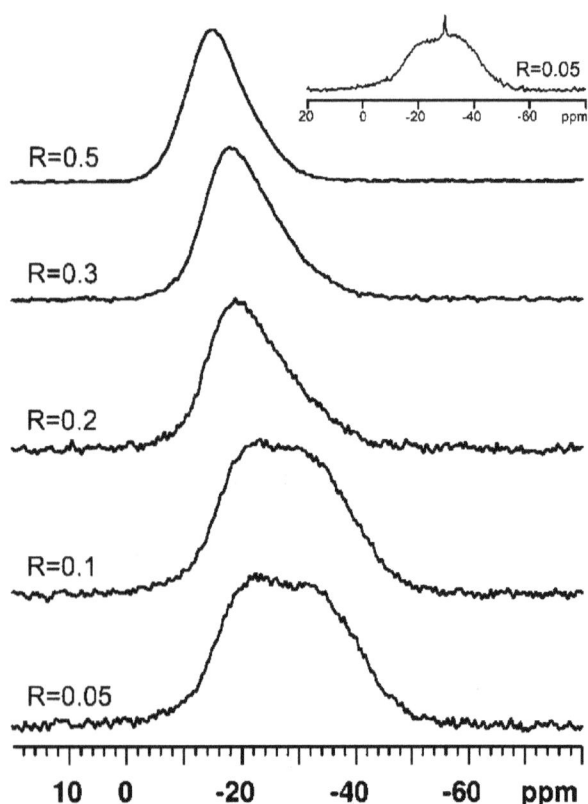

Figure 7. ^{31}P MAS NMR spectra of caesium borophosphate glasses recorded at 14·1 T. Inset shows low alkali rubidium borophosphate glass with devitrified BPO_4

relative number of bonded phosphate units, these peaks may, in principle, be assigned to $^{[4]}B$ bonded to 0, 1 and 2 phosphate units at a minimum, and 2, 3 and 4 phosphate units at a maximum, or possibly some combination thereof. However, previous work in sodium borophosphate glasses has established that these $^{[4]}B$ chemical shifts correspond to 2, 3 and 4 bonded P.[1,14] Using these assignments, mass balance calculations indicate that there are not enough phosphorus atoms available in the composition to allow for the observed number of $^{[4]}B(3P/4P)$ units unless P sharing (so that a phosphorus Q^4 unit is bonded to more than one $^{[4]}B$ unit) is invoked. Hence, the four-coordinated borons bonded to multiple phosphate tetrahedra are not isolated, but form clusters of (presumably) charge ordered BO_4^- and PO_4^+, similar to crystalline BPO_4.

Further evidence of such clustering comes from the ^{11}B and ^{31}P chemical shifts of crystalline BPO_4, which coincide with the most shielded $^{[4]}B$ (−3·3 ppm) and Q^4 peaks (−30 ppm),[9,11,12] suggesting that the local structural units in the low alkali glasses resemble the strictly alternating "order" of BPO_4. Finally, the ^{31}P MAS NMR spectrum of one sample (rubidium borophosphate, R=0·05) shows a sharp peak at −30 ppm superimposed upon the broader glassy signals (Figure 7), clearly indicative of a small fraction of crystallized BPO_4. This strongly suggests that the

glass structure is such that crystallization of BPO_4 is favourable under conditions of insufficiently rapid quenching.

Clustering of BPO_4-like regions in low alkali borophosphates is also consistent with the observation of B_2O_3-like ring/non-ring fractions at these compositions. With a sizeable proportion of $^{[4]}B$ and Q^4 units sequestered into nanoscale domains of a separate phase, the remaining glass matrix consists of a relatively small fraction of anionic $^{[4]}B$ and neutral Q^3 units which would barely perturb the ring/non-ring fraction.

The NMR spectra betray significant structural differences between the low alkali (4 and 8 mol%) and higher alkali compositional regimes, implying that the structural model described for the former does not apply to the latter. For example, the ^{11}B MAS NMR spectra of higher alkali glasses (Figure 1) show little evidence of multiple $^{[4]}B$, and the $^{[3]}B$ spectral region no longer resembles that of B_2O_3.[18,19,29,34] In the ^{31}P MAS NMR spectra (Figure 7), the glasses with alkali content greater than 14·5 mol% have little intensity in the Q^4 region (−30 ppm). The N_4 plot (Figure 4) exhibits a crossover of the ternary borophosphate glasses with the binary borates around R=0·3, above which the curves begin to "fan out" in a more complex manner related to the distribution of network modifiers between $^{[4]}B$ and other anionic species. In the high alkali compositional range there is no evidence of clustering, with all data pointing toward simple charge compensation of $^{[4]}B$ by alkali cations and progressive depolymerization of phosphate groups in the network, also by alkali cations. For these compositions, charge repulsion between the anionic $^{[4]}B$ and Q^2 phosphate species leads to preferential connectivity of phosphate units with $^{[3]}B$, resulting in the ^{11}B MAS NMR detection of a single $^{[4]}B$ environment (Figure 1). Accordingly, $^{11}B\{^{31}P\}$ REDOR experiments on a high alkali glass (R=0·5) show significantly weaker dephasing behaviour for $^{[4]}B$ signals than in the low alkali glasses (data not shown). This observation can be understood both by the avoidance of anionic [4]B and Q^2 phosphate species at high alkali levels principally leaving [3]B adjacent to [4]B units, and by the presence of BPO_4-like clustering at low alkali loadings bringing a high proportion of [4]B into close proximity to phosphate units. The ^{31}P MAS NMR results (Figure 7) support this connectivity arrangement in the high alkali region, with spectral intensity concentrated from −10 to −30 ppm, a region previously assigned to $Q^3(3B)$ and $Q^2(2B)$.[9,11,12]

One of the aims of this research is to discern whether heavy alkali cations have a different impact on the glass structure than the oft-studied lithium and sodium. Caesium borophosphate glasses are displayed throughout because these are expected to yield the greatest contrast; however, differences in the NMR spectra for different alkali modifier cations are subtle (Figure 2). The ^{31}P NMR peak intensities are nearly identical and the ^{11}B N_4 data are essentially indistinguishable in the low alkali region. However, similar to the binary alkali borates,[18,19] evidence of different boron speciation can be observed above $R=0.3$, implying that network modifier size and electropositive character exert a non-negligible influence upon the balance of three- and four-coordinated boron. This is presumably related to the preferential stabilization of nonbridging oxygens by heavier alkali cations inferred from the behaviour of binary alkali borates.[18,19] Moreover, small distinctions amongst the relative contributions of the different [4]B peaks are observed in the phase-separated compositional region (Figures 2 and 5).

DFT calculations were successful in assigning the [4]B peaks according to the clear shielding trend exhibited by the number of bonded phosphate groups, and agree with $^{11}B\{^{31}P\}$ REDOR experiments and previous work.[11,13–15] However, attempts to compare the numerical values of the calculated chemical shifts with the actual number of bonded phosphates failed due to the inaccuracy of the calculated values. Although the quantitative correlation is imperfect, it should be noted that very high level basis sets (6311++g(d,p) and aug-cc-pvtz) and geometry optimized clusters were necessary to produce a roughly linear trend. Finally, there is no reason to presume that the trend *should* be linear, especially as previous experimental work has suggested that the chemical shift difference between [4]B(1P) and [4]B(2P) may be smaller than that between [4]B(3P) and [4]B(4P).[11,17]

Conclusion

We have focused on the structural role of heavy alkali modifiers in a mixed network former borophosphate glass rich in borate, finding small but measurable size effects in high alkali glasses. At alkali oxide loadings less than about 10 mol%, clustering of BPO_4-like

amorphous regions segregate from a major glassy phase comprising principally [3]B species and a small fraction of anionic [4]B and neutral Q^3 phosphate species. At higher alkali loadings, the apparently homogeneous glass consists mainly of [4]B and Q^2 phosphate groups interspersed amongst intervening [3]B. The presence of two distinct compositional regimes, phase separation and the complexity introduced by the mixed network former environment precludes a comprehensive account of the speciation and connectivity beyond the first coordination sphere. However, it is clear that any potential applications of these materials will be contingent on careful compositional and morphological control.

Acknowledgements

The authors are grateful to Dr Kirk Marat (University of Manitoba) and Dr Victor Terskikh (National Ultrahigh-Field NMR Facility for Solids) for technical assistance. SK thanks the Natural Sciences and Engineering Research Council of Canada, the Canada Foundation for Innovation, and the Manitoba Research Innovation Fund for infrastructure and operational support. VKM acknowledges NSERC for a post-graduate scholarship (PGSD3). Access to the 900 MHz NMR spectrometer was provided by the National Ultrahigh-Field NMR Facility for Solids (Ottawa, Canada), a national research facility funded by CFI, the Ontario Innovation Trust, Recherche Québec, the National Research Council Canada and Bruker BioSpin, and managed by the University of Ottawa (www.nmr900.ca). NSERC is acknowledged for a Major Resources Support grant.

References

1. Zielniok, D., Cramer, C. & Eckert, H. *Chem. Mater.*, 2007, **19**, 3162.
2. Saranti, A., Koutselas, I. & Karakassides, M. A. *J. Non-Cryst. Solids*, 2006, **352**, 390.
3. Lee, S., Kim, J. & Shin, D. *Solid State Ionics*, 2007, **178**, 375.
4. Karan, N. K., Natesan, B. & Katiyar, R. S. *Solid State Ionics*, 2006, **177**, 1429.
5. Shelby, J. E. *Introduction to Glass Science and Technology.* 2nd ed., Royal Society of Chemistry, Cambridge, 2005, p 290.
6. Kumar, S., Vinatier, P., Levasseur, A. & Rao, K. J. *J. Solid State Chem.*, 2004, **177**, 1723.
7. Takebe, H., Harada, T. & Kuwatara, M. *J. Non-Cryst. Solids*, 2006, **352**, 709.
8. Videau, J. J., Ducel, J. F., Suh, K. S. & Senegas, J. *J. Alloy Compounds*, 1992, **188**, 157.
9. Carta, D., Qiu, D., Guerry, P., Ahmed, I., Abou Neel, E. A., Knowles, J. C., Smith, M. E. & Newport, R. J. *J. Non-Cryst. Solids*, 2008, **354**, 3671.
10. Ducel, J. F., Videau, J. J., Suh, K. S. & Senegas, J. *Phys. Chem. Glasses*, 1994, **35**, 10.
11. Elbers, S., Strojek, W., Koudelka, L. & Eckert, H. *Solid State Nucl. Magn.*, 2005, **27**, 65.
12. Mackenzie, K. J. D. & Smith, M. E. *Multinuclear Solid-State NMR of Inorganic Materials*, Pergamon, Oxford, 2002, p 727.
13. Raskar, D. B., Eckert, H., Ewald, B. & Kniep, R. *Solid State Nucl. Magn.*, 2008, **34**, 20.
14. Rinke, M. T. & Eckert, H. *Phys. Chem. Chem. Phys.*, 2011, **13**, 6552.
15. Strojek, W., Felise, C. M., Eckert, H., Ewald, B. & Kniep, R. *Solid State Nucl. Magn.*, 2007, **32**, 89.
16. Martineau, C., Michaelis, V. K., Schuller, S. & Kroeker, S. *Chem. Mater.*,

2010, **22**, 4896.

17. Zeyer-Dusterer, M., Montagne, L., Palavit, G. & Jäger, C. *Solid State Nucl. Magn.*, 2005, **27**, 50.

18. Kroeker, S., Aguiar, P. M., Cerquiera, A., Clarida, W. J., Doerr, J., Olesiuk, M., Ongie, G., Affatigato, M. & Feller, S. A. *Phys. Chem. Glasses: Eur. J. Glass Sci. Technol. B*, 2006, **47**, 393.

19. Michaelis, V. K., Aguiar, P. M. & Kroeker, S. *J. Non-Cryst. Solids*, 2007, **353**, 2582.

20. Michaelis, V. K. & Kroeker, S. *Phys. Chem. Glasses: Eur. J. Glass Sci. Technol. B*, 2009, **50**, 249.

21. Frisch, M. J., Trucks, G. W., Schlegel, H. B., Scuseria, G. E., Robb, M. A., Cheeseman, J. R., Montgomery, J.A., Vreven, T., Kudin, K. N., Burant, J. C., Millam, J. M., Iyengar, S. S., Tomasi, J., Barone, V., Mennucci, B., Cossi, M., Scalmani, G., Rega, N., Petersson, G. A., Nakatsuji, H., Hada, M., Ehara, M., Toyota, K., Fukuda, R., Hasegawa, J., Ishida, M., Nakajima, T., Honda, Y., Kitao, O., Nakai, H., Klene, M., Li, X., Knox, J., Hratchian, H. P., Cross, J. B., Bakken, V., Adamo, C., Jaramillo, J., Gomperts, R., Stratmann, R. E., Yazyev, O., Austin, A. J., Cammi, R., Pomelli, C., Ochterski, J. W., Ayala, P. Y., Morokuma, K., Voth, G. A., Salvador, P., Dannenberg, J. J., Zakrzewski, V. G., Dapprich, S., Daniels, A. D., Strain, M. C., Farkas, O., Malick, D. K., Rabuck, A. D., Raghavachari, K., Foresman, J. B., Ortiz, J. V., Cui, Q., Baboul, A. G., Clifford, S., Cioslowski, J., Stefanov, B. B., Liu, G., Liashenko, A., Piskorz, P., Komaromi, I., Martin, R. K., Fox, D. J., Keith, T., Al-Laham, M. A., Peng, C. Y., Nanayakkara, A., Challacombe, M., Gill, P. M. W., Johnson, B., Chen, W., Wong, M. W., Conzalez, C. & Pople, J. A. *Gaussian 03*, Gaussian: 2003.

22. Becke, A. D. *Phys. Rev. A*, 1988, **38**, 3098.

23. Becke, A. D. *J. Chem. Phys.*, 1993, **98**, 5648.

24. Lee, C., Yang, W. & Parr, R. G. *Phys. Rev. B*, 1988, **37**, 785.

25. Dunning, T. H. *J. Chem Phys.*, 1989, **90**, 1007.

26. Kendall, R. A., Dunning, T. H. & Harrison, R. J. *J. Chem. Phys.*, 1992, **96**, 6796.

27. Krishnan, R., Binkley, J. S., Seeger, R. & Pople, J. A. *J. Chem. Phys.*, 1980, **72**, 650.

28. McLean, A. D. & Chandler, G. S. *J. Chem. Phys.*, 1980, **72**, 5639.

29. Kroeker, S. & Stebbins, J. F. *Inorg. Chem.*, 2001, **40**, 6239.

30. Youngman, R. E., Haubrich, S. T., Zwanziger, J. W., Janicke, M. T. & Chmelka, B. F., *Science*, 1995, **269**, 1416.

31. Massiot, D., Bessada, C., Coutures, J. P. & Taulelle, F. *J. Magn. Reson.*, 1990, **90**, 231.

32. Gullion, T. & Schaefer, J. *J. Magn. Reson.*, 1989, **81**, 196.

33. Schulze, G. E. R. *Z. Phys. Chem.*, 1934, **24**, 215.

34. Aguiar, P. M. & Kroeker, S. *J. Non-Cryst. Solids*, 2007, **353**, 1834.

Phys. Chem. Glasses: Eur. J. Glass Sci. Technol. B, February 2013, **54** (1), 42–51

Spectroscopic study of manganese-containing borate and borosilicate glasses: cluster formation and phase separation

Doris Möncke,*[a,b] *Doris Ehrt*[a] *& Efstratios I. Kamitsos*[b]

[a] *Otto-Schott-Institut für Glaschemie, Friedrich-Schiller-Universität Jena, Fraunhoferstr. 6, 07743 Jena, Germany*
[b] *Theoretical and Physical Chemistry Institute, National Hellenic Research Foundation, 48 Vassileos Constantinou Avenue, 11635 Athens, Greece*

Manuscript received 30 November 2011
Revised version received 22 March 2012
Accepted 23 July 2012

The coordination and bonding of Mn^{2+} ions in glasses can be probed sensitively and selectively by electron paramagnetic resonance (EPR) and photoluminescence spectroscopy. These methods also give information on Mn–Mn ion interactions and cluster formation. Mn^{2+} ions were found to be tetrahedrally coordinated in borosilicate glasses of high optical basicity, and octahedrally coordinated in low alkaline borosilicate glasses (Duran-type) as well as in binary borate glasses. Broad emission bands and multicomponent fluorescence decay curves in Duran glasses indicate very strong Mn–Mn ion interactions and the presence of multiple Mn^{2+} sites, even at low Mn-levels. The EPR spectra show exchange narrowing with increasing Mn content in the Duran series, which is caused by a decrease in the Mn–Mn distances as edge sharing MnO_6 octahedra are formed. The network structure of Mn-containing binary and ternary borate glasses is discussed on the basis of their infrared spectra. Addition of MnO to Duran glasses is found to cause the preferential transformation of $[BO_3]^0$ to $[BO_4]^-$ groups, and to a lesser extent of silicate Q^4 to Q^3 units. An increase of the relative population of homopolar B–O–B or Si–O–Si bonds, with the simultaneous decrease in the number of mixed B–O–Si bonds, is also observed, and this explains the visible phase separation of Duran glasses when MnO is added in excess of 4 mol%.

1. Introduction

Manganese occurs in glasses as divalent or trivalent ions. Both Mn^{2+} and Mn^{3+} have been extensively studied spectroscopically with regards to their coordination and bonding environment. The different valence states can be distinguished by optical spectroscopy.[1–6] However, the presence of Mn^{3+} can easily obscure the transitions of Mn^{2+} ions in the optical spectra, especially for higher manganese levels. Photoluminescence and EPR (electron paramagnetic resonance) spectroscopy on the other hand allow the study of a wide concentration range without interference from Mn^{3+} ions, due to their selectivity towards the divalent ions. For low Mn^{2+} contents, EPR spectroscopy provides information on site symmetry, via the hyperfine splitting parameters of the sextet resonance,[6–11] and on Mn-bonding characteristics.[12] Dipole–dipole interactions between Mn^{2+} ions can be studied at higher Mn-levels, as this may either cause linewidth broadening,[13,14] or exchange narrowing of the main EPR resonance.[14,15] Tetrahedrally and octahedrally coordinated Mn^{2+} ions can be distinguished by the position of the emission maximum in the fluorescence spectra.[5,7,16–18] The position and form of

the emission maximum depends also on clustering effects and the existence of multiple Mn-sites. Cluster formation and concentration quenching also lead to decreased luminescence intensities and to shorter fluorescence lifetimes.

A previous paper discussed in detail the role of Mn^{2+} as a structural probe ion for EPR and luminescence spectroscopic studies in very different glass systems.[7] The role of the divalent manganese ion not only as a structure indicator, but also as an intermediate glass former was explored in strontium metaphosphate glasses in detail by Konidakis *et al*,[6] who showed that increasing MnO content leads to the formation of clusters consisting of an increasing number of corner sharing MnO_6 octahedra with a considerable covalent character in the Mn–O bonding.

Cluster formation in borosilicate glasses behaves quite differently from phosphate glasses. EPR spectroscopy, for example, shows exchange narrowing with increasing Mn-levels, but not the previously observed linewidth broadening of the $g=2$ resonance.[6,7] Furthermore, phase separation is already observed at a relatively low doping level of only 4 mol% MnO. Fluorescence spectroscopy gives very short lifetimes with multi-exponential lifetime curves for the borosilicate glasses, which is indicative of a complex site distribution. The opposite was observed for

Corresponding author. Email doris.moencke@uni-jena.de
Original version presented at VII Int. Conf. on Borate Glasses, Crystals and Melts, Dalhousie University, Halifax, Nova Scotia, Canada, 21–25 August 2011

phosphate glasses for which mono-exponential fits described the decay curves very well, as do Förster fits which consider the effect of concentration quenching that is expected to occur at higher Mn levels.[7,17,18]

Infrared spectroscopy is employed in this study as an additional investigative tool to obtain more information on MnO_6 cluster formation in Duran-like borosilicate glasses. For comparison, different Mn containing borate glasses were also studied, such as manganese, sodium and zinc borates. Mn-doped zinc borate glass was studied due to the similarities between Mn^{2+} and Zn^{2+} ions, including their preference for either four- or six-fold coordination, similar ionic radii of ca. 0·6 Å (CN=4) and ca. 0·8 Å (CN=6), and the fact that both Mn^{2+} and Zn^{2+} ions show photoluminescence. In addition, information is already available for the zinc borate system, such as the phase diagrams from Hummel & Harrison.[19,20]

2. Glasses and methods

2.1. Glass samples

Glass compositions and values of selected physical properties, including density, ρ, glass transition temperature, T_g, and optical basicity, Λ, are listed in Table 1. The Mn-dopant concentration of the glasses varies from 10^{19} to 10^{22} Mn^{2+}/cm^3. Most samples were melted under reducing conditions, resulting in colourless to weak yellow coloured glasses. The binary manganese borate glasses were prepared at 1100–1300°C. Liquid–liquid phase separation of the $MnO.4B_2O_3$ and $MnO.5B_2O_3$ melts[20] resulted in an upper layer of partially crystallized H_3BO_3,[20] and a lower layer of vitreous manganese borate. This glass was transparent, with a light orange colour, and in

analogy to the $ZnO–B_2O_3$ phase diagram its composition was assumed to correspond to ca. $50MnO.50B_2O_3$, which is consistent with the measured physical properties listed in Table 1[20] (where this glass is denoted "MnB phase"). Mn-doped zinc borate glasses were similarly prepared at 1100°C. The borosilicate glasses were melted at 1650°C, and all samples were carefully annealed. More details on preparation, properties and structure of the different glasses studied can be found elsewhere.[6,7,16–18,20–22] High purity raw materials were used for glass preparation to ensure that EPR signals due to Fe^{3+} or other paramagnetic impurities are absent in all samples.

2.2. Spectroscopy

2.2.1. Optical absorption spectroscopy and optical basicity

Optical absorption spectra were recorded on a conventional double beam spectrometer (UV–VIS–NIR Perkin Elmer, Lambda 19), on polished sample plates of thickness from 0·5 mm to 10 mm. For details on the quantitative analysis of Mn^{3+} in the Mn/Sr metaphosphate series, for which a lower molar extinction coefficient of $\varepsilon(Mn^{3+})$=25 L $mol^{-1}cm^{-1}$ was used, see Refs 1, 2 and 6, and references therein.

The optical absorption spectra displayed in Figure 1 show d–d electronic transitions of Mn^{2+} (d^5) with a dominant peak at ca. 415 nm.[1–7,16–18] All transitions of the d^5 Mn^{2+} ion are spin forbidden, and the molar extinction coefficient of these transitions is about 250 times lower than that of the spin allowed transitions of the d^4 ion Mn^{3+}.[6] As a result, even traces of Mn^{3+} ions can mask the transitions of Mn^{2+} ions and bestow the often observed deep purple colour on glass sam-

Table 1. Compositions (mol%) and selected properties of the studied glasses; glass transition temperature, T_g, refractive index, n_e (546 nm), Abbe number, v_e, density, ρ, optical basicity, Λ (Λ_{th} is the calculated average basicity, and Λ_{Mn} is the site basicity determined from the Mn-absorption), fluorescence maxima for extinction, Ex-λ_{max}, and emission, Em-λ_{max}, luminescence lifetime, τ_e, and Mn^{2+} coordination number, Mn-CN

Glass sample	NCS	NBS1	NBS2	Duran series	MnNB	MnB-phase	ZnB
Glass composition in mol%							
SiO_2	74	74	74	83	-	-	
B_2O_3	–	10	20·7	12	83·3	50	50
Na_2O	16	16	4·3	4	12·7	-	
Other oxides	10 CaO	-	1 Al_2O_3	1 Al_2O_3	-	-	50 ZnO
MnO	(ppm)	(ppm)	(ppm)	+0–4%	4	50	$1\times10^{20} Mn^{2+}/cm^3$
Thermal and optical properties							
T_g (°C) ±5	530	553	442	530	455	580	570
n_e±0·0001	1·52	1·51	1·47	1·473	n.m.	1·654	1·646
v_e±0·6	60	63	65	66	n.m.	55	52
ρ (g/cm³)±0·01	2·49	2·45	2·18	2·22	2·19	2·97	3·37
Optical basicity							
Λ_{th}	0·57	0·55	0·49	0·49	0·46	0·57	0·55
Λ_{Mn}	0·64	0·64	0·54	0·61	0·48	0·57	0·53
Photoluminescence data for Mn^{2+} doping level of 5000 ppm							
Mn^{2+}/cm^3	$1·4\times10^{20}$	$1·3\times10^{20}$	$1·2\times10^{20}$	1×10^{20}	$0·8\times10^{21}$	1×10^{22}	1×10^{20}
Ex-λ_{max}(nm)	420+430	420+430	412	417	407	625	411
Em-λ_{max} (nm)	525	525	640	640	647	730	605
τ_e	8 ms	8 ms	2 ms	2 ms	5 ms	10 µs+100 µs	7 ms
Mn-CN	4	4	6	6	6	6	6

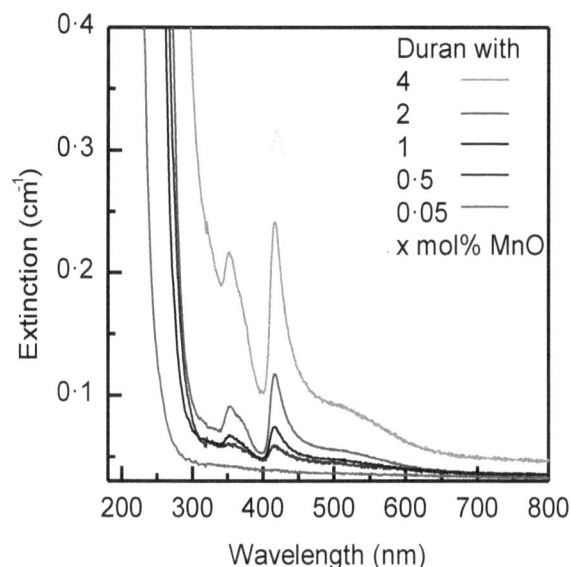

Figure 1. Optical absorption spectra of Mn-doped Duran-like glasses melted under reducing conditions. The arrow shows the slight shift to longer wavelengths in the position of the $^6A_{1g}(S) \rightarrow {}^4E_g(G)$ and $^6A_{1g}(S) \rightarrow {}^4A_{1g}(^4G)$ transitions of Mn^{2+} with increasing Mn concentration. The same transition is used for the determination of the local optical basicity of Mn sites, Λ_{Mn} [In colour online]

ples.[1–7] If not stated otherwise, all samples discussed in this paper were melted under reducing conditions and so the effect of trivalent Mn ions can be ignored.

The optical basicity concept, as introduced by Duffy & Ingram,[3,4] is used here to express the average electron donor power in Mn-doped glasses, and to explore its effect on the Mn bonding characteristics. Individual oxide basicity values are available from Duffy & Ingram for a wide range of oxides,[3] and the average optical basicity of glass, Λ_{th}, was calculated accordingly from the equivalent fractions of the oxides, X_A ..., and the corresponding basicities, $\Lambda(A)$ ..., of the glass constituents:

$$\Lambda = X_A \Lambda(A) + X_B \Lambda(B) + \dots \quad (1)$$

with $\Lambda_{SiO_2} = 0.48$, $\Lambda_{B_2O_3} = 0.42$, $\Lambda_{Al_2O_3} = 0.4$, $\Lambda_{Na_2O} = 1.15$, $\Lambda_{CaO} = 1.00$, $\Lambda_{ZnO} = 1.03$, and $\Lambda_{MnO} = 1.0$.[3,4,6]

The Mn^{2+} ion can also be used as a probe ion for the determination of the local Mn site basicity. The energy of the band corresponding to the $^6A_{1g}(S) \rightarrow {}^4E_g(G)$ and $^6A_{1g}(S) \rightarrow {}^4A_{1g}(^4G)$ transitions is related to a change in the electron density provided by the anions of the glass matrix as depicted in Equation (2).[1–7] The parameters needed are: $\nu_{S \rightarrow G}$, the measured absorption maximum of the $^6S \rightarrow {}^4G$ transition; $\nu^f_{S \rightarrow G}$, the value of the "free" Mn^{2+} ion frequency of 26 846 cm^{-1}; k, which is specific to the metal ion with $k = 0.0688$ for Mn^{2+}; and the value 2.56 which is specific for the oxygen ligands:[4]

$$\Lambda_{site} = \frac{\nu^f_{S \rightarrow G} - \nu_{S \rightarrow G}}{2.56 k \nu^f_{S \rightarrow G}} \quad (2)$$

2.2.2. Photoluminescence spectroscopy

Luminescence spectra were obtained on a single beam spectrometer (Shimadzu RF-5301PC) in reflection mode, using comparable slit widths for all glasses measured in this work. The emission intensities and maxima were obtained after excitation in the maximum around 400 nm. Measurements were performed on 10 mm thick polished glass plates with dimensions of 15×20 mm^2. Fluorescence lifetimes were obtained for the emission maxima, employing a home-made system with excitation at 337 nm provide by focused N_2-laser pulses (<500 ps).[17,18] The lifetime, τ_e, corresponds to the time required for the normalized fluorescence intensity induced by the laser pulses to drop to $1/e$ (i.e. 36.8%) of its initial value.[17,18]

2.2.3. Electron paramagnetic resonance spectroscopy

EPR spectra were recorded at X-band frequency (Bruker, ~10 GHz) at room temperature with the addition of the spin standard diphenylpicrylhydrazyl (dpph) for spectral calibration; for details see Ref. 7. The linewidth, ΔH_{pp}, is defined as the peak-to-peak distance of the derivative spectra. The hyperfine coupling constant, A_{hfs}, was estimated from the measured peak-to-peak distance of the Hfs (hyperfine splitting) signal and was taken as the average value, A_{av} of the differences of peak-to-peak, Δ_{pp}, and trough-to-trough, Δ_{tt}, distances

$$A_{av} = \frac{1}{6} \left[\frac{\Delta_{Opp} + \Delta_{Ott}}{5} + \frac{\Delta_{Mpp} + \Delta_{Mtt}}{3} + \frac{\Delta_{Ott} + \Delta_{Itt}}{1} \right] \quad (3)$$

where Δ_O, Δ_M and Δ_I refer to the differences of the outer (1st and 6th), middle (2nd and 5th) and inner (3rd and 4th) lines, respectively.[7,10] The peak-to-peak distance increases with increasing magnetic field, and differs when measured either peak-to-peak or trough-to-trough, due to strain broadening of the individual hyperfine lines. However, since similar values are found for each average of outer, middle and inner distances, it can be said that the hyperfine splitting is well behaved in a statistical sense, and that the increased differences in the distances with the magnetic field are due to instrumental strain broadening, and not to site asymmetry effects.[10] An attempt to simulate the spectra with EasySpin software was not successful because of the overlap of the singlet and the sextet spectrum.

2.2.4. Vibrational spectroscopy

Infrared (IR) measurements were performed on polished glass samples using a vacuum Fourier transform spectrometer. The IR spectra were measured in reflectance mode at quasi-normal incidence

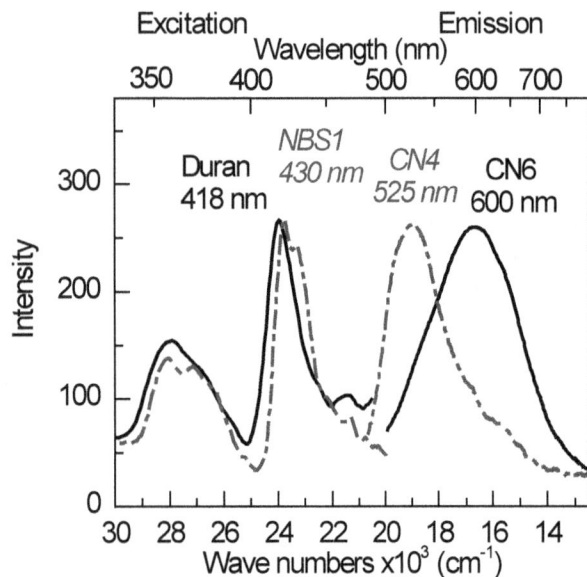

Figure 2. Photoluminescence excitation and emission spectra of two Mn-doped borosilicate glasses. The emission maximum shifts from 525 nm in the high alkaline borosilicate glasses NBS1 ($CN_{Mn}=4$) to 600 nm in the low alkaline borosilicate Duran-like glass ($CN_{Mn}=6$) [In colour online]

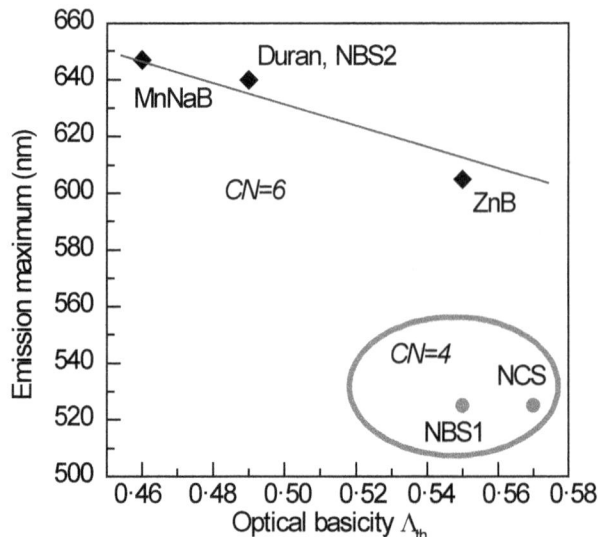

Figure 3. Photoluminscence emission maxima versus theoretical optical basicity for borosilicate glasses containing 1×10^{20} Mn^{2+} ions/cm^3 (1×10^{22} Mn^{2+} ions/cm^3 for the manganese metaborate glass phase, MnB). The line is a guide to the eye, for glasses with predominantly six-fold coordinated Mn^{2+} ions (black diamonds). Glasses with predominantly four-fold coordination for Mn^{2+} ions are depicted by red circles [In colour online]

(~11°) in the range 30–7000 cm^{-1} with 2 cm^{-1} resolution, and each spectrum represents the average of 200 scans. The reflectance spectra were analyzed by Kramers–Kronig transformation[23] to yield the absorption coefficient spectra, $a(v)$, from the expression $a(v)=4\pi vk(v)$, where $k(v)$ is the imaginary part of the complex refractive index and v is the infrared frequency in cm^{-1}. Raman spectroscopy is very helpful in analyzing the structure of Mn-doped metaphosphate glasses.[6] However, the borate and borosilicate glasses studied in this work were found to exhibit very strong fluorescence backgrounds from Mn^{2+} ions when excited with the available laser lines (488, 514, 633 and 785 nm), and this effect prevented the acquisition of useful Raman spectra.

3. Results and discussion

3.1. Photoluminescence measurements

Figure 2 shows the excitation and emission spectra of two borosilicate glasses. The emission of Mn^{2+} consists of one broad band attributable to the transition from the excited $^4T_1(G)$ level to the $^6A_1(S)$ ground state. With increasing optical basicity of the glass matrix (Table 1) the excitation maximum shifts to longer wavelengths, similar to the shift observed in the absorption maximum. The energy of the $^4T_1(G)$ state depends strongly on the bonding environment of Mn^{2+} ions, and consequently the emission energy varies with coordination number, or generally with the electron donor power of the ligands.

Alkaline borosilicate glasses of higher optical basicity provide a significant number of nonbridging

oxygen atoms to act as ligands for coordination with Mn^{2+} ions, and show fluorescence in the green, which is typical for four-fold coordinated Mn^{2+} ions.[7,16–18] A strong crystal field, as for octahedral coordination, results in an orange to red emission.[5–7,16–18] The glasses NCS and NBS1 show green emission around 525 nm, and the low alkaline borosilicate glass Duran shows orange to red emission around 600 nm. The borate glasses gave also an orange emission indicative of octahedral Mn^{2+} coordination. For glasses of comparable Mn concentration, the emission maximum shifts to longer wavelengths as the optical basicity decreases (Figure 3). The wavelength of the emission maximum tends to increase with increasing Mn concentration within the Duran glass series. In the same series, the emission intensity at lower Mn levels increases with increasing Mn^{2+} concentration, but decreases again for higher Mn levels as the Mn^{2+} concentration increases further.

The broadness of the emission band of Duran glasses (Figure 2) indicates a broad distribution of Mn^{2+} sites,[7,16,18] which is also supported by the fluorescence decay curves (not shown here; see for example Ref. 7). The bend form of these curves, which need multi-exponential fits, the relative short lifetimes, and the fact that two distinct lifetime curves are obtained for different excitation wavelengths, are all indicative of a diverse site distribution and of strong Mn–Mn interactions, such as found in manganese clusters.[7,16–18] Fluorescence lifetimes of high alkaline borosilicate glasses are somewhat higher than those of low alkaline borosilicate glasses, while the rate of

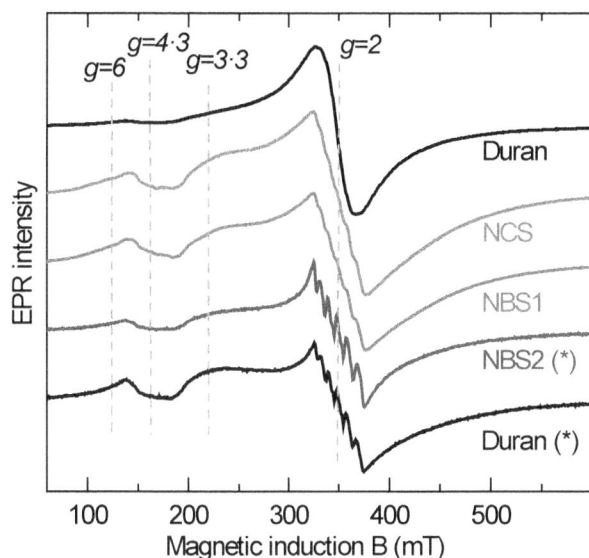

Figure 4. EPR spectra of borosilicate glasses doped with 5000 ppm MnO; the NBS2 and Duran-like samples (*) contain only 500 ppm MnO [In colour online]

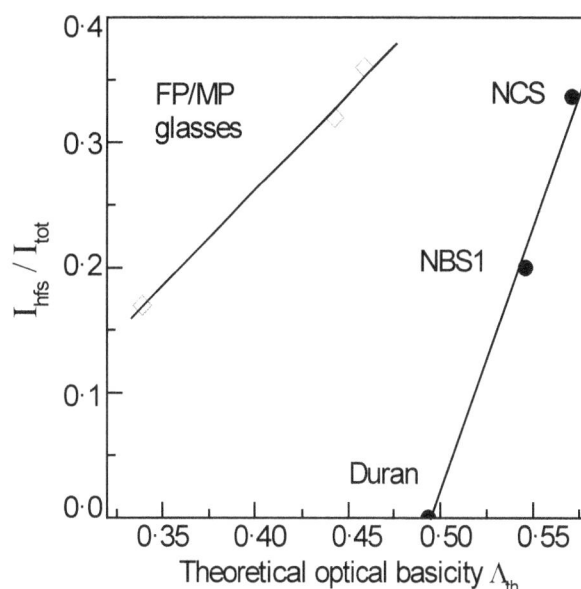

Figure 5. Ratio of the intensity of the single EPR hfs lines, I_{hfs}, over the total intensity, I_{tot}, of the g=2 resonances for borosilicate glasses (black, circles). Data for fluoride phosphate glasses (diamonds) are included for comparison.[7] All glasses contain 5000 ppm, ~$1×10^{20}$ Mn^{2+}/cm^3

concentration quenching in NCS and NBS1 samples is found to be lower than the rate observed for Duran-type glasses.[16–18] The fluorescence emission of Mn^{2+} is found in the red for both the manganese borate and the Mn-doped zinc borate glasses, indicating that Mn^{2+} ions are six-fold coordinated with oxygen atoms.

3.3. EPR spectroscopy

EPR spectra and diverse assignments of the observed resonances are discussed in detail in Refs 6 and 7, and references therein. All Mn containing glasses studied here show the characteristic spectra of Mn^{2+} with a strong signal centred close to the free electron value of g=2·0023, which is indicative of Mn^{2+} ions in nearly perfect cubic symmetry.[1,6–10] Additional much weaker signals are apparent at lower magnetic fields at g=3·3 and between g=4·3 and g=6 (Figure 4).

The main resonance at g=2 is due to allowed m_s ½→−½ magnetic dipolar transitions from the $^6S_{5/2}$ ground state of Mn^{2+}. The signature pattern, i.e. hyperfine splitting, arises from the interaction between the spin of the unpaired $3d^5$ electrons ($S=^5/_2$) and the spin of the ^{55}Mn nucleus, $I(^{55}Mn, 100\%)=^5/_2$.[1,6–13] The Hfs structure arises from highly symmetric sites in which isolated Mn^{2+} ions are well separated from each other, without magnetic interactions. The downward slope, which is apparent even for very low Mn levels (see for example Figure 4), results from the superposition of the sextet with a broad singlet signal due to sites in which Mn^{2+} ions are positioned close enough to experience magnetic interaction.

The low field signals are associated with Mn^{2+} ions in tetragonal (g=3·3), or rhombic distorted (g=4·3) symmetries.[1,6–11] Relative to the g=2 resonance these

signals are more pronounced for glasses in which Mn^{2+} is four-fold coordinated than in glasses in which Mn^{2+} is six-fold coordinated. The glasses of this study generally show only weak low field signals compared to the main resonance at g=2, indicating that manganese ions are predominantly present in undistorted cubic sites.

As already mentioned, the hfs structure is indicative of isolated Mn ions. Plotting the ratio of the hfs intensity, I_{hfs}, to the overall signal intensity, I_{tot}, at g=2 for glasses of comparable Mn levels, it is found that I_{hfs}/I_{tot} increases with increasing average optical basicity of glass (Figure 5). However, this trend may not reflect directly the impact of the average glass basicity, but rather the preference of Mn^{2+} ions to coordinate to certain ligands, resulting in slightly different local optical basicity values. For example, for low average basicity fluoride phosphate glasses (FP) with low phosphate levels, the local Mn site basicities are significantly higher than the theoretical basicity values, indicating that Mn^{2+} ions prefer coordination to phosphate rather than fluoride ligands. The higher the phosphate content of FP glasses, the higher the Mn levels at which the hfs lines start to disappear.[7] Thus the preference of Mn^{2+} ions for phosphate ligands in FP glasses, or for borate ligands in borosilicate glasses, results in a heterogeneous Mn^{2+} ion distribution, as more Mn^{2+} ions concentrate in glass regions that contain more of the preferred ligands. Because of the closer proximity of the Mn^{2+} ions and the subsequent onset of magnetic interactions, the hfs structure already starts to disappear at relatively low overall Mn concentrations. Even at doping levels of

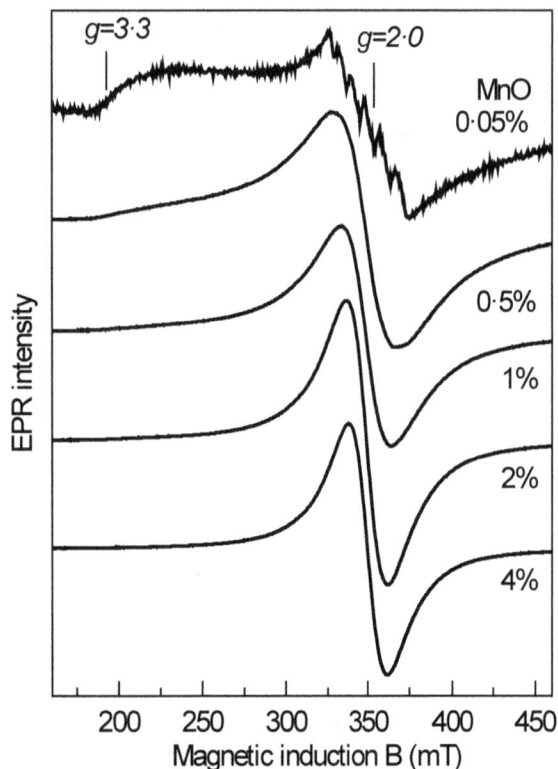

Figure 6. The effect of manganese concentration on the EPR spectra of Duran-like glasses (intensity not to scale)

only 500 ppm, Duran glass, with only 12 mol% B_2O_3, shows a less well resolved hfs than NBS2, which has 21 mol% B_2O_3. This signifies that only a few isolated Mn^{2+} sites are present in Duran. Mn^{2+} sites are also more evenly distributed in NBS1 and NCS glasses, since the coordinating ligands provided by the nonbridging oxygen atoms of the silicate groups are spread over the whole glass volume. NMR studies of NBS1 glass showed that nonbridging oxygen atoms are bonded to silicate and not to borate groups. Therefore, in comparison to the soda–lime–silica glass NCS, the Mn^{2+} ions in NBS1 exhibit a lower I_{hfs}/I_{tot} ratio (Figure 5) because they are restricted to a smaller portion of the glass structure.[21]

EPR spectra of the concentration series of MnO-doped Duran glasses are shown in Figure 6. The intensity of the main signal at $g=2$ increases with increasing Mn levels. The typical sextet due to isolated Mn^{2+} ions disappears for higher dopant levels, and merges with the underlying broad singlet resonance from sites which encounter Mn–Mn interactions.[1,6–13]

The phenomenon of exchange narrowing is clearly evident, as the overall linewidth of the resonance at $g=2$ falls with the addition of MnO from 50 to 25 mT. This is contrary to observations made for the metaphosphate series MnO_x–SrO_{1-x}–P_2O_5 which showed a doubling of the linewidth from 50 to 100 mT with progressive replacement of SrO by MnO.[6,7]

The linewidth ΔH_{pp} depends on several factors:

$$\Delta H_{pp}=\Delta H_{total}=\Delta H_{nat}+\Delta H_{dd}+\Delta H_{ex}+\Delta H_o \qquad (4)$$

where ΔH_{nat} is the natural linewidth, ΔH_{dd} arises from dipole–dipole interaction, ΔH_{ex} from exchange interactions, and ΔH_o from other factors, such as the distribution of distorted sites in glasses or a non-resolved hyperfine splitting. The natural linewidth ΔH_{nat} can be obtained by linear extrapolation of ΔH to zero Mn^{2+} concentration, and depends only on the spin–lattice and spin–spin relaxation. The natural linewidth of Mn^{2+} in glasses may be close to $\Delta H_{nat}=50$ mT, the value found for very low Mn concentration (100–500 ppm) in both phosphate and Duran glasses.[6,7]

Unlike dipolar coupling, which causes linewidth broadening, exchange coupling leads to a narrowing of the linewidth. Exchange forces are effective only at relative short distances, and are therefore expected to occur only at very high dopant concentrations.[8,14,15] In metaphosphate glass, $MnO.P_2O_5$, clusters of corner sharing MnO_6-octahedra develop, and the linewidth of the corresponding EPR spectrum increases as the number of participating octahedra increases.[6,7] The Mn–Mn distances for corner sharing MnO_6 octahedra do not decrease below a value of about 4.2 Å, and only dipole–dipole interactions are experienced by the Mn^{2+} ions.[13] According to Figure 6, exchange narrowing already occurs in Duran glasses at very low doping levels. Because exchange narrowing is indicative of very short Mn–Mn distances, it must be assumed that the Mn–Mn distances fall below the value 4 Å observed in the cluster type that forms in metaphosphate glasses. Shorter Mn–Mn distances might, for example, be realized by the formation of edge sharing MnO_6 octahedra, in analogy with the nanodomains of corner and edge sharing NiO_6 octahedra described previously by Galoisy et al for Ni-doped low alkaline borosilicate glasses.[24] Other examples of exchange narrowing can be found in the study of Komatsu et al for Fe ion clusters which may consist of similar structures.[15]

Figure 7 shows the change in g-values for two glass series with variable Mn concentration. In the literature, a positive shift of the g-value is associated with the formation of an internal magnetic field which can be produced by magnetocrystalline anisotropy or interparticle dipolar interaction.[15] Fidone et al suggested that increasing covalence in Mn–O interactions reduces the positive charge on the ion and gives a positive contribution to g.[25] Values above the free electron value of $g=2.0023$ are indicative of more covalent, and those below of more ionic Mn–O bonding character.[25] Both Duran and metaphosphate glass have similar optical basicity values, and the g-values for low Mn concentrations for both glasses is close to the free electron value. The g-value increases with increasing Mn concentrations in the metaphosphate series, indicating a more covalent character of the Mn–O–Mn bonds within the clusters compared to the Mn–O–P bonds of the glass matrix.[6,7] The

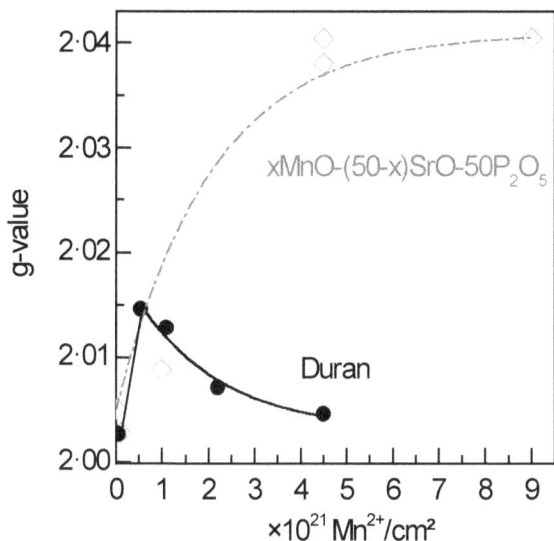

Figure 7. Comparison of the shift in the g-values with increasing MnO content of the metaphosphate and the Duran-like glass series. The concentration values of the Duran series have been reduced to the borate phase[7] [In colour online]

concentration of Mn^{2+} ions in the Duran glass series in Figure 7 has been adjusted to concentration levels in the borate phase where Mn^{2+} ions are preferably dissolved. The g-values of the Duran glasses for low

Mn^{2+} concentrations (up to 5×10^{20} Mn^{2+}/cm^3) fall on the same line as those of the metaphosphate glass series. Contrary to the phosphate glass series, however, the g-values decrease for Mn^{2+} concentrations in Duran over 5×10^{20} Mn^{2+}/cm^3. This decrease may be indicative of more ionic bonds within the clusters, with edge or even face sharing MnO_6 octahedra, for which oxygen atoms have to be shared between more than two Mn^{2+} ions.

3.4. Infrared spectroscopy

The infrared spectra of the Mn-doped Duran glasses (0·05 to 4 mol% MnO) are shown in Figure 8. The spectra appear typical of a low alkaline borosilicate glass with only bridging oxygen atoms. The dominant absorption band at 1095 cm⁻¹ and its shoulder at ca. 1200 cm⁻¹ are due to asymmetric stretching vibrations of Si–O–Si bridges in a three dimensional network of connected Q^4 silicate groups, that is SiO_4 tetrahedra with four bridging oxygen atoms.[26,27] The band at ca. 1390 cm⁻¹ arises from B–O stretching vibrations in neutral trigonal borate units, $[BO_3]^0$, while B–O stretching vibrations in tetrahedral $[BO_4]^-$ groups contribute to the absorption at ca. 925 cm⁻¹.[23,28] The bending and rocking motions of Si–O–Si bridges are

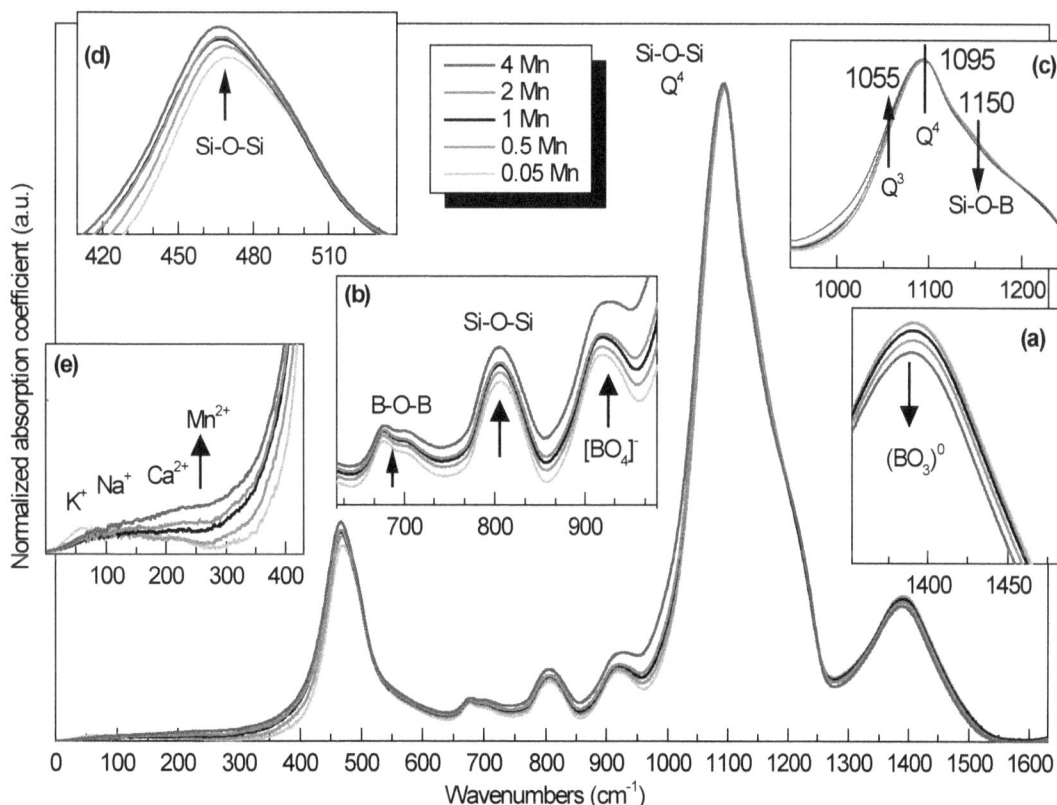

Figure 8. Infrared absorption coefficient spectra derived by Kramer–Kronig transformation of the reflectance spectra measured on Duran-like glasses doped with increasing amounts of MnO (in mol%). The insets (a) to (e) show selected spectral regions expanded to highlight the effect of MnO on the Duran glass structure: (a) neutral borate triangles [BO_3]^0, (b) bridging bonds B–O–B and Si–O–Si and charged borate tetrahedral units [BO_4]^-, (c) mixed Si–O–B bonds, Q^4 and Q^3 silicate species, (d) Si–O–Si bridges, and (e) cation motion bands. See text for discussion of the vibrational modes giving rise to the IR bands of Duran glass [In colour online]

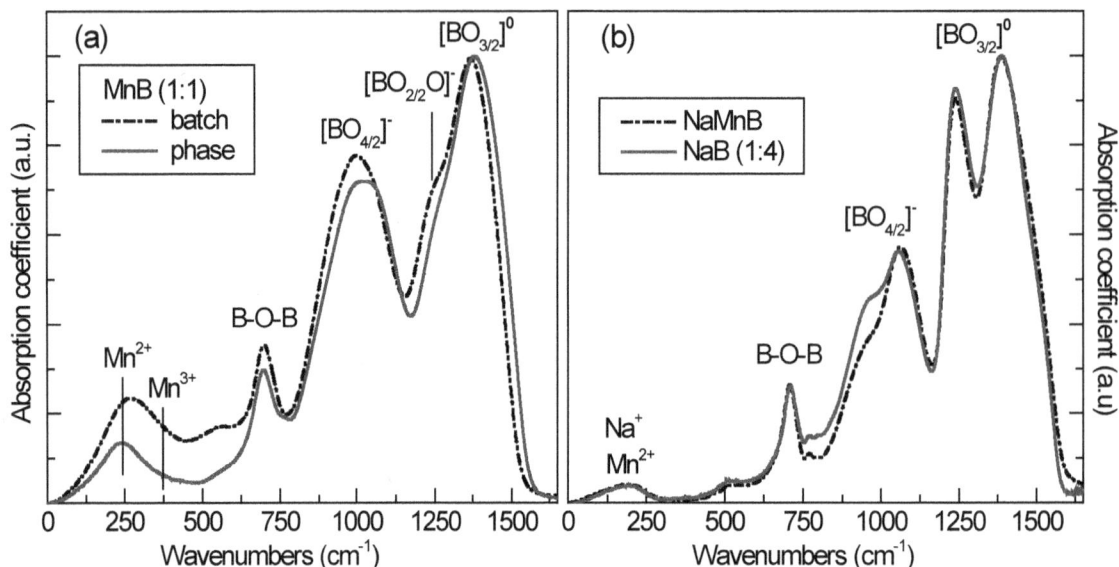

Figure 9. Infrared absorption coefficient spectra after Kramer–Kronig transformation of the reflectance spectra of borate glasses: (a) as prepared glass with batch composition $50MnO.50B_2O_3$ and with a significant amount of Mn^{3+} (black, dot-dash), and the $50MnO.50B_2O_3$ phase obtained from the liquid–liquid phase separated melt containing only Mn^{2+} (blue, solid line); (b) binary $20Na_2O.80B_2O_3$ glass (blue, solid line) in comparison with $4MnO.12\cdot7Na_2O.83\cdot3B_2O_3$ glass (black, dot-dash). All compositions are in mol% [In colour online]

associated with the bands at 805 cm^{-1} and ca. 465 cm^{-1}, respectively,[26,27] while bending vibrations of B–O–B bridges in various bonding environments give rise to bands at ca. 675 and 702 cm^{-1}.[23]

The addition of MnO to Duran glass influences the IR spectrum in several ways, including the clear change in the ratio of four-fold to three-fold coordinated borate species. This is shown in Figure 8, where the relative intensity of the band around 1390 cm^{-1} decreases (inset a) while that at 925 cm^{-1} increases (inset b) upon increasing MnO content. Corresponding changes in the silicate part of the spectrum are less obvious. Nevertheless, inset (c) shows that the addition of MnO causes a slight down shift of the main 1095 cm^{-1} silicate band by ca. 5 cm^{-1} and the gradual development of a shoulder at ca. 1055 cm^{-1}. These changes indicate transformation of some neutral Q^4 silicate species into charged Q^3 units.[26,27] Judging from the extent of the observed spectral changes, we suggest that MnO modifies the borate part to a larger degree than the silicate part of Duran glass, in agreement with previous studies on the distribution of divalent transition metal ions in phase separated borosilicate glasses.[29] Thus, manganese ions appear to have a preference for incorporation into borate domains, where they cause the conversion of neutral BO_3 groups into ligand providing $[BO_4]^-$ units. This process facilitates the gradual phase separation of the Duran glass into modified borate-rich and silica-rich domains. Additional evidence for this is provided by the increase in relative intensity of bands at ca. 805 cm^{-1} (inset b) and ca. 465 cm^{-1} (inset d), both bands being related to Si–O–Si bridges in silica-rich domains, and also by the slight intensity decrease at

ca. 1150 cm^{-1}, where the asymmetric stretching mode of mixed Si–O–B bridges is observed.[22]

The introduction of Mn^{2+} ions into the glass matrix is also evident in the far-IR region where cation site motion bands are active and are known to shift to higher wavenumbers with increasing cation field strength, i.e. with decreasing mass and increasing cation valence. The maxima of the cation motion bands for low and high basicity sites in binary borates are found in the ranges 100–185 cm^{-1} for K^+,[29,31] 130–240 cm^{-1} for Na^+[29,31] and 150–300 cm^{-1} for Ca^{2+}.[32] For Mn^{2+} ions in metaphosphate glasses the cation motion band was measured at 240 cm^{-1}.[6] Even though the cation motion bands of K^+, Na^+ and Ca^{2+} ions are not separated in the spectrum of Duran glass (inset e of Figure 8), MnO addition causes a distinct increase in the absorption between 200 and 300 cm^{-1}, where the Mn^{2+} motion band is expected.

Phase separation also affects binary $MnO–B_2O_3$ and $ZnO–B_2O_3$ glasses, though in a quite different way. Liquid–liquid phase separation is observed when the B_2O_3 content exceeds the MnO content; for example, the $20MnO.80B_2O_3$ batch composition separates in the melt into a light B_2O_3 phase and a heavier manganese borate phase.[20] The phase diagram of $ZnO–B_2O_3$ shows for melting temperatures of 1100°C exactly such a liquid–liquid phase separation, when the B_2O_3 exceeds the ZnO fraction. Batch compositions with 50 or 56 mol% ZnO result in homogenous, transparent glasses, while a melt with only 44 mol% ZnO separates into two layers, a light B_2O_3-rich phase with B_2O_3 crystals on the surface, and a clear transparent zinc borate phase.[20] As mentioned before, a composition $50MnO.50B_2O_3$ was assumed for the manganese borate phase obtained from the phase separated $20MnO.80B_2O_3$ melt. Infrared spectra of the manganese borate phase and the as prepared $50MnO.50B_2O_3$ glass were compared in order to verify this assumption (Figure 9(a)). The overlaying B_2O_3 phase prevented the oxidation of Mn^{2+} to Mn^{3+} in the underlying manganese borate phase, while the 50:50 melt was not protected in a similar fashion, and therefore it contains a considerable amount of Mn^{3+}. This is clearly evident from the resulting nearly black colour of the as-melted glass compared to the light orange colouring of the manganese borate phase. While Figure 9(a) shows considerable similarities of the two spectra, it also manifests key differences both in the mid- and far-IR spectral regions. The mid-IR spectrum of the manganese borate phase is very similar to that reported for calcium metaborate glass,[32] indicating similar borate structures consisting of $[B\emptyset_4]^-$ tetrahedral units (ca. 1000 cm^{-1}), charged $B\emptyset_2O^-$ triangles (ca. 1250 cm^{-1}) and neutral $[B\emptyset_3]^0$ triangles (ca. 1375 cm^{-1}). The far-IR region of the manganese borate phase is dominated by the band characteristic of Mn^{2+} ions at ca. 240 cm^{-1}.[6] The spectrum of the as-prepared $50MnO.50B_2O_3$ glass (denoted as 'batch' in Figure 9(a)) clearly shows a higher degree of modification in comparison to the manganese borate phase. This is manifested by larger relative intensities of the bands characteristic of $[B\emptyset_4]^-$ and $B\emptyset_2O^-$ units, as well as by an upshift of the Mn^{2+} motion band to ca. 270 cm^{-1}. The far-IR profile of the 'batch' is also consistent with the partial oxidation of manganese to Mn^{3+}. For the cation motion band of Mn^{2+} at 240–270 cm^{-1}, it is expected that trivalent Mn^{3+} ions will be active between 370–400 cm^{-1}.[33,34] Indeed, the spectrum of the $50MnO.50B_2O_3$ 'batch' glass exhibits a shoulder at ca. 370 cm^{-1} on the high-frequency side of the Mn^{2+} band. It is of note that the bending and stretching vibrational modes of crystalline Mn_2O_3 are found between 200 and 700 cm^{-1}.[35,36]

Infrared spectra of glass $20Na_2O.80B_2O_3$, and of the very similar NaMnB, where 4 mol% Na_2O are replaced by MnO, are shown in Figure 9(b). It is observed that replacement of Na^+ by Mn^{2+} affects the spectra only marginally. The borate network structure, and consequently the type and number of the different borate entities are very much alike. Nevertheless, the small change in the form of the $[B\emptyset_4]^-$ envelope is indicative of different connectivities of tetrahedral borate groups, with penta- and triborate units exhibiting two bands at 960 and 1060 cm^{-1}, and diborate units exhibiting a band at 1000 cm^{-1}.[23,37]

4. Conclusions

This paper was motivated by the aspiration of explaining the different linewidth trends observed in the EPR spectra of metaphosphate and Duran glasses with increasing Mn concentration. The linewidth

doubled in metaphosphate glasses with increasing MnO content, while the opposite effect is observed in the Duran series, for which it drops to half the initial value. Even more, phase separation inhibits the preparation of Duran glasses with more than 4 mol% MnO.

It was shown previously that MnO_6 clusters in metaphosphate glasses consist of corner sharing MnO_6 octahedra, with nearest Mn–Mn distances around 4 Å.[6] Linewidth broadening due to dipole–dipole interactions is a consequence of cluster formation in which an increasing number of corner sharing MnO_6 octahedra participate. Exchange narrowing, however, is indicative of even stronger magnetic interactions, which are typically observed only for ions at close proximity, i.e. only after a further decrease in Mn–Mn distances.

IR spectroscopy revealed the transformation of $[B\emptyset_3]^0$ to $[B\emptyset_4]^-$ groups upon addition of MnO to the Duran glasses, and simultaneously a relative increase in the number of Si–O–Si bridges. Mn ions create their own preferred bonding environment by accumulating borate ligands at the sites of the MnO_6 clusters, and as a consequence induce phase separation in the glass. EPR and fluorescence data indicate the presence of several Mn sites with strong interactions between them. These findings could be explained by the formation of edge or even face sharing MnO_6 octahedra next to the commonly observed corner sharing octahedra, similar to the nanodomains of corner and edge sharing NiO_6 octahedra seen in low alkaline borosilicate glasses,[24] for which Ni–Ni distances decrease from $4·08$ Å for corner sharing octahedra to $2·94$ Å for edge sharing octahedra. Mn–Mn distances for the slightly bigger Mn^{2+} ions in analogous corner and edge sharing MnO_6 octahedra would fall in the same range, $4·2$ Å for corner sharing, $3·1$ Å for edge sharing, and $2·5$ Å for face sharing MnO_6 octahedra. The latter two values are taken from Menzel et al,[13] or were calculated using average Mn–O distances of $2·2$ Å, which are typical Mn–O distances for octahedral Mn sites in crystalline samples.[38]

Acknowledgement

DM gratefully acknowledges EU support under grant no. MTKD-CT-2006-042301, and the financial contribution from the Otto-Schott-Institut of the Friedrich-Schiller-Universität, Jena for her participation at the Seventh International Conference on Borate Glasses, Crystals, and Melts.

References

1. Wong, J. & Angell, C. A. *Glass Structure by Spectroscopy*, M. Dekker Inc, New York, 1976, ch. 6, pp. 191–407, and ch. 9, pp. 555–668.
2. Bates, T. *Modern Aspects of the Vitreous State*, Vol. 2, Ed. J. D. Mackenzie, Butterworth Inc, Washington, 1962, pp. 195–254.
3. Duffy, J. A. *Bonding, Energy Levels & Bands in Inorganic Solids*, (first

ed.), Longman Scientific and Technical, Essex, 1990.

4. Duffy, J. A., Ingram, M. D. & S. Fong, *Phys. Chem. Chem. Phys.*, 2000, **2**, 1829–1833.

5. Kawano, M., Takebe, H. & Kuwabara, M. *Opt. Mater*, 2009, **32**, 277–280.

6. Konidakis, I., Varsamis, C., Kamitsos, E. I., Möncke, D. & Ehrt, D. *J. Phys. Chem. C*, 2010, **114**, 9125–9138.

7. Möncke, D., Kamitsos, E. I., Herrmann, A., Ehrt, D. & Friedrich, M. *J. Non-Cryst. Solids*, 2011, **357**, 2542–2551.

8. Pilbrow, J. R. *Transition Ion Electron Paramagnetic Resonance*, Oxford University Press, USA, 1991.

9. Kliava, J. *Phys. Status Solidi B*, 1986, **134**, 411–455.

10. Warne, J. O., Pilbrow, J. R. & MacFarlane, D. R. *J. Non-Cryst. Solids*, 1992, **140**, 314–318.

11. Misra, S. K. *Appl. Magn. Reson.*, 1996, **10**, 193–216.

12. Simanek, E. Miller, K. A. *J. Phys. Chem. Solids*, 1970, **31**, 1027–1040.

13. Menzel, E. R., Vincent, W. R. & Wasson, J. R. *J. Magn. Reson.*, 1976, **21**, 165–172.

14. Kumagai, H. *J. Phys. Soc. Jpn*, 1954, **9**, 369–375.

15. Komatsu, T., Saga, N. & Kunugi, M. *J. Appl. Phys.*, 1979, **50**, 6469–6474.

16. Ehrt, D. *J. Non-Cryst. Solids*, 2004, **348**, 22–29.

17. Herrmann, A. & Ehrt. D. *Glass Sci. Technol.*, 2005, **78**, 99–105.

18. Ehrt, D. & Herrmann. A. *Verre*, 2005, **11**, 13–18.

19. Harrison, D. E. & Hummel, F. A. *J. Electrochem. Soc.*, 1956, **103**, 491–498.

20. Ehrt, D. *Phys. Chem. Glasses: Eur. J. Glass Sci. Technol. B*, 2013, **54** (2), 65.

21. Möncke, D. Ehrt, D. Eckert, H. & Mertens, V. *Phys. Chem. Glasses*, 2003, **44**, 113–116.

22. Möncke, D., Ehrt, D., Varsamis, C.-P. E., Kamitsos, E. I. & Kalampounias, A. G. *Glass Tech.: Eur. J. Glass. Sci. Technol. A*, 2006, **47**, 133–137.

23. Kamitsos, E. I., Patsis, A. P., Karakassides, M. A. & Chryssikos, G. D.

J. Non-Cryst. Solids, 1990, **126**, 52–67.

24. Galoisy, L., Calas, G. & Cormier, L. *J. Non-Cryst. Solids*, 2003, **321**, 197–203.

25. Fidone, W. H. & Stevens, K. W. H. *Proc. Phys. Soc.*, 1959, **73**, 116–117.

26. Kamitsos, E. I., Patsis, A. P. & Kordas, G. *Phys. Rev. B*, 1993, **48**, 12499–12505.

27. Ingram, M. D., Davidson, J. E., Coats, A. M., Kamitsos, E. I. & Kapoutsis, J. A. *Glastech. Ber. Glass Sci. Technol.*, 2000, **73**, 89–104.

28. Kamitsos, E. I., Kapoutsis, J. A., Jain, H. & Hsieh, C. H. *J. Non-Cryst. Solids*, 1994, **171**, 31–45.

29. Ehrt, D., Reiß, H. & Vogel, W. *Silikattechnik*, 1976, **27**, 304–3098, *ibid*, 1977, **28**, 359–364. (In German.)

30. Kamitsos, E.I. Chryssikos, G.D. Karakassides, M.A. *J. Phys. Chem.*, 1987, **91**, 1067–1073.

31. Kamitsos, E. I., Chryssikos, G. D., Patsis, A. P. & Duffy, J. A. *J. Non-Cryst. Solids*, 1996, **196**, 249–254.

32. Yiannopoulos, Y. D., Chryssikos, G. D. & Kamitsos, E. I. *Phys. Chem. Glasses*, 2001, **42**, 164–172.

33. Kamitsos, E. I., Varsamis, C. P. E. & Vegiri, A. *XIX Int. Congr. on Glass – Invited Papers*, Society of Glass Technology, Sheffield, UK, 2001, pp. 234–246.

34. Nelson, B. N. & Exarhos, G. J. *J. Chem. Phys.*, 1979, **71**, 2739–2747.

35. Chen, Z. W., Lai, J. K. L. & Shek, C. H. *J. Non-Cryst. Solids*, 2006, **352**, 3285–3289.

36. Julien, C. M., Massot, M. & Poinsignon, C. *Spectrochim. Acta A*, 2004, **60**, 689–700.

37. Kamitsos, E. I., Karakassides, M. A. & Chryssikos, G. D. *J. Phys. Chem.*, 1987, **91**, 1073–1079.

38. McKeown, A., Kot, W. K., Gan, H. & Pegg, I. L. *J. Non-Cryst. Solids*, 2003, **328**, 71–89.

Phys. Chem. Glasses: Eur. J. Glass Sci. Technol. B, February 2013, **54** (1), 52–59

Formation of an outer borosilicate glass layer on Late Bronze Age Mycenaean blue vitreous relief fragments

D. Möncke,[*,1,2] *D. Palles,*[1] *N. Zacharias,*[3] *M. Kaparou,*[3,4] *E. I. Kamitsos*[1] & *L. Wondraczek*[2]

[1] *Theoretical and Physical Chemistry Institute, National Hellenic Research Foundation, 48 Vassileos Constantinou Avenue, 11635 Athens, Greece*

[2] *Otto-Schott-Institut, Friedrich-Schiller-Universität Jena, Fraunhoferstr. 6, 07743 Jena, Germany*

[3] *Laboratory of Archaeometry, Department of History, Archaeology and Cultural Resources Management, University of Peloponnese, 24100 Kalamata, Greece*

[4] *Laboratory of Archaeometry, Institute of Materials Science, N.C.S.R. Demokritos, 15310 Aghia Paraskevi, Attiki, Greece*

Manuscript received 6 December 2011
Revised version received 21 April 2012
Accepted 16 November 2012

Ancient glass samples from Greece were studied by a combination of SEM/EDX to determine their chemical composition and infrared (IR) and Raman spectroscopy for their glass structure. The archaeological samples consisted of three blue vitreous Mycenaean relief fragments from the Late Bronze Age. The chemical composition of the samples is consistent with typical soda–lime–silica glasses, with the possible use of plant ash instead of soda in preparation. The deep blue colour of the samples is due to tetrahedrally coordinated Co^{2+} ions. The infrared spectra of the Mycenaean relief fragments deviate from all other previously measured ancient samples as the spectra resemble those of highly polymerized silica or highly polymerized low alkaline borosilicate glasses. Only spectra taken on relatively fresh cuts are consistent with the analyzed soda–lime–silica glass composition. Measurements on cuts on the sides of the fragments show that only a very thin layer (<1 µm) on the front and back of the samples has a fully polymerized glass network. Strong background fluorescence hindered Raman spectroscopy in the highly polymerized surface layer. Boron could not be detected by SEM/EDX (scanning electron microscopy/energy dispersive x-ray spectroscopy) because of the thinness of the outer layers, whereas reflectance infrared measurements were not impacted by the bulk glass composition. Possible chemical reaction mechanisms, conditions and likely boron sources leading to the formation of this outer layer are discussed.

Introduction

Crucial questions in the archaeology of ancient glasses concern provenance, technology and trade routes. How and where did the technology of glass making evolve, and how did the knowledge spread throughout the Mediterranean to the rest of Europe and Asia? By far the most common investigations of ancient glasses focus on the quantitative evaluation of glass composition. Scanning electron microscopy (SEM) or x-ray-fluorescence (XRF) are the methods of choice, although other techniques, such as inductively coupled plasma optical emission spectroscopy (ICP-OES) or laser ablation mass spectroscopy are also employed. Quantitative analysis of minor components, trace elements and even isotopic ratios give important clues for the identification of raw materials used for glass preparation and therefore on the provenance of vitreous artifacts. However, mixing of raw materials from different sources or recycling of glasses creates additional difficulties in the examination of ancient glasses.[1]

Techniques which focus on the glass structure are much less frequently employed than those used for the analysis of chemical composition alone. Raman and infrared (IR) spectroscopy allow for the assessment of the structure of glasses nondestructively. These techniques are also sensitive to structural changes induced over time by weathering,[2,3] or even structural variations caused by different melting and cooling temperatures.[4,5] Vibrational spectroscopy may thus provide new information and help to gain more insights into the provenance, technology and trade routes of ancient glasses.

We report in this work results on Mycenaean glass fragments from Greece by combining SEM/EDX to analyze the chemical composition, and IR and Raman spectroscopy to investigate the glass structure of these ancient samples. Infrared spectra of the Mycenaean relief fragents revealed a very thin surface layer of different glass structure. Such a layer had escaped detection by other analytical tech-

* Corresponding author. Email doris.moencke@uni-jena.de
Original version presented at VII Int. Conf. on Borate Glasses, Crystals and Melts, Dalhousie University, Halifax, Nova Scotia, Canada, 21–25 August 2011

#37   #32   #33

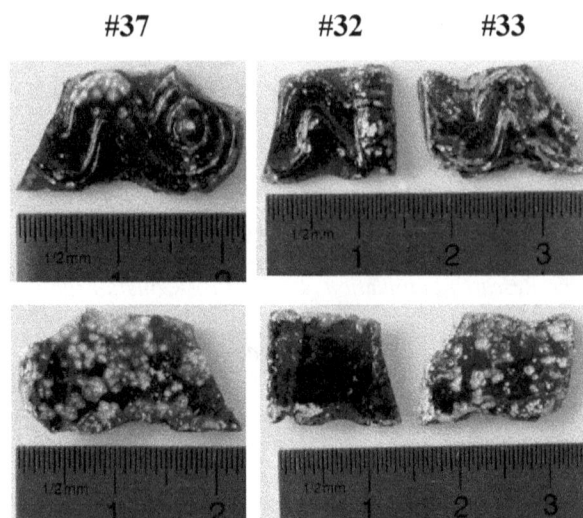

Figure 1. Photographs of the Mycenaean relief fragments. The top row shows the front and the bottom row the back of the samples (Photographs taken by DM)

niques which probe deeper into the bulk glass than reflectance infrared spectroscopy. Possible chemical reaction mechanisms and reaction conditions leading to the formation of this surface layer are discussed.

Experimental

Three blue vitreous relief fragments from the Mycenaean Period (LBA IIIB, 1330–1190 BC) were found as part of burial offerings in a tomb which was excavated in 1989 near Palaia Epidavros, Peloponnese, Greece.[6] The samples showed signs of corrosion in the form of white intrusions on the surfaces, which were more pronounced on the back than on the front of the samples (see Figure 1). The fragments were otherwise of deep blue colour, and showed wide areas of 'healthy glass.' Spectra were taken from the surface, front and back, as well as from more recent cuts on the sides of the fragments. Raman spectra were also obtained from cuts on the sides.

Raman spectra were collected on a Renishaw confocal micro-Raman spectrometer, using different excitation lines, including the 488 and 514 nm lines of an argon ion laser and the 633 nm line of a He–Ne laser. Infrared measurements were performed on the glass samples using a vacuum Fourier transform spectrometer (Vertex 80v, Bruker). The IR spectra were measured in reflectance mode at quasinormal incidence (~11°) in the range 30–7000 cm^{-1} with a resolution of 2 cm^{-1}, each spectrum representing the average of 200 scans. The reflectance spectra were analyzed by Kramers–Kronig (KK) transformation, as detailed elsewhere,[2,7,8] to yield the absorption coefficient spectra, $\alpha(\nu)$, from the expression $\alpha(\nu)=4\pi\nu k(\nu)$, where $k(\nu)$ is the imaginary part of the complex refractive index and ν is the infrared frequency in cm^{-1}. Reflectance measurements on small sample areas or when pinpointing weathered or coloured areas were

Table 1. Average chemical composition of the studied Mycenaean ancient glass samples according to SEM/EDX measurements (±3 to 5%)

Sample no	#32		#33		#37	
	wt%	mol%	wt%	mol%	wt%	mol%
SiO$_2$	64·2	64·5	63·9	63·8	65·2	65·9
Na$_2$O	16·6	16·2	17·0	16·5	19·7	19·3
CaO	8·4	9·1	8·5	9·1	5·7	6·1
MgO	3·8	5·6	3·7	5·5	3·1	4·7
Al$_2$O$_3$	2·7	1·6	2·5	1·5	1·6	1·0
K$_2$O	1·6	1·0	1·4	0·9	1·0	0·6
CoO*	-	-	0·12	0·1	0·16	0·13
CuO*	0·52	0·39	0·31	0·23	0·21	0·16

Other traces include: Fe$_2$O$_3$, TiO$_2$, MnO, Cr$_2$O$_3$, NiO, ZnO, SO$_3$, Cl$_2$O

Calculated	7% Q^2	9% Q^2	
network	93% Q^3	91% Q^3	99·5% Q^3
connectivity			0·5% Q^4

*later performed prompt gamma activation analysis gave values around 0·07–0·09 wt% CoO (±5%) and 0·09–1·1 wt% CuO (±10%) for all three samples

also obtained in the range from 500 to 7000 cm^{-1} using an IR microscope (Hyperion 2000, coupled to an Equinox 55 FT-IR system by Bruker). The resolution was about 4 cm^{-1}, with each spectrum representing an average of 100 scans.

Quantitative analysis was obtained with a scanning electron microscope (SEM) equipped with an energy dispersive x-ray analyzer, model FEI (type Inspect, with Super Ultra Thin Window). Backscattered Electron Imaging was employed to provide both topographic and compositional evidences. The EDX system was operated with an accelerating voltage of 25 kV and a beam current of 1·5 nA, and by using a net counting time of 400 Live sec. An internal to the set-up ZAF correction mode was used to normalize the standardless analysis.

Results

All samples displayed the deep blue colour typical for tetrahedrally coordinated Co^{2+} ions, and chemical analysis confirmed the presence of up to 0·16 wt% of CoO in these samples. It is possible that the higher concentration of CuO found by analysis might shift the colour perception from cobalt blue to marine blue. However, due to the much higher extinction coefficient of Co^{2+} compared to Cu^{2+}, especially in the visible wavelength region, cobalt is the main colourant, even if the Co content of #32 was below the detection limit of analysis. Chemical analysis by SEM/EDX characterized all samples as typical soda–lime–silica glasses; the composition for the three fragments is listed in Table 1. (Recently performed prompt gamma activation analysis (PGAA) indicates for all three Mycenaean samples values of 0·07–0·09 wt% CoO and 0·09–1·1 wt% CuO.) SiO$_2$ levels were found to be around 64–66 mol% (64–65 wt%), and the Na$_2$O content around 16–19 mol% (17–20 wt%). The relatively high MgO content (3–4 wt%) and K$_2$O levels (1–2 wt%) indicate that plant ash and not soda

Table 2. Assignments of infrared and Raman bands measured in this work on Mycenaean ancient glass samples, following Refs. 2–5,7,8,13–21

Structural unit	Infrared band (cm^{-1})	Raman band (cm^{-1})
silicate Q^4	1090, 1200(shoulder); ν_{as}(Si–O–Si)	
silicate Q^3	1050; asymmetric stretching	1093; symmetric stretching
silicate Q^2	970; asymmetric stretching	945; symmetric stretching
Si–OH	900–980; ν(Si–O)	
	3730; ν(O–H) of free Si–OH	
$[BO_4]^-$	930; asymmetric stretching	
$[BO_3]^0$	1300–1500; asymmetric stretching	
Si–O–Si	785–805; bending	780; bending
	465–475; rocking	598; ν_s(Si–O–Si)
Si–O–B	655; bending	
B–O–B	585; bending	
Cation-site	100–400; cation site vibration	
H_2O	2200–3500; ν(O–H) in H-bonded H_2O	
Si–OH	and Si–OH	
	1600; bending of H_2O	

was most probably used as the Na_2O source.[9]

IR spectra of silicate glasses with SiO_2 content around 65 mol% would display their main band around 1055 cm^{-1}, which results from the asymmetric stretching vibration of Si–O bonds in SiO_4 tetrahedra with three bridging and one nonbridging oxygen atoms (Table 2) , also called a Q^3 unit.[8] Q^n denotes SiO_4 tetrahedra with n bridging and $4-n$ nonbridging oxygen atoms. Theoretical considerations, based on the chemical glass composition, confirm that Q^3 units will be by far the most common silicate unit in these glasses, with also a small fraction of Q^2 units:

$$65SiO_{4/2}(Q^4) + 35M_2O$$
$$\rightarrow 60SiO_{3/2}O^-M^+(Q^3) + 5SiO_{2/2}O_2^{2-}2M^+(Q^2)$$

and for modifier oxides of divalent metals:

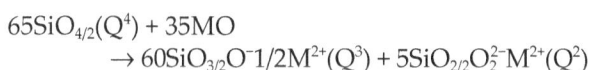

$$65SiO_{4/2}(Q^4) + 35MO$$
$$\rightarrow 60SiO_{3/2}O^-1/2M^{2+}(Q^3) + 5SiO_{2/2}O_2^{2-}M^{2+}(Q^2)$$

Figure 2(a) to (c) shows a comparison of the infrared spectra of several different historic glass samples. All glasses are of the soda–lime–silica composition, but in different states of weathering. The Roman samples (Figure 2(b)) were polished prior to measuring, and thus, represent the spectra of the bulk glass structure rather than of a surface structure modified by weathering.[10,11] The glass beads from the Classical Hellenic Period (Figure 2(c)) showed visible signs of weathering in the form of very thin whitish coatings on the untreated surface which were removed for one of the samples.[10,12] SEM analysis of the glass composition of the samples from the Roman and Classical Hellenic Period resulted in SiO_2 contents of ca. 70 mol%, which signifies a glass network consisting predominantly of Q^3 silicate groups with only few Q^2 and Q^4 units from the disproportionation $2Q^3 \rightarrow Q^2+Q^4$.[7,10,13,14] The IR spectra of these samples should have been very similar if they had not undergone weathering effects. It is clearly evident from Figure 2(c) that weathering of the Classical Hellenic glass beads caused the depolymerization of the glass network. In addition to the dominant band at 1065 cm^{-1}, which arises from asymmetric stretching vibrations of (Si–O) bonds

in Q^3 units, there is a strong shoulder at 950 cm^{-1}, caused by analogous vibrations of Q^2 units. However, in contrast to the glass beads, the Mycenaean relief fragments show a highly polymerized silicate glass network. The dominant band is shifted to 1100 cm^{-1} and displays a shoulder at ca 1200 cm^{-1} (see Figure 2(a) and Table 2), both features being characteristic of

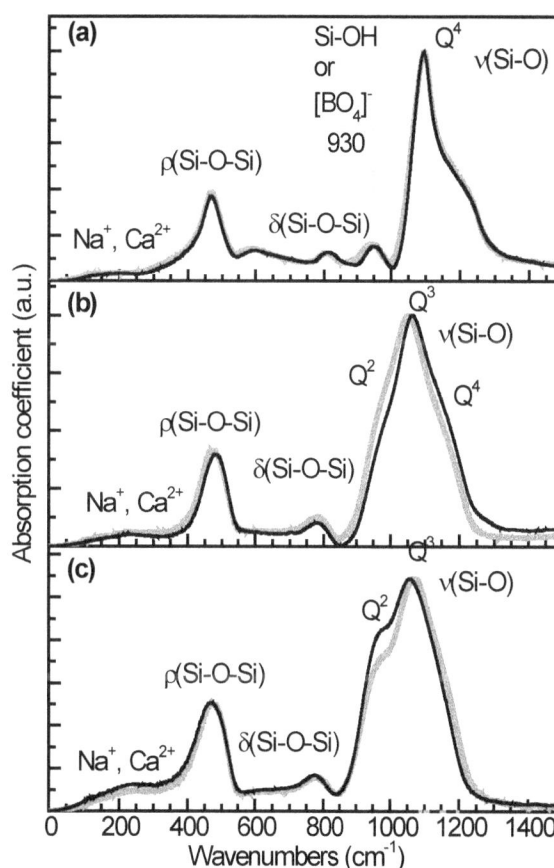

Figure 2. Infrared absorption coefficient spectra after KK transformation of the reflectance spectra taken from (a) the outer blue surface of the Mycenaean relief fragment #32 (thin black line) and #33 (thick grey line); (b) the polished surface of two vitreous samples from the Late Roman Period [10,11]; and (c) the untreated surface (thin black line) and the slightly polished surface (thick grey line) of two different blue glass beads from the Classical Hellenic Period[10,12]

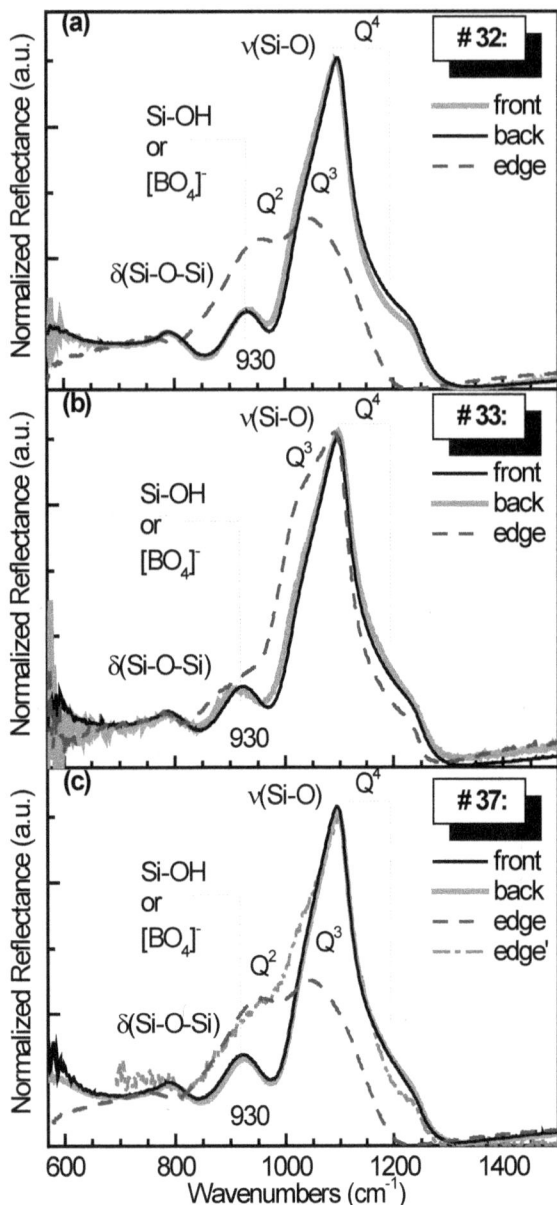

Figure 3. Micro-infrared reflectance spectra of three Mycenaean relief fragments taken from the front (thin black line), the back (thick grey line) and from relatively fresh cuts on the edges (dashed lines) of sample: (a) #32, (b) #33 and (c) #37. The red spectrum of #37 (dash dot) was measured on the edge close to the surface layer [Colour available online]

asymmetric stretching vibrations of Si–O–Si bridges in a three dimensional network of Q^4 silicate units.[15] Variations in the position of the rocking vibrations of Si–O–Si linkages from 465 cm^{-1} for the Mycenaean fragments (Figure 2(a)) to 475 cm^{-1} for the weathered glass beads (Figure 2(b,c)) are also indicative of a depolymerization of the silicate network in the weathered beads, because a frequency increase for this mode is associated with the presence of more nonbridging oxygen atoms.[7]

On the other hand, IR measurements on the Mycenaean samples taken on relatively fresh cuts on the sides of the samples resulted in spectra expected for glasses of soda–lime–silica composition. Due to the small probed areas of the cuts, only reflectance measurements by micro-FT-IR spectroscopy was possible. Figure 3(a) to (c) presents the spectra taken from the front, back and the cuts of the three Mycenaean fragments. Clearly, the spectra from fresh cuts on the sides of fragments show bands for the stretching vibrations of both Q^2 and Q^3 silicate groups. The only exception is sample #33 where the edge also shows a certain degree of polymerization. For sample #37, two spectra were taken on the edge; one taken in the centre of the edge shows bands for Q^2 and Q^3 groups, while one taken on the edge close to the surface layer indicates a combination of bulk and surface spectra, and thus points to a transition in glass structure.

Taking into account the chemical composition of the Mycenaean glasses, the bulk spectra from the sides of the fragments in Figure 3 indicate some degree of depolymerization of the glass network due to weathering, which is reflected by strong bands for Q^2 units. Nonetheless, it should be pointed out that relative band intensities in reflectance spectra (Figure 3) are not directly comparable to those depicted in the absorption coefficient spectra of (Figure 2). This is because the wavenumber factor (v) in the expression $\alpha(v)=4\pi v k(v)$ for the absorption coefficient enhances higher frequency bands relative to those at lower frequencies, i.e. the Q^2 band is relatively less pronounced than Q^3 in the absorption coefficient spectra.

Also of interest is the band at ca. 930 cm^{-1} in Figure 3, which is not typical for vitreous silica. Bands in the region 900–980 cm^{-1} may be related to the Si–O stretching vibration of free silanol groups (Si–OH), or to Si–O_{nb} vibrations where the non-bridging oxygen atoms, O_{nb}, are charge compensated by metal cations, and have been observed in hydrated silicates including leached silicate glasses,[2,3] sol-gel SiO_2[15] and raw perlite materials.[16] It is noted though that B–O stretching vibrations in $[BO_4]^-$ tetrahedra are also observed by IR in the range 900–1000 cm^{-1} for borate[7] and borosilicate glasses.[8,10,13,17] The spectra of vitreous silica and Duran-type low alkaline borosilicate glasses[13–15,17] are very similar to the spectra of the Mycenaean relief fragments, as all give evidence of only bridging oxygen atoms in Q^4 silicate groups, and in the case of low alkaline borosilicate glasses the addition of $[BO_4]^-$ tetrahedra.

Raman spectra for fragments #32 and #37 are displayed in Figure 4(a) and (b). The spectra taken on the outer surface on the front or back of the samples do not provide any structural information due to a strong fluorescent background (Figure 4(b)), which is much lower in the spectra taken from the cuts on the sides of fragments (Figure 4(a)). Such an impeding fluorescence is known for glasses with a high amount of fluorescent dopants,[10,17] and in silicate glasses seems to be more enhanced when the number

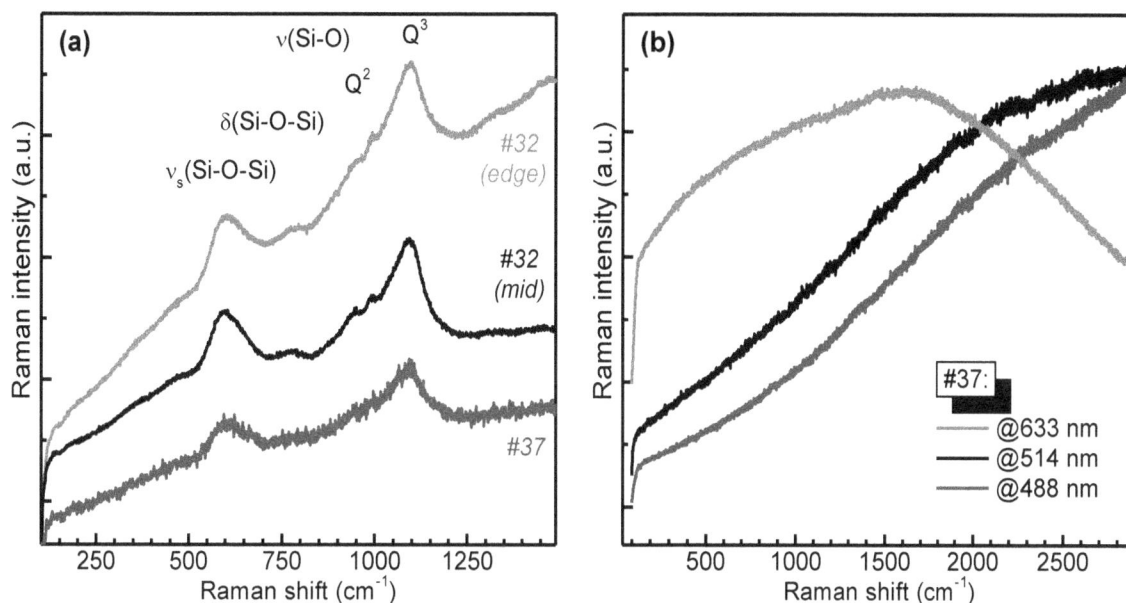

Figure 4. Raman spectra from (a) the sides of fragments #32 and #37 with excitation at 514 nm, and (b) the outer surface layer of fragment #37 with excitation at 488, 514 and 633 nm, respectively, from bottom to top [Colour available online]

of nonbridging oxygen atoms is low. The spectra from the sides of the fragments in Figure 4(a) are consistent with the Raman spectrum of bulk soda–lime–silica glass,[18] and with the IR spectra taken from the fragment sides. The strongest bands are caused by symmetric stretching vibrations of Si–O bonds in Q^3 groups (1090 cm^{-1}) and the symmetric stretching vibrations of Si–O–Si bonds around 600 cm^{-1} (Table 2). A weak band at ca. 950 cm^{-1} can be assigned to the symmetric stretching vibration of Q^2 groups.

Discussion

Two possible scenarios leading to the formation of a polymerized outer glass layer on the Mycenaean glass fragments will be discussed in this section.

Scenario 1

Scenario 1 follows the hypothesis of a profound leaching and subsequent re-polymerization of the network. Several authors have studied leaching experiments on alkali silicate glasses by infrared reflectance spectroscopy: For example McDonald *et al*[2] analyzed sodium silicate glasses and showed that the initial depolymerization of the glass network is followed at longer leaching times by the formation of a gel layer, which transforms into a silica-type network. Lynch *et al*[3] pursued this research, including drying experiments, thus proving that a silica-like gel layer can form on the surface of a leached sodium silicate glass, and that drying will eventually lead to condensation of silanol groups (Si–OH) and the formation of a silica-type network. Lynch *et al*[3] further mentioned that volume shrinkage between the gel and the new glass layer resulted in a cracked

surface upon drying.

No visible surface degradation or cracks are observed by the naked eye in the blue areas of the Mycenaean samples, even though volume shrinkage should be higher for glasses with an initial SiO_2 content of only 65 mol%, compared to 80 mol% in the experiments by MacDonald and Lynch. Another discrepancy between these experiments and the Mycenaean samples is the fact that small monovalent ions, such as Na$^+$ or Li$^+$, are known to leach easily from glasses, but divalent ions, such as Mg^{2+} or Ca^{2+}, often remain in the glass matrix. Therefore, the IR spectra of the Mycenaean glasses should still show the presence of silicate groups with nonbridging oxygen atoms, charge balanced by these divalent modifier cations. Furthermore, leaching of all modifier cations would also include the colouring agent Co^{2+}, and only a colourless or white surface should remain. It is worthwhile to consider at this point that the impeding fluorescent background observed in the Raman spectra from the samples surface would be hard to explain in the absence of fluorescent Co^{2+} ions.

White corroded spots are indeed present on the surfaces of the Mycenaean fragments. However, the IR reflectance spectra of these white corroded spots are typical of fully polymerized SiO_2, with weak shoulders at ca 930 cm^{-1} (Figure 5). Well resolved bands around 930 cm^{-1}, which could be due to Si–OH, Si–O$_{nb}$ and/or [BO$_4$]$^-$ groups, are observed only in the spectra recorded on blue surface areas. As also shown in Figure 5, several of the spectra measured on white corroded spots exhibit very broad bands around 2200–3500 cm^{-1} due to O–H stretching vibrations in a variety of environments, including hydrogen bonded Si–OH groups and molecular water. It is of note that the O–H stretching vibration of free silanol groups

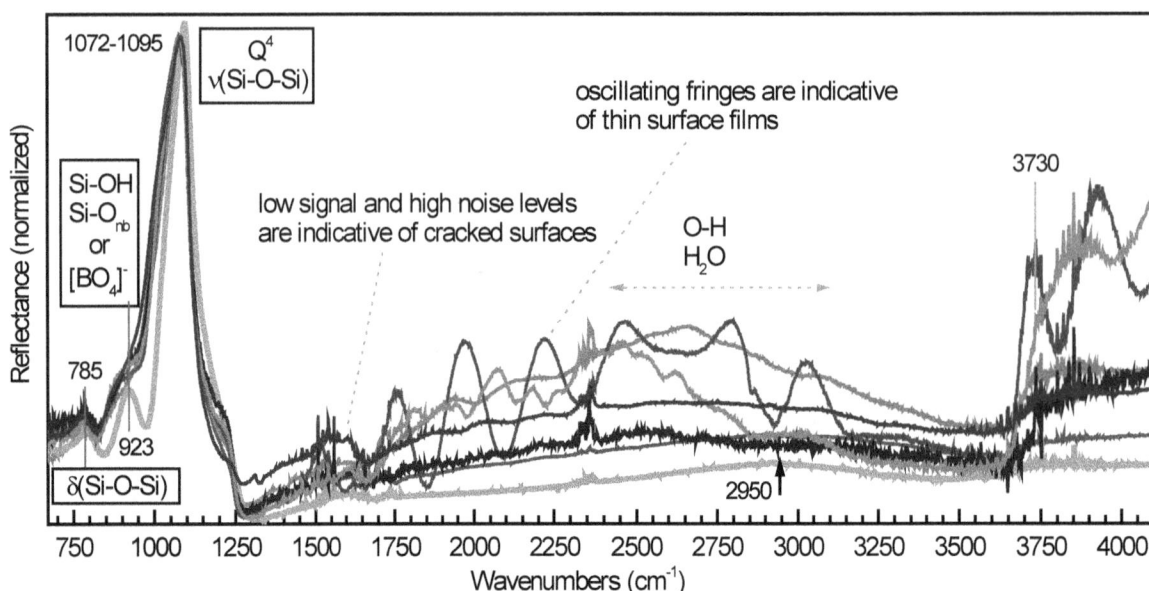

Figure 5. Micro-infrared reflectance spectra of white corroded areas on the surface of the Mycenaean fragments. The spectrum from a blue surface area is included (grey line) for comparison. Spectra with oscillating fringes are from iridescent spots (purple and red lines), and the spectrum with the highest noise level is from a greenish white area (dark blue line) [Colour available online]

(Si–OH) usually gives sharp bands above ca. 3600 cm^{-1}. For example, such bands were measured from 3620 to 3745 cm^{-1} for diatomite minerals,[19] at 3712 cm^{-1} for palygorskite[20] and at 3740 and 3750 cm^{-1} for silanol groups on silica.[21] Inspection of Figure 5 shows that a sharp band at 3730 cm^{-1} was measured for only one white corroded spot. However, this particular 3730 cm^{-1} band cannot be unambiguously attributed to free Si–OH groups because of the presence of strong oscillating fringes in such weathered films. The weak shoulder at ca 930 cm^{-1} in Figure 5 may be related to SiO$_{nb}$ bonds of Q^2 species probed in the bulk of glass, because the infrared beam can penetrate the thin silica-type layer of the white corroded spots. Generally, such spots give IR signals and signal to noise ratios much lower than those in spectra taken from blue areas, indicating much smoother surfaces in the blue regions of the Mycenaean fragments. It seems that the leaching scenario is very helpful for explaining the occurrence of a silica-like network in the weathered white spots, where even Co^{2+} ions are leached from the matrix and the shrinkage of the network upon re-polymerization resulted in a heterogeneous surface.

Scenario II

The observed spectra of the smooth blue surface areas, however, may have resulted from a quite different mechanism. As observed in Figure 5, the spectra of blue surface areas exhibit weak and broad band envelops peaking at ca 2950 cm^{-1} (O–H stretching of hydrogen bonded water[16]) and 1600 cm^{-1} (bending of water[16]), and lack sharp bands due to free Si–OH groups. Both these observations and the preceding

discussion suggest that the well defined band for blue surface areas at 923 cm^{-1} should rather be attributed to [BO$_4$]$^-$ groups. This proposition is strengthened by the fact that the observed spectra bear a high resemblance to those of duran and other low alkaline borosilicate glasses (see Figure 6).[10,13,17] This raises some very obvious questions; namely, how exactly such a borosilicate layer was formed or, even more

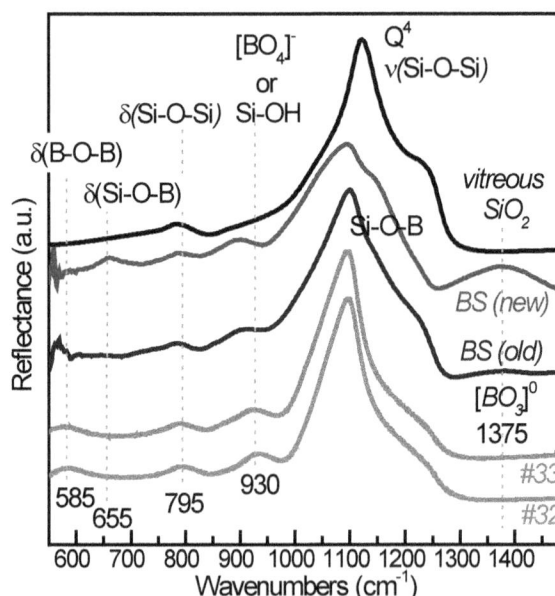

Figure 6. Specular reflectance spectra from blue surface areas of Mycenaean relief fragments (#32 and #33), in comparison with the specular reflectance spectra of vitreous silica (top) and micro-infrared reflectance spectra of low alkaline borosilicate glasses (BS), new: as prepared sample, old: 10 year old sample, partially hydrolyzed [Colour available online]

important, how and why was boron introduced into the glass in the first place? A possible answer, though at this point purely speculative, is that the blue relief fragments were possibly "glued" as decorative elements onto the surface of a gilded artifact, and were, at least partially, gilded themselves. This would result in a quite attractive gold-blue colour scheme. Following this line of thought, borax could have been used as flux in the goldworks. Borax has been known since antiquity; for example the Babylonians used borax for soldering and later, since about 300 BC, borax was also used in India as flux for goldsmiths.[22] Schliemann wrote in 1878, "While speaking of soldering, I may mention that Professor Landerer informs me that the Mycenean goldsmiths soldered gold with the help of borac (borate of soda), which is still used at the present day for the same purpose..." and "... that this was imported from Persia and India under the name of *Baurac-Pounxa-Tinkal*."[23]

Assuming that borax was used in a gilding process, the borate components could have reacted easily at elevated temperatures with the silicate glass, a reaction well known when melting borate glasses in silica crucibles. Diffusion of borate into the glass would be very limited; restricting the borosilicate layer to the outer µm or even less, which corresponds to the probing depth of reflectance IR measurements. Through this reaction of borax with the silicate glass, the borate component would be incorporated into the silicate matrix mainly in the $[BO_4]^-$ tetrahedral form, without excluding the presence of some trigonal $[BO_3]^0$ units. This process would generate a more polymerized network compared to the original soda–lime–silica glass matrix. In order to charge balance the $[BO_4]^-$ tetrahedra, a certain number of modifier cations would also be retained in the modified surface layer. This is in agreement with the absorption coefficient spectra in Figure 2(a), where cation motion activity of different modifier cations[24] is clearly evident in the far-IR region (below 400 cm^{-1}). Weathering on the surface leads in a final step to leaching of some of the modifier ions and of trigonal $[BO_3]^0$ groups, resulting in a silica-like network of SiO_4 and $[BO_4]^-$ tetrahedra. Volume shrinking due to weathering would be less than in scenario 1, which might explain the good surface quality of the Mycenaean samples. Moreover, while transition metal ions are not very soluble in pure silica glass, the tetrahedral borate groups offer good ligands for the coordination of Co^{2+} ions, or generally for divalent transition metal ions.[17,25]

Chemical analysis should be able to distinguish between both scenarios from the composition of the surface layer. However, the typically applied methods probe too deep to analyze the outer layer alone, and the results include all elements of the underlying bulk composition. Another problem is the insensitivity of many methods for the light element boron, including SEM or XRF. On the other hand, Raman and infrared spectroscopy are surface probing methods and sensitive to both trigonal $[BO_3]^0$ and tetrahedral $[BO_4]^-$ groups. For example, Figure 6 shows that $[BO_3]^0$ groups are easily probed at ca. 1375 cm^{-1} in a freshly prepared low alkaline borosilicate glass ('BS new'), and that such groups are almost lost from the probed glass surface within a time span of ten years of storing the glass under normal laboratory atmosphere. This high susceptibility of trigonal $[BO_3]^0$ groups towards hydrolysis easily explains why the broad 1300–1500 cm^{-1} borate band is absent from the spectra of the more than 3000 year old Mycenaean glass fragments. Further experiments are necessary in order to explain the formation of the highly polymerized glass layer on the surface of the Mycenaean relief fragments. To this aim, laboratory experiments on model glasses with the same composition are currently planned.

Conclusions

Infrared and Raman spectroscopy are useful tools for studies of historic glasses, as they provide valuable information on the glass structure. The analysis of Mycenaean relief glass fragments from the late Bronze Age suggest different weathering effects on glass surfaces. On broken edges caused by weathering a depolymerization of the glass network was observed, as evident by an increased number of Q^2 units in the IR spectra. Raman spectroscopy probes deeper into the bulk glass and shows a much stronger prevalence of Q^3 units, as expected from the bulk glass composition. On the other hand, a very thin surface layer of a highly polymerized glass network was found to form on the front and back surfaces of the Mycenaean relief fragments, and to correspond spectroscopically to either silica or duran-type compositions. The high fluorescent background of the highly polymerized outer layer prevented the use of Raman spectroscopy, but the same technique was found to be useful for the analysis of bulk glass samples with higher levels of nonbridging oxygen atoms. The IR spectra of white corroded spots on the fragment surfaces were found to be consistent with a SiO_2-type composition which is formed by leaching and is accompanied by considerable O–H and water content. Corresponding IR spectra of the blue surface areas suggested a combined silica-type and Duran-type composition with considerably less water content. It is suggested that the use of borax as a flux in the process of gilding artifacts may explain the presence of a borate component in the blue relief fragments.

Acknowledgements

The authors want to thank the 27th Ephorate of Prehistoric and Classical Antiquities, Corinth, the

25th Ephorate of Byzantine Antiquities, Corinth, the American School of Classical Studies (Corinth Excavations), the 13th Ephorate of Prehistoric and Classical Antiquities, Volos, the 4th Ephorate of Prehistoric and Classical Antiquities, Nafplion, Argolida, and the Society for Messenian Archaeological Studies all in Greece for permission to study the samples. DM wants to thank the Otto-Schott-Institute of the Friedrich-Schiller-University, Jena and NANONLO project of the Theoretical and Physical Chemistry Institute of NHRF in Athens for financial support.

References

1. Rehren, T. *J. Archaeol. Sci.*, 2008, **35**, 1345–1354.
2. MacDonald, S. A., Schardt, C. R., Masiello, D. J. & Simmons, J. H. *J. Non-Cryst. Solids*, 2000, **275**, 72–82.
3. Lynch, M. L., Folz, D. C. & Clark, D. E. *J. Non-Cryst. Solids*, 2007, **353**, 2667–2674.
4. Kamitsos, E. I. & Karakassides, M. A. *Phys. Chem. Glasses*, 1989, **30**, 235–236.
5. Raffaëlly-Veslin, L., Champagnon, B. & Lesage, F. *J. Raman Spectrosc.*, 2008, **39**, 1120–1124.
6. Kaparou, M., Zacharias, N. & Murphy, J. *Archaeological glass weathering and resulting implications to analytical glass studies*, Int. Symp.: History, Technology and Conservation of Ancient Metal, Glasses and Enamels, 16–19 November 2011, Athens, Greece, submitted to Proceedings)
7. Kamitsos, E. I., Patsis, A. P., Karakassides, M. A. & Chryssikos, G. D. *J. Non-Cryst. Solids*, 1990, **126**, 52–67.
8. Kamitsos, E. I., Kapoutsis, J. A., Jain, H. & Hsieh, C. H. *J. Non-Cryst. Solids*, 1994, **171**, 31–45.
9. Walton, M. S. Shortland, A., Kirk, S. & Degryse, P. *J. Archaeol. Sci.*, 2009, **36**, 1496–1503.
10. Möncke, D., Palles, D., Zacharias, N., Kaparou, M., Papageorgiou, M., Oikonomou, A., Kamitsos, E. I. & Wondraczek, L. *Chemical and Spectroscopic Investigation of a Greek Glass Archaeological Collection Spanning the Mycenaean to Roman Period as Probed by SEM/EDS, IR and Raman Techniques*, Int. Symp.: History, Technology and Conservation of Ancient Metal, Glasses and Enamels, 16–19 November 2011, Athens, Greece, submitted to Proceedings).
11. Papageorgiou, M. & Zacharias, N. *A comparative and analytical study of late antiquity glass vases: Collections from Peloponnese and Magnesia, Greece*, In Second ARCH_RNT Symp. Proc., Ed N. Zacharias, University of Peloponnese Publications, 2012, in print. (In Greek.)
12. Zacharias, N., Beltsios, K., Oikonomou, A., Karydas, A., Bassiakos, Y., Michael, C. T. & Zarkadas, C. *Opt. Mater.*, 2008, **30**, 1127–1133.
13. Möncke, D., Ehrt, D., Varsamis C.-P. E., Kamitsos, E. I. & Kalampounias, A. G. *Glass Tech.: Eur. J. Glass Sci. Technol. A*, 2006, **47**, 133–137.
14. Ingram, M. D., Davidson, J. E., Coats, A. M., Kamitsos, E. I. & Kapoutsis, J. A. *Glastech. Ber. Glass Sci. Technol.*, 2000, **73**, 89–104.
15. Kamitsos, E. I., Patsis, A. P. & Kordas, G. *Phys. Rev. B*, 1993, **48**, 12499–12505.
16. Roulia, M., Chassapis, K., Kapoutsis, J. A., Kamitsos, E. I. & Savvidis, T. *J. Mater. Sci.*, 2006, **41**, 5870–5881.
17. Möncke, D., Ehrt, D. & Kamitsos, E. I. Spectroscopic Study of Manganese-Containing Borate and Borosilicate Glasses: Cluster Formation and Phase Separation, Presentation at the 7th Borate Conference, Halifax 2011, *Phys. Chem. Glasses.: Eur. J. Glass Sci. Technol. B*, 2013, **54** (1), 42–51.
18. Dussauze, M., Rodriguez, V., Lipovskii, A., Petrov, M., Richardson, K. Smith, C., Cardinal, T., Fargin, E. & Kamitsos, E. I. *J. Phys. Chem. C*, 2010, **114**, 12754–12759.
19. Yuan, P., Wu, D. Q., He, H. P. & Lin, Z. Y. *Appl. Surf. Sci.*, 2004, **227**, 30–39.
20. Gionis, V., Kacandes, G. H., Kastritis, I. D. & Chryssikos, G. D. *Am. Miner.*, 2006, **91**, 1125–1133.
21. Morrow, B. A. & McFarlan, A. J. *J. Phys. Chem.*, 1992, **96**, 1395–1400.
22. Speel, E. & Bronk, H. *Berliner Beiträge zur Archäometrie*, 2001, **18**, 43–100, pp.75.
23. Schliemann von, H. *Mycenae: A Narrative of Researches and Discoveries at Mycenae and Tiryns*, 1878. (Digitally printed version Cambridge University Press, 2010, pp. 231.)
24. Kamitsos, E. I., Chryssikos, G. D., Patsis, A. P. & Duffy, J. A. *J. Non-Cryst. Solids*, 1996, **196**, 249–254.
25. Ehrt, D., Reiß, H. & Vogel, W. *Silikattechnik*, 1977, **28**, 359–364. (In German.)

Phys. Chem. Glasses: Eur. J. Glass Sci. Technol. B, February 2013, **54** (1), 60–63

Radiation pattern control by the sidewall angle of Bi₂O₃–B₂O₃ based glass phosphor doped with Yb³⁺ and Nd³⁺

Shingo Fuchi,* Shunichi Kobayashi, Koji Oshima & Yoshikazu Takeda

Department of Crystalline Materials Science, Graduate School of Engineering, Nagoya University, Furo-cho, Chikusa-ku, Nagoya 464-8603, Japan

Manuscript received 25 October 2011
Revised version received 7 March 2012
Accepted 4 April 2012

The dependence of the radiation pattern on the sidewall angle of Bi₂O₃–B₂O₃ based glass phosphors doped with Yb³⁺ and Nd³⁺ has been investigated. Batwing-like radiation patterns were observed for sidewall angles of 0° and 15°. On the other hand, Lambertian-like radiation patterns were observed for sidewall angles of 30° and 45°. The highest near-infrared (NIR) light intensity was obtained for a sidewall angle of 30°. Calculated radiation patterns were close to experimental results. The calculated results reveal that the change of radiation patterns was due to the change of average path length of the NIR light in the glass phosphor. Therefore, it is concluded that, for practical uses, glass phosphor design should include the sidewall angle to obtain the appropriate radiation pattern.

1. Introduction

Wideband light sources have been used in a fibre gyroscope,[1,2] for optical coherence tomography (OCT – a novel cross-sectional imaging technique for biological tissues),[3] and for absorption spectrometry for an agricultural applications.[4] Since OCT is based on the Michelson interferometer, a low coherence wideband light source has several advantages for a high depth resolution. In general, a wideband light source is desirable for absorption spectrometry in order to measure a broader region of an absorption spectrum. The central wavelength in the near-infrared (NIR) region (around 1000 nm) is suitable for the above applications, and leads to penetration depths in biological tissues larger than at other wavelengths.[3] A recent study has shown that imaging biological tissues at the central wavelength yields minimal degradation, due to water dispersion, of the OCT axial resolution.[5] Therefore, the central wavelength should be ~1000 nm.

Super luminescent diodes (SLDs) and light emitting diodes (LEDs) in the NIR region and a halogen lamp are usually used as NIR wideband light sources. SLDs and LEDs are small and have a long lifetime. However, the spectral width of SLDs and LEDs is approximately 50 nm, which is insufficient for the above applications. Zhang *et al* have reported a method for synthesis of several LEDs to broaden the spectral width.[6] However, synthesis of several LEDs and control of the light intensity of individual LEDs are complicated for practical uses. On the other

hand, a halogen lamp has a very wide spectral width. Therefore, absorption spectrometry uses a halogen lamp as a NIR wideband light source. However, the size of the halogen lamp is much bigger than that of SLDs and LEDs. Moreover, the lifetime of halogen lamps is much shorter than that of SLDs and LEDs.

Therefore, we propose a new type of light source by combining an LED and a NIR emitting phosphor in one package. Since this light source is similar to a white LED that is composed of a blue LED and a yellow phosphor, it is simple and useful for practical uses. To realize this light source, we synthesized a phosphor that emits at around 1000 nm using Bi₂O₃–B₂O₃ based glasses doped with Yb³⁺ and Nd³⁺.[7,8] We reported that a high power (over 1 mW) NIR light source with a Gaussian-like shape spectrum was successfully achieved by using this glass phosphor and a high power amber LED.[9] The spectral width was from 950 nm to 1100 nm (with a central wavelength of 1014 nm, and a full width at half maximum (FWHM) of 98 nm). As is well known, the radiation pattern of a light source is important for practical uses. Since the refractive index of Bi₂O₃–B₂O₃ based glass is ~2, total internal reflection has an influence on the optical properties of the light source. Therefore, in this paper, we discuss the dependence of the radiation patterns on the sidewall angle of Yb³⁺ and Nd³⁺ co-doped Bi₂O₃–B₂O₃ based glass phosphor.

2. Experimental results

Samples used in this study were synthesized by a melt quenching method. Powders of Yb₂O₃, Nd₂O₃, Bi₂O₃, H₃BO₃, and Sb₂O₃ were mixed in a nominal mo-

*Corresponding author. Email fuchi@numse.nagoya-u.ac.jp
Original version presented at VII Int. Conf. on Borate Glasses, Crystals and Melts, Dalhousie University, Halifax, Nova Scotia, Canada, 21–25 August 2011

Figure 1. Photographs of $1Yb_2O_3$–$4Nd_2O_3$–$47Bi_2O_3$–$47B_2O_3$–$1·0Sb_2O_3$ glass phosphor with sidewall angles of $0°$, $15°$, $30°$ and $45°$ [Colour available online]

lar composition of $1Yb_2O_3$–$4Nd_2O_3$–$47Bi_2O_3$–$47B_2O_3$–$1Sb_2O_3$. The mixed powders were melted in an Al_2O_3 crucible in an electric furnace at 1250°C. Sb_2O_3 was used to suppress reduction of Bi^{3+}. After 10 min at the high temperature, the molten liquid was poured into stainless steel mould plates kept at room temperature for formation of the glass phosphor. The samples contained Al_2O_3 (~10 mol%) from the crucible. However, there was no degradation of luminescence intensity. The sidewall angles of the stainless steel mould plates were set at $0°$, $15°$, $30°$, and $45°$ (a sidewall angle of $0°$ corresponds to a column shape). The thickness of the glass phosphor was set at 3 mm. The samples were not annealed.

Figure 1 shows photographs of the samples used in this study. The volume of all the samples was the same. The yellow colour is due to Bi_2O_3. The variation in sidewall angle is clearly observable in Figure 1.

Figure 2 shows the experimental results for the dependence of radiation patterns on sidewall angle. The radiation pattern depends strongly on the sidewall angle. Batwing-like radiation patterns are observed for sidewall angles of $0°$ and $15°$. On the other hand, Lambertian-like (i.e. isotropic) radiation patterns are observed for sidewall angles of $30°$ and $45°$. The NIR light intensity in the angular ranges from $-90°$ to $-30°$

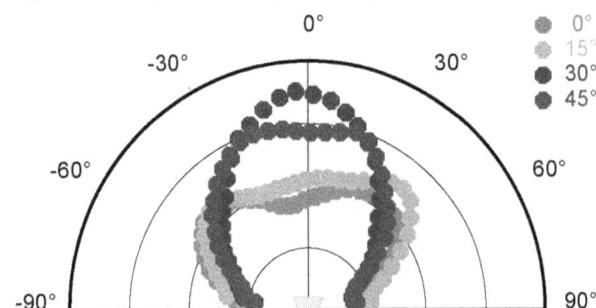

Figure 2. The experimental dependence of the radiation patterns on sidewall angle. The experimental error is within the point size [Colour available online]

and from $30°$ to $90°$ is almost the same for all samples. However, for angles in the range from $-30°$ to $30°$, the NIR light intensity for sidewall angles of $30°$ and $45°$ is higher than for sidewall angles of $0°$ and $15°$. Thus, it is in the angular range from $-30°$ to $30°$ that the radiation patterns for the NIR light intensity is affected by the sidewall angle.

As shown in Figure 2, the radiation pattern can be controlled by varying the sidewall angle. The maximum NIR light intensity (with the Lambertian-like radiation pattern) is realized for a sidewall angle of $30°$. Therefore, a sidewall angle of $30°$ is suitable for practical uses of $1Yb_2O_3$–$4Nd_2O_3$–$47Bi_2O_3$–$47B_2O_3$–$1Sb_2O_3$ glass phosphor. Moreover, the sidewall angle is easily controlled by changing the sidewall angle of the mould plates.

3. Calculated results

Figure 3 shows a schematic drawing of a two-dimensional calculation model for the radiation pattern. 55 light source points were arranged in the glass phosphor. The NIR light radiates in all directions from each individual light source point. The refractive index of the glass phosphor was assumed to be 2. The glass phosphor is irradiated by excitation light from below. The calculation model includes total internal reflection and refraction of the NIR light at the surface of the glass phosphor. However, NIR light that was reflected more than two times in the glass phosphor was ignored. The reflection and scattering of the excitation light was ignored. The absorption of the excitation light and the self absorption of the NIR light were included in the calculation model. The effects of the absorption of the excitation light and the self absorption of the NIR light have already been reported in Ref. 10.

Figure 4 shows the calculated radiation patterns at sidewall angles of $0°$, $15°$, $30°$ and $45°$. The calculated radiation patterns are similar to the experimental

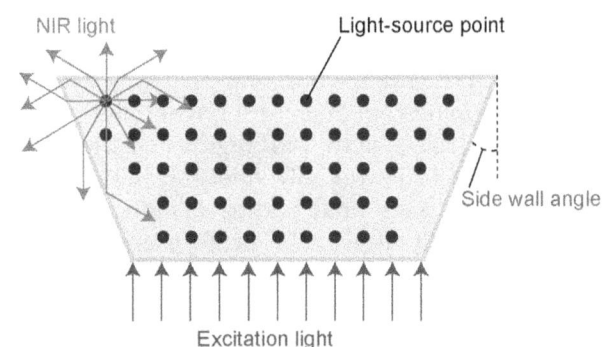

Figure 3. Schematic drawing of the calculation model. 55 light source points were arranged in the glass phosphor. Total internal reflection (up to a maximum of two reflections), the absorption of the excitation light, and the self absorption of the NIR light are included in the calculation model [Colour available online]

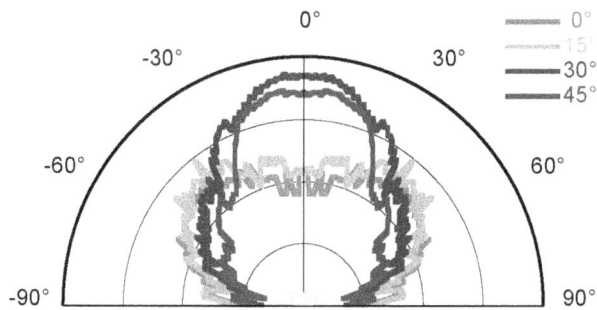

Figure 4. Calculated radiation patterns for the glass phosphor with sidewall angles of 0°, 15°, 30° and 45° [Colour available online]

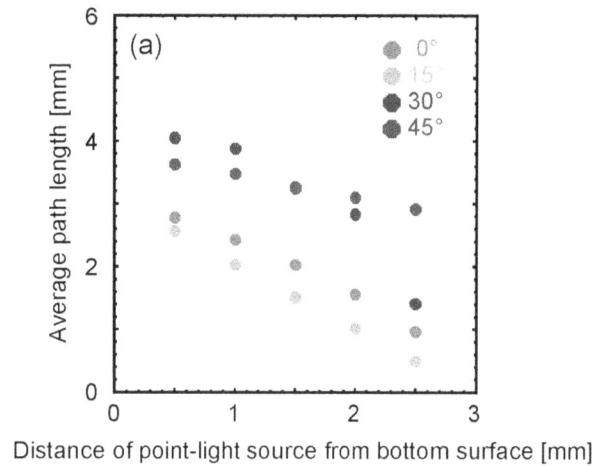

Distance of point-light source from bottom surface [mm]

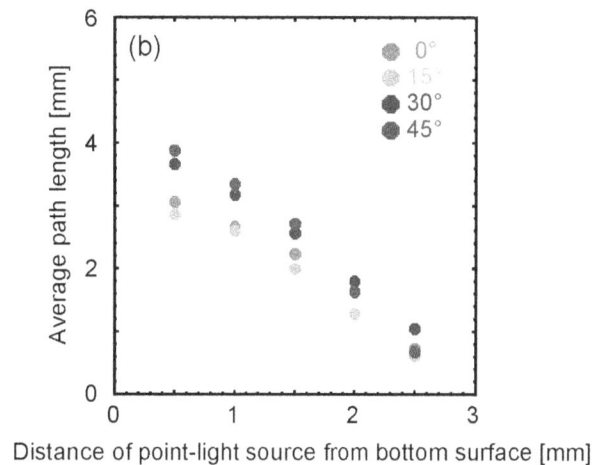

Distance of point-light source from bottom surface [mm]

Figure 5. The average path length of the NIR light in the glass phosphor at angles of (a) 0° and (b) 30°, for sidewall angles of 0°, 15°, 30° and 45°. At an angle of 0°, the average path lengths for sidewall angles of 30° and 45° are longer than for sidewall angles of 0° and 15° [Colour available online]

radiation patterns, as shown in Figure 2. The calculated radiation patterns have batwing-like patterns for sidewall angles of 0° and 15°. On the other hand, Lambertian-like radiation patterns are predicted for sidewall angles of 30° and 45°. For angles in the ranges from −90° to −30° and from 30° to 90°, the NIR light intensity is almost the same for all samples. However, for angles in the range from −30° to −30°, the NIR light intensity for sidewall angles of 30° and 45° is higher than for sidewall angles of 0° and 15°.

Since the calculated radiation patterns are similar to the experimental radiation patterns, it is concluded that the simple light behaviours included in the calculation model occur in the glass phosphor. These simple light behaviours are easily included in the design of the glass phosphor.

4. Discussion

As shown in Figures 2 and 4, the NIR light intensity in the angular range from −30° to −30°depends strongly on the sidewall angle. On the other hand, for angles in the ranges from −90 to −30° and from 30 to 90°, the NIR light intensity is almost the same for all sidewall angles. Therefore, we focus our attention on the behaviour for angles of 0° and 30°. Figure 5 shows the calculated dependence of the average path length of the NIR light on the distance from the bottom surface of the glass phosphor. The average path length decreases with increasing distance from the bottom surface both the angles of 0° and 30°. This tendency is well understood, since the NIR light mainly comes from the top surface, not from the bottom surface.

For an angle of 0°, the average path lengths for sidewall angles of 30° and 45° are significantly longer than for sidewall angles of 0° and 15°. On the other hand, for an angle of 30°, the average path lengths for sidewall angles of 30° and 45° are only slightly longer than those for sidewall angles of 0° and 15°. For example, at an angle of 0° and a distance 0·5 mm from the bottom surface, the average path lengths are 2·79 mm (sidewall angle of 0°), 2·57 mm (sidewall angle of 15°), 4·05 mm (sidewall angle of 30°), and 3·63 mm (sidewall angle of 45°), respectively. Therefore, the

maximum difference of the average path length is 1·48 mm. This value is almost half of the glass phosphor thickness. On the other hand, at an angle of 30° and at a distance 0·5 mm from the bottom surface, the average path lengths are 3·06 mm (sidewall angle of 0°), 2·86 mm (sidewall angle of 15°), 3·67 mm (sidewall angle of 30°), and 3·88 mm (sidewall angle of 45°), respectively. Therefore, the maximum difference of the average path length is 1·02 mm. This value is smaller than for an angle of 0°.

We have already reported that the NIR light intensity strongly depends on the absorption of the excitation light, the self-absorption of the NIR light, the luminescence efficiency, and the thickness of the glass phosphor.[10] The NIR light intensity I is represented by

$$I = \frac{\eta I_{ex}\alpha_{ex}\exp(-\alpha_{lumi}d)\left(\exp\left((\alpha_{lumi}-\alpha_{ex})d\right)-1\right)}{(\alpha_{lumi}-\alpha_{ex})} \quad (1)$$

where η is the luminescence efficiency, I_{ex} is the excitation light intensity at the bottom surface, α_{ex} is the absorption coefficient of the excitation light, α_{lumi} is the absorption coefficient of the NIR light, and d is the glass phosphor thickness.[10] In this paper, we assume that the average path length of the NIR light should be substituted for d in Equation (1). Since the composition of samples used in this study is the same, η, α_{ex}, and α_{lumi} should be the same for all samples. Therefore, the NIR light intensity depends only on the average path length of the NIR light. We measured the absorption coefficient (without surface reflection) at 590 nm and at 980 nm as α_{ex} and α_{lumi}, respectively. Substituting the measured absorption coefficient at 590 nm (0·847 mm^{-1}) and at 980 nm (0·043 mm^{-1}) in Equation (1), we can estimate the NIR light intensity in units of $I/(\eta I_{ex})$. The estimated NIR intensity at an angle of 0° is 0·986 (sidewall angle of 0°), 0·973 (sidewall angle of 15°), 1·005 (sidewall angle of 30°), and 1·007 (sidewall angle of 45°), respectively. Although detailed analyses are needed to clarify the quantitative dependence of the NIR intensity on the average path length, this rough estimation indicates that at an angle of 0°, the NIR intensity for sidewall angles of 30° and 45° is higher than for sidewall angles of 0° and 15°. From these results and discussions, we can conclude that the change of radiation patterns is due to the change of the average path length of the NIR light in the glass phosphor.

5. Conclusions

The dependence of radiation pattern on the sidewall angle of Bi_2O_3–B_2O_3 based glass phosphors doped with Yb^{3+} and Nd^{3+} has been investigated. Batwing-like radiation patterns were observed for sidewall angles of 0° and 15°. On the other hand, Lambertian-like radiation patterns were observed for sidewall angles of 30° and 45°. We have calculated the radiation patterns using a simple model which includes the total internal reflection, the absorption of the excitation light, and the self absorption of the NIR light, and the calculated radiation patterns are close to the experimental results. Moreover, it was calculated that the average path length of the NIR light at an angle of 0° in the glass phosphors with sidewall angles of 30° and 45° is longer than for sidewall angles of 0° and 15°. It was revealed that the change of radiation patterns is due to the change of the average path length of the NIR light in the glass phosphor. From these results, we conclude that glass phosphor design for practical uses should include the sidewall angle to control the radiation pattern.

Acknowledgements

This work was supported in part by the Technology Development Program for Advanced Measurement and Analysis of the Japan Science and Technology Agency (JST).

References

1. Cutler, C. C., Newton, S. A. & Shaw, H. J. *Opt. Lett.*, 1980, **5**, 488.
2. Blin, S., Kim, H. K., Digonnet, M. J. F. & Kino, G. S. *J. Lightwave Technol.*, 2007, **25**, 861.
3. Fercher, A. F., Drexler, W., Hitzenberger, C. K. & Lasser, T. *Rep. Progr. Phys.*, 2003, **66**, 239.
4. Siesler, H. W., Ozaki, Y., Kawano, S. & Heise, H. M. *Near-Infrared Spectroscopy-Principles, Instruments, Applications*, WILEY-VCH, Weinheim, Germany, 2002.
5. Wang, Y., Nelson, J., Chen, Z., Reiser, B., Chuck, R. & Windeler, R. *Opt. Express*, 2003, **11**, 1411.
6. Zhang, Y., Sato, M. & Tanno, N. *Opt. Lett.*, 2001, **26**, 205.
7. Fuchi, S., Sakano, A. & Takeda, Y. *Jpn. J. Appl. Phys.*, 2008, **47**, 7932.
8. Fuchi, S., Sakano, A., Mizutani, R. & Takeda, Y. *Glass Technol.: Eur. J. Glass Sci. Technol. A*, 2009, **50**, 319.
9. Fuchi, S., Sakano, A., Mizutani, R. & Takeda, Y. *Appl. Phys. Express*, 2009, **2**, 32102.
10. Fuchi, S., Sakano, A., Mizutani, R. & Takeda, Y. *Appl. Phys. B*, 2011, **105** (4), 877–881.

Phys. Chem. Glasses: Eur. J. Glass Sci. Technol. B, June 2013, **54** (3), 109–114

Structure and properties of lithium borate glass electrolytes synthesised by a mechanochemical technique

Akitoshi Hayashi,[1] *Daisuke Furusawa, Yuuki Takahashi, Keiichi Minami & Masahiro Tatsumisago*

Department of Applied Chemistry, Graduate School of Engineering, Osaka Prefecture University, 1-1 Gakuen-cho, Naka-ku, Sakai, Osaka 599-8531, JAPAN

Manuscript received 30 November 2011
Revision received 25 August 2012
Manuscript accepted 18 December 2012

Lithium borate glasses with high Li_2O content (up to 75 mol% Li_2O) were prepared by mechanical milling using a planetary ball mill apparatus. The local structure and conductivity of the glasses were examined, and the lithium orthoborate glass showed a conductivity of $3 \cdot 4 \times 10^{-7}$ S cm^{-1} at room temperature as a powder-compressed pellet. Glasses at the ortho-oxosalt compositions in the systems $Li_2O–M_xO_y$ (M=Si, Ge or Al) were also prepared. For these systems, the glass forming regions for mechanical milling are similar to or wider than for rapid quenching, but for the $Li_2O–P_2O_5$ system, glasses could only be prepared by milling for compositions that contained less than 70 mol% Li_2O. A hot pressing technique was effective in decreasing the grain boundary in milled glass pellets. A hot pressed lithium orthosilicate glass pellet showed an ambient temperature conductivity of $2 \cdot 1 \times 10^{-6}$ S cm^{-1}, which is similar to the bulk conductivity of the corresponding quenched glass.

1. Introduction

Ion conducting glasses are promising solid electrolytes for solid state battery applications. Glassy electrolytes prepared by melt quenching, including rapid quenching, have been widely studied in the oxide and sulphide systems with high lithium ion concentrations.[1–4] Oxide glasses with high Li_2O contents have been prepared by twin roller rapid quenching and the conductivity of these glasses increases with increasing Li_2O content in the lithium silicate system.[2] Amorphous LiPON (*Lithium Phosphorus OxyNitride*) thin films show a conductivity of 10^{-6} S cm^{-1} at room temperature and they have been applied as a solid electrolyte in thin film lithium secondary batteries.[5] Recently, it was reported that amorphous $Li_3PO_4–Li_4SiO_4$ thin films are useful as a buffer material for decreasing interfacial resistance at the $LiCoO_2$ electrode and $Li_2S–P_2S_5$ electrolyte in bulk-type solid state rechargeable lithium batteries.[6] Those thin film electrolytes were prepared by gas phase techniques such as RF magnetron sputtering and pulsed laser deposition. Development of simple and convenient techniques to prepare bulk oxide glass electrolytes with high lithium ion concentration is desirable.

We have previously prepared several sulphide-based glass electrolytes with high lithium ion concentration by a mechanochemical technique with a planetary ball mill apparatus.[7–10] This preparation technique has several advantages, such as a wide glass forming region, room temperature process and direct preparation of fine glass powders. Sulphide glass electrolytes in the system $Li_2S–Al_2S_3$, which were difficult to prepare by melt-quenching, were prepared via mechanochemistry.[10] For oxide systems, SnO-based oxide glasses were prepared by ball milling;[11,12] these glasses are attractive as a negative electrode for lithium secondary batteries and as a sealing glass with a low melting point. In the SnO–B_2O_3 system, glasses were prepared at compositions up to 80 mol% SnO, which could not be prepared by a conventional melt quenching technique.[11] The mechanochemical technique is thus useful for finding new glassy systems with high lithium ion concentrations.

In this study, lithium borate glass electrolytes with high Li_2O contents were prepared by the mechanochemical technique; reagent grade chemicals of Li_2O and B_2O_3 were reacted to prepare glasses using a planetary ball mill apparatus. The glass formation region, local structure and conductivity of the lithium borate glasses were investigated. The structure and conductivity of the ball milled glasses were compared with those of melt quenched glasses. The glass forming regions for other oxide systems $Li_2O–M_xO_y$ (M=Si, P, Ge, or Al) were also examined, and the conductivities of lithium ortho-oxosalt glasses were compared to those of lithium orthoborate glasses.

2. Experimental

Reagent grade chemicals of crystalline Li_2O (Furuuchi Chem., 99·9%) and amorphous B_2O_3 (Sigma-Aldrich,

Corresponding author. Email hayashi@chem.osakafu-u.ac.jp
Original version presented at VII Int. Conf. on Borate Glasses, Crystals and Melts, Dalhousie University, Halifax, Nova Scotia, Canada, 21–25 August 2011

99·99%) powders were used as starting materials for sample preparation. Reagent grade chemicals, SiO_2 (Kojundo Chem., 99·99%), P_2O_5 (Sigma-Aldrich, 99·90%), GeO_2 (Wako Pure Chem., 99·9999%) and Al_2O_3 (Kojundo Chem., 99·999%) were also used as other glass formers. The mechanical milling treatment was carried out for batches (0·7 g) of the mixed materials of $xLi_2O.(100-x)M_xO_y$ in a zirconia pot (45 mL volume) with 160 zirconia balls (5 mm in diameter) using a high energy planetary ball mill apparatus (Fritsch Pulverisette 7). The rotation speed was fixed at 370 rpm and the milling periods were from 0 to 40 h. All the processes were conducted at room temperature in a dry Ar-filled glove box to avoid the participation of water molecules in the mechanochemical reaction process.

X-ray diffraction (XRD) measurements (Cu K_α) were performed using a Shimadzu XRD-6000 diffractometer. Differential scanning calorimetry (DSC) was carried out using a Seiko Instruments DSC 6200 with the powder samples sealed in an Al pan in a dry Ar atmosphere. The heating rate was 10°C min⁻¹. Raman spectra of the milled samples were measured using a Raman spectrometer (HORIBA Jobin Yvon, LabRAM HR-800) using the 325 nm line of a He–Cd laser. ²⁹Si MAS (magic angle spinning) NMR measurements were conducted using a Varian Unity Inova 300 spectrometer. The observed frequency, 90° pulse length, and recycle pulse delay were 59·59 MHz, 2·5 µs, and 30 s, respectively.

Electrical conductivities were measured for a pellet obtained by cold pressing (360 MPa) of the milled powders; the diameter and thickness of the pellets were 10 mm and ca. 1 mm, respectively. The morphology of the cross section of a powder compressed pellet was examined by a JEOL JSM-5300 scanning electron microscope. Gold electrodes were formed on both faces of the pellet by vacuum evaporation, and AC impedance measurements were then carried out in a dry Ar atmosphere using a Solartron Analytical 1260A impedance/gain phase analyser in a frequency range from 0·1 Hz to 8 MHz.

3. Results and discussion

3.1. Glass preparation in the system Li_2O–B_2O_3

Lithium borate glasses at compositions of 50, 66·7 and 75 mol% Li_2O were prepared by mechanical milling. Figure 1 shows XRD patterns of the lithium pyroborate samples ($66·7Li_2O.33·3B_2O_3$) prepared by mechanical milling for different periods of time. A starting mixture of crystalline Li_2O and amorphous B_2O_3 was used. The intensity of the peak attributable to Li_2O decreased with increasing milling time, and an amorphous halo pattern was observed after milling for 40 h. A DSC curve for the $66·7Li_2O.33·3B_2O_3$ sample milled for 40 h is shown in Figure 2. The amorphous sample prepared by milling showed an

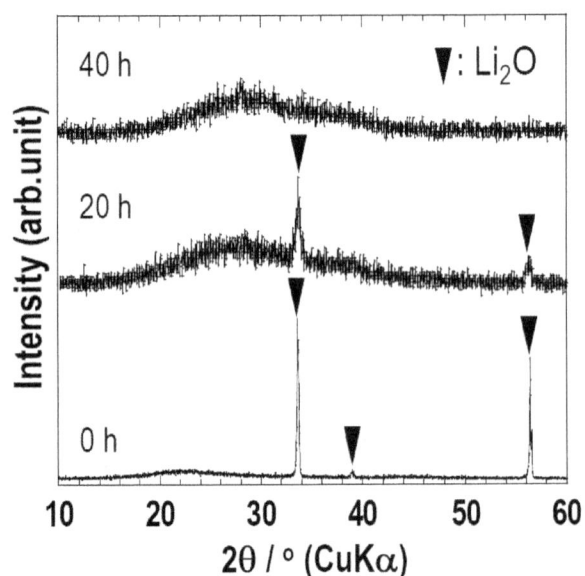

Figure 1. XRD patterns of lithium pyroborate $66·7Li_2O.33·3B_2O_3$ (mol%) samples prepared by mechanical milling for different periods of time

endothermic change attributable to a glass transition at ca. 260°C, and an exothermic peak due to crystallisation at ca. 280°C. The glass transition temperature, T_g, of $66·7Li_2O.33·3B_2O_3$ glass prepared by melt quenching has been reported to be 266°C,[2] which is similar to T_g for the milled glass. Thus the amorphous $66·7Li_2O.33·3B_2O_3$ sample prepared by milling was in the glassy state. The lithium borate glasses at compositions of 50 and 75 mol% Li_2O were prepared by mechanical milling for 20 and 40 h, respectively. The DSC curves of the milled glasses suggest that both the glass transition and crystallisation temperatures decrease with increasing Li_2O content, although the glass transition phenomenon for the $75Li_2O.25B_2O_3$ glass was vague.

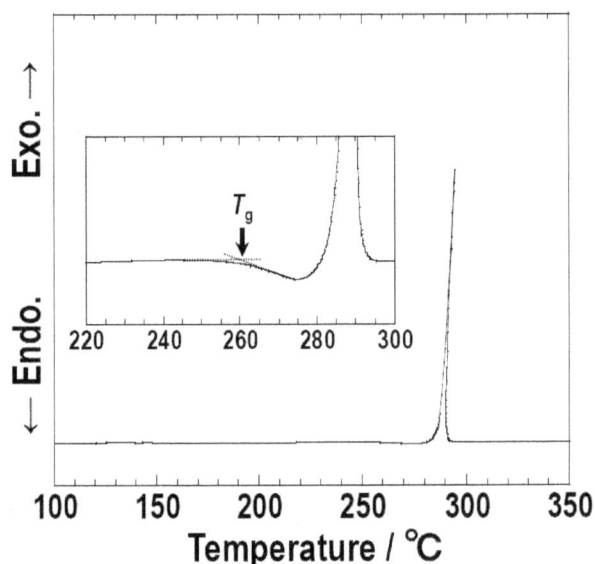

Figure 2. DSC curve for the $66·7Li_2O.33·3B_2O_3$ sample milled for 40 h

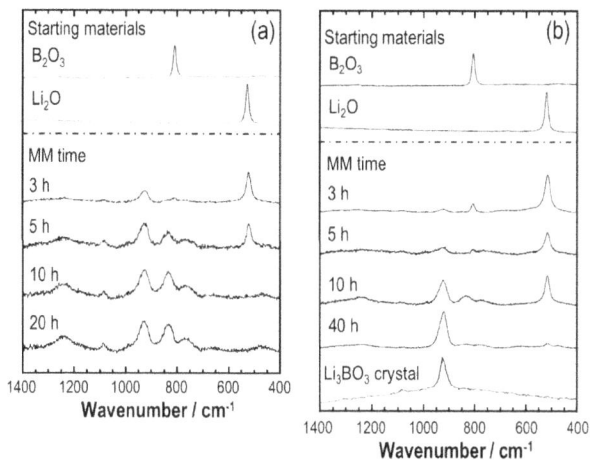

Figure 3. Raman spectra for (a) the lithium pyroborate ($66 \cdot 7Li_2O.33 \cdot 3B_2O_3$) samples, and (b) the lithium orthoborate ($75Li_2O.25B_2O_3$) samples, prepared by mechanical milling for different periods of time. Also shown are the Raman spectra of the starting materials, Li_2O and B_2O_3, and of crystalline Li_3BO_3

The mechanism for the formation of glasses by mechanical milling has not been clarified in detail. To examine the reaction process between Li_2O and B_2O_3 during milling, the local structure of the samples prepared with shorter milling periods of time was analyzed by Raman spectroscopy. Figure 3 shows the Raman spectra for the lithium pyroborate samples ($66 \cdot 7Li_2O.33 \cdot 3B_2O_3$) (a) and the lithium orthoborate samples ($75Li_2O.25B_2O_3$) (b) prepared by mechanical milling (MM) for different periods of time. The spectra of the starting materials of crystalline Li_2O and amorphous B_2O_3 are also shown for comparison.

As shown in Figure 3 (a), a new band at about 930 cm^{-1} appeared for the $66 \cdot 7Li_2O.33 \cdot 3B_2O_3$ sample after milling for 3 h. Two bands at about 840 and 770 cm^{-1}, in addition to the band at 930 cm^{-1}, were observed after milling for 5 h, and their intensities increased with increasing milling time. The band due to the starting material Li_2O vanished after milling for 10 h. The spectrum for the milled sample for 20 h was similar to both that for 10 h, and that for the melt quenched glass at the same composition.[13] For the $75Li_2O.25B_2O_3$ sample as shown in Figure 3(b), a similar change was observed during milling. The band at 930 cm^{-1} was newly observed in the sample milled for 3 h. The bands due to the starting materials almost vanished after milling for 40 h, and only one band at 930 cm^{-1} remained. The spectrum of crystalline Li_3BO_3 is also shown for comparison in Figure 3(b), and this also exhibits a band at 930 cm^{-1}.

The bands at 930 and 840 cm^{-1} are attributable to orthoborate (BO_3^{3-}) and pyroborate ($B_2O_5^{4-}$) groups, respectively.[13,14] The band at 770 cm^{-1} is attributable to six-membered borate rings with one BO_4 unit (triborate groups); because of the broadness of the band, it also includes the band (745 cm^{-1}) attributed

to six-membered borate rings with two BO_4 units (di-triborate groups). The Raman spectra for both the $66 \cdot 7Li_2O.33 \cdot 3B_2O_3$ and $75Li_2O.25B_2O_3$ samples indicated that isolated orthoborate groups without any bridging oxygen atoms were formed by initial mechanical milling of the starting mixture of Li_2O and B_2O_3. After the formation of the BO_3^{3-} groups, pyroborate and triborate groups with bridging oxygen atoms were formed after an increase in milling time. Lithium borate glasses with local structure corresponding to the nominal compositions were thus prepared by mechanical milling.

The electrical conductivity was determined for the pelletised glasses by the ac impedance technique. Figure 4 shows the temperature dependence of the conductivities of the $xLi_2O.(100-x)B_2O_3$ glasses prepared by mechanical milling. The conductivities were calculated from the total resistances including bulk and grain boundary components, because the two components were not distinguished in Nyquist plots. The conductivities for all the glasses obeyed the Arrhenius equation. The conductivity of the glasses increased with increasing Li_2O content. The lithium orthoborate ($75Li_2O.25B_2O_3$) glass exhibited the highest conductivity of $3 \cdot 4 \times 10^{-7}$ S cm^{-1} at room temperature. The activation energy for conduction for the glass was 52 kJ mol^{-1}. The lithium orthoborate glass was not prepared by melt quenching, and this paper is the first report of the conductivity of lithium orthoborate glass. The lithium pyroborate ($66 \cdot 7Li_2O.33 \cdot 3B_2O_3$) and lithium metaborate ($50Li_2O.50B_2O_3$) glasses prepared by milling showed conductivities of $1 \cdot 1 \times 10^{-8}$ and $3 \cdot 6 \times 10^{-9}$ S cm^{-1} at room temperature, which is lower than the conductivities of 6×10^{-7} and 3×10^{-7} S cm^{-1} for lithium pyroborate and lithium metaborate glasses prepared by melt

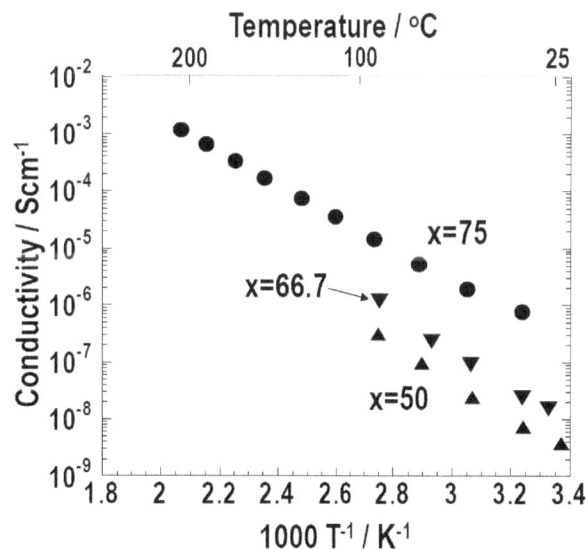

Figure 4. Temperature dependence of the conductivities of the $xLi_2O.(100-x)B_2O_3$ glasses prepared by mechanical milling. The error on the conductivity values is included within the symbol size

quenching.[2] The conductivity of the melt quenched glass is a bulk conductivity, which does not include a grain boundary resistance, and thus the difference in conductivity may be based on the sample shape for conductivity measurements. This suggests that the bulk conductivity of lithium orthoborate glass would be larger than 3.4×10^{-7} S cm^{-1} at room temperature.

3.2. Glass preparation in the systems Li$_2$O–M$_x$O$_y$

Several oxide glasses in the systems Li$_2$O–M$_x$O$_y$ (M=Si, P, Ge or Al) were prepared by mechanical milling. Lithium phosphate glasses with 50 and 66.7 mol% Li$_2$O were obtained by milling, but it was difficult to prepare glasses with more than 70 mol% Li$_2$O. The glasses in the systems Li$_2$O–M$_x$O$_y$ (M=Si, Ge or Al) were not mechanochemically prepared from crystalline Li$_2$O and M$_x$O$_y$ as starting materials. In the Li$_2$O–SiO$_2$ system, XRD peaks attributable to crystalline SiO$_2$ still remained after milling for 60 h. On the other hand, Li$_2$O–SiO$_2$ glasses were prepared using amorphous SiO$_2$ as a starting material, which had been obtained by milling of crystalline SiO$_2$ for 25 h in advance. Glasses in the systems Li$_2$O–GeO$_2$ and Li$_2$O–Al$_2$O$_3$ were also mechanochemically synthesized using amorphous GeO$_2$ and Al$_2$O$_3$, which

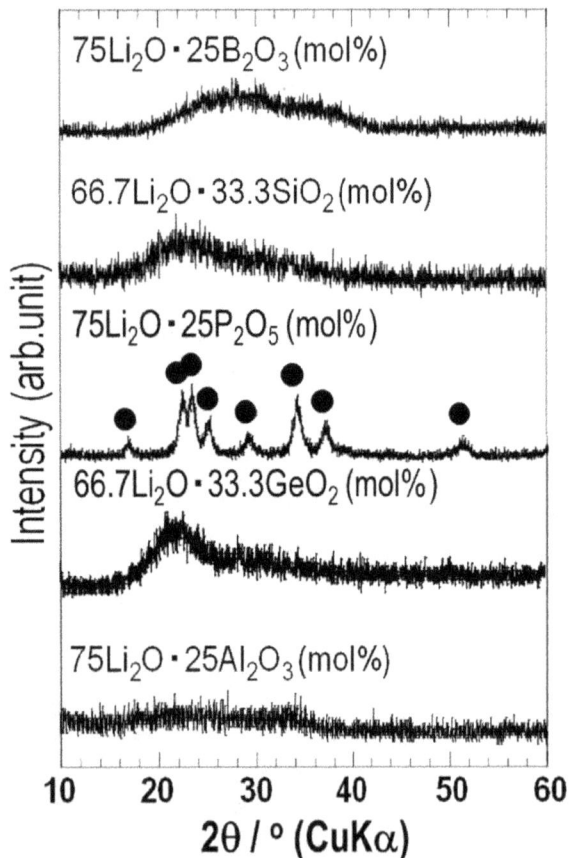

Figure 6. Glass forming region for the systems Li$_2$O–M$_x$O$_y$ (M=B, Si, P, Ge or Al) by conventional melt quenching (white area), rapid quenching (shaded area) and mechanical milling (symbols). Open circles denote the compositions where a glass was formed by mechanical milling, and a closed circle denotes a crystallised composition. The ortho-oxosalt composition in each system is indicated by an arrow

had been milled in advance.

Figure 5 shows the XRD patterns of Li$_2$O–M$_x$O$_y$ samples at ortho-oxosalt compositions prepared by mechanical milling. Except for the phosphate system, amorphous halo patterns were observed for the borate, silicate, germanate and aluminate systems. In the phosphate system, crystalline Li$_3$PO$_4$ was directly formed after milling and glass was not obtained at this composition.

Figure 6 shows the glass forming region for the Li$_2$O–M$_x$O$_y$ systems by conventional melt quenching (white area), twin roller rapid quenching (shaded area) and mechanical milling (symbols). Open circles denote a composition where glass was formed by mechanical milling, and a closed circle denotes A crystallised composition. The ortho-oxosalt composition in each system is indicated by an arrow. The glass forming region for mechanical milling is similar to, or wider than, that for rapid quenching. In particular, glasses at the ortho-oxosalt compositions were prepared in the borate, germanate and aluminate systems, which have not previously been prepared by quenching methods. The ^{29}Si MAS NMR spectrum of the lithium orthosilicate glass (66.7Li$_2$O.33.3SiO$_2$) is shown in Figure 7. A single peak at −65 ppm was observed which is attributable to Q^0 units, i.e. a silicon atom coordinated to no bridging oxygen atoms.[15]

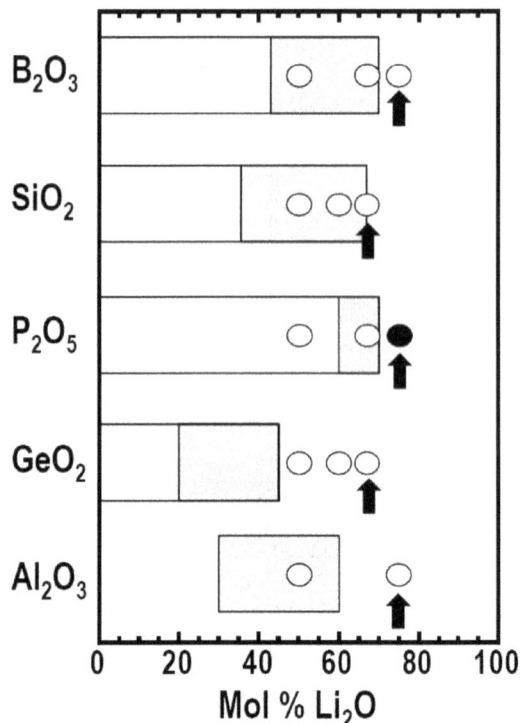

Figure 5. XRD patterns of Li$_2$O–M$_x$O$_y$ (M=B, Si, P, Ge or Al) samples at the ortho-oxosalt compositions prepared by mechanical milling. Closed circles indicate the positions of the Li$_3$PO$_4$ Bragg peaks

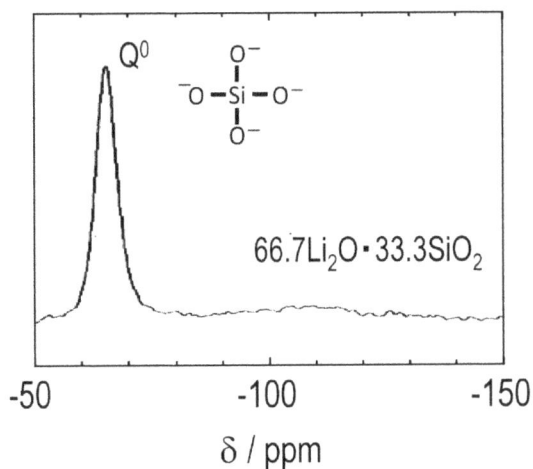

Figure 7. ^{29}Si MAS NMR spectrum of the lithium ortho-silicate (66·7Li$_2$O.33·3SiO$_2$) glass

Thus, in the lithium orthosilicate glass prepared by mechanical milling, the silicons are only in isolated SiO$_4^{4-}$ ions.

Figure 8 shows the conductivity for pelletised Li$_2$O–M$_x$O$_y$ with (M=B, Si, Ge or Al) glasses at the ortho-oxosalt compositions prepared by mechanical milling. The glasses with M=B, Si or Ge exhibited almost the same conductivity, 10^{-7} S cm^{-1}, at room temperature. On the other hand, the 75Li$_2$O.25Al$_2$O$_3$ glass showed one order of magnitude lower conductivity than the other glasses.

The conductivity of a pellet of the milled 66·7Li$_2$O.33·3SiO$_2$ glass was 3·0×10^{-7} S cm^{-1} at room temperature, which is one order of magnitude lower than that of a flake-like glass sample prepared by rapid quenching.[2] The difference in conductivity may be due to a grain boundary resistance in the milled pellet. A glass transforms into supercooled

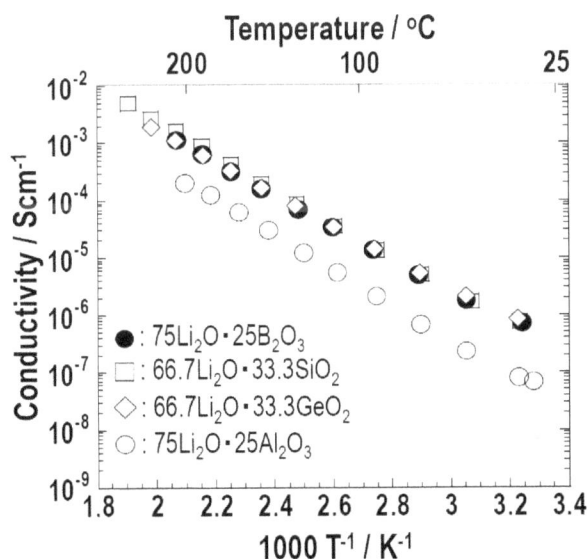

Figure 8. Conductivities for pelletized Li$_2$O–M$_x$O$_y$ (M=B, Si, Ge or Al) glasses at the ortho-oxosalt compositions prepared by mechanical milling

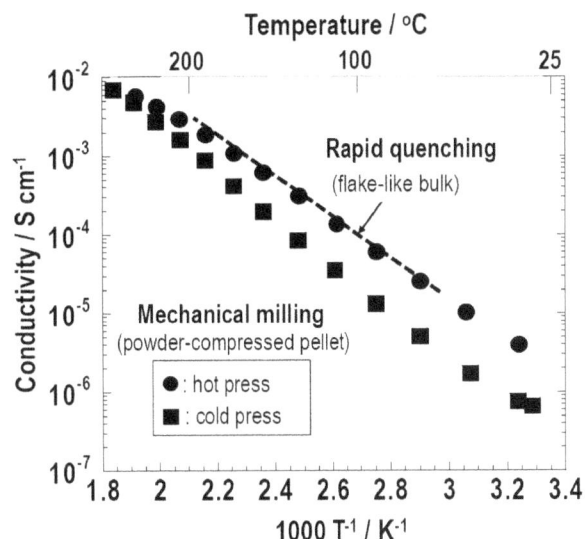

Figure 9. Temperature dependence of the conductivity of 66·7Li$_2$O.33·3SiO$_2$ glasses prepared by mechanical milling and rapid quenching. Closed squares and circles respectively denote the conductivities of the milled glass pellets prepared by cold press and hot press, whilst the dashed line denotes the conductivity of rapidly quenched glass

liquid and softens at around or beyond the glass transition temperature; the grain boundary in a glass pellet can thus be decreased by using viscous flow of the supercooled liquid at around T_g.[16]

Figure 9 shows the temperature dependence of the conductivity of the 66·7Li$_2$O.33·3SiO$_2$ glasses prepared by mechanical milling and rapid quenching.

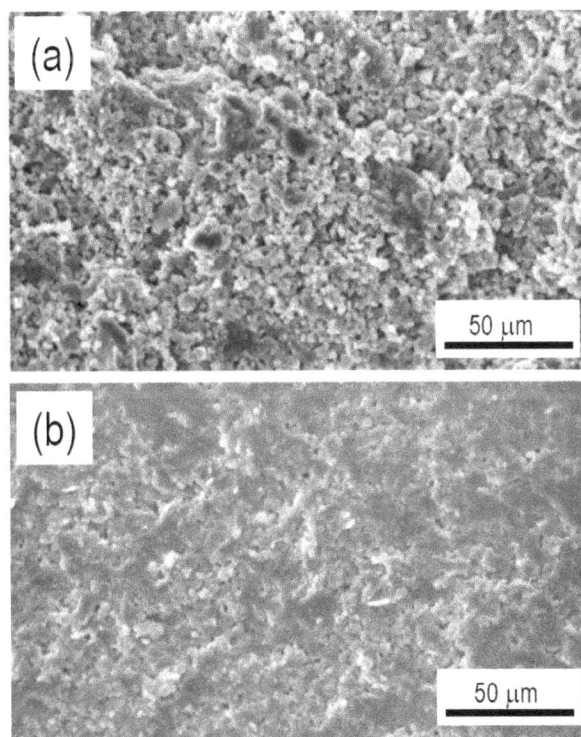

Figure 10. Cross-sectional SEM images of the 66·7Li$_2$O.33·3SiO$_2$ glass pellets prepared by (a) cold pressing, and (b) hot pressing

Closed squares denote the conductivities of a milled glass pellet prepared by a conventional cold press at 370 MPa for 5 min at room temperature. On the other hand, closed circles denote the conductivities of a milled glass pellet obtained by a hot press at 370 MPa for 5 h at 290°C, which is between the glass transition and crystallisation temperatures. The hot pressed pellet shows a conductivity of $2 \cdot 1 \times 10^{-6}$ S cm^{-1}, which is one order of magnitude higher than for the cold pressed pellet, and is almost the same as the bulk conductivity of the rapidly quenched glass.[2] The activation energy for conduction for mechanically milled $66 \cdot 7Li_2O.33 \cdot 3SiO_2$ glass decreased from 55 to 47 kJ mol^{-1} as a result of hot pressing. Figure 10 shows the cross-sectional SEM images of the $66 \cdot 7Li_2O.33 \cdot 3SiO_2$ glass pellets prepared by cold pressing (a), and hot pressing (b). Softening adhesion among the glass particles decreased the grain boundaries and voids in the pellet. The decrease of grain boundary resistance by hot pressing brought about an enhancement of conductivity of the lithium orthosilicate glass prepared by mechanical milling.

4. Conclusions

The glass forming region in the $Li_2O–B_2O_3$ system for mechanical milling is wider than that for rapid quenching, and a glass at the orthoborate composition (Li_3BO_3) was prepared by milling. The local structure of lithium borate glasses prepared by mechanical milling and melt quenching was almost the same. The conductivity of the lithium borate glasses increased with increasing Li_2O content, and the Li_3BO_3 glass showed the highest conductivity of $3 \cdot 4 \times 10^{-7}$ S cm^{-1}.

The conductivity of Li_4SiO_4 and Li_4GeO_4 glasses prepared by milling exhibited almost the same conductivity. A hot press technique was effective in increasing the conductivity of pelletised glass electrolytes prepared by milling. Mechanochemical synthesis is useful for developing oxide glass electrolytes with high lithium contents, which are difficult to obtain by conventional melt quenching.

References

1. Tatsumisago, M., Narita, H., Minami, T. & Tanaka, M. *J. Am. Ceram. Soc.*, 1983, **66**, C210.
2. Tatsumisago, M., Yoneda, K., Machida, N. & Minami, T. *J. Non-Cryst. Solids*, 1987, **95–96**, 857.
3. Pradel, A. & Ribes, M. *Solid State Ionics*, 1986, **18–19**, 351.
4. Minami, T., Hayashi, A. & Tatsumisago, M. *Solid State Ionics*, 2000, **136–137**, 1015.
5. Bates, J. B., Dudney, N. J., Neudecker, B., Ueda, A. & Evance, C. D. *Solid State Ionics*, 2000, **135**, 33.
6. Sakurai, Y., Sakuda, A., Hayashi, A. & Tatsumisago, M. *Solid State Ionics*, 2011, **182**, 59.
7. Morimoto, H., Yamashita, H., Tatsumisago, M. & Minami, T. *J. Am. Ceram. Soc.*, 1999, **82**, 1352.
8. Tatsumisago, M., Yamashita, H., Morimoto, H. & Minami, T. *Solid State Ionics*, 2000, **136**, 483.
9. Hayashi, A., Hama, S., Morimoto, H., Tatsumisago, M & Minami, T. *J. Am. Ceram. Soc.*, 2001, **84**, 477.
10. Hayashi, A., Fukuda, T., Hama, S., Yamashita. H., Morimoto, H., Minami, T. & Tatsumisago, M. *J. Ceram. Soc. Jpn.*, 2004, **112**, S695.
11. Hayashi, A., Nakai, M., Morimoto, H., Minami, T. & Tatsumisago, M. *J. Mater. Sci.*, 2004, **39**, 5361.
12. Hayashi, A., Konishi, T., Nakai, M., Morimoto, H., Tadanaga, K., Minami, T. & Tatsumisago, M. *J. Ceram. Soc. Jpn.*, 2004, **112**, S713.
13. Tatsumisago, M., Takahashi, M., Minami, T., Tanaka, M., Umesaki, N. & Iwamoto, N. *J. Ceram. Soc. Jpn.*, 1986, **94**, 12.
14. Konijnendijk, W. L. & Stevels, J. M. *J. Non-Cryst. Solids*, 1975, **18**, 307.
15. Grimmer, A. R., Magi, M., Hahnert, M. Stade, H., Samosono, A., Wieker, W. & Lippmaa, E. *Phys. Chem. Glasses*, 1984, **25** (2), 105–8.
16. Kitaura, H., Hayashi, A., Ohtomo, T., Hama, S. & Tatsumisago, M. *J. Mater. Chem.*, 2011, **21**, 118.

Phys. Chem. Glasses: Eur. J. Glass Sci. Technol. B, October 2013, **54** (5), 195–198

Luminescence efficiency of B$_2$O$_3$–Sb$_2$O$_3$–Bi$_2$O$_3$ glass phosphor doped with Sm^{3+}

Koji Oshima,* Shingo Fuchi, Shunichi Kobayashi & Yoshikazu Takeda

Department of Crystalline Materials Science, Graduate School of Engineering, Nagoya University, Furo-cho, Chikusa-ku, Nagoya 464-8603, Japan

Manuscript received 9 November 2011
Revised version received 22 January 2012
Accepted 28 May 2013

The dependence of the output power of Sm^{3+}-doped glass phosphor on Sm$_2$O$_3$ concentration and thickness has been investigated. The output power increased with increasing Sm$_2$O$_3$ concentration up to 1 mol%. However, the output power decreased with Sm$_2$O$_3$ concentration more than 1 mol% due to concentration quenching. The output power increased with increasing glass phosphor thickness. The highest output power was achieved for a thickness of 10 mm. The luminescence efficiency of a 10 mm thick glass phosphor doped with 1 mol% Sm$_2$O$_3$ was 0·5%. Moreover, the luminescence efficiency of Sm^{3+} was estimated to be 12%.

1. Introduction

Wideband light sources are used in absorption spectrometry for agricultural applications,[1] fibre gyroscopes,[2,3] and for optical coherence tomography (OCT) – a novel cross-sectional imaging technique for biological tissues.[4] In general, a wideband light source is desirable for absorption spectrometry in order to measure a broader region of the absorption spectrum. Since OCT is based on the Michelson interferometer, a low coherence, wideband light source has several advantages for high depth resolution. Moreover, a central wavelength in the near-infrared (NIR) region is suitable for the above applications, leading to penetration depths in biological samples larger than that at other wavelengths.[4]

Super luminescent diodes (SLDs) and light emitting diodes (LEDs) in the NIR region and halogen lamps are usually used as NIR wideband light sources. SLDs and LEDs are small and have a long lifetime. However, the spectral widths of SLDs and LEDs are 50 nm at maximum, which is insufficient for most spectroscopic applications. This spectral width corresponds to an OCT depth resolution of around 10 μm, which is insufficient for some medical applications. Zhang *et al* have reported a method for the assembly of several LEDs to shorten the coherent length.[5] However, the assembly of several LEDs and control of the light intensity of individual LEDs are complicated for practical uses. On the other hand, the halogen lamp has a very wide spectral width. Therefore, absorption spectrometry uses the halogen lamp as the light source. However, the size of the halogen lamp is much larger than that of SLDs and LEDs. Moreover, the lifetime of the halogen lamp is

much shorter than that of SLDs and LEDs.

Therefore, we propose a novel wideband NIR light source by combining a wideband NIR phosphor and a LED as an excitation light source in one package. To realize this light source, we have previously synthesised a phosphor using Bi$_2$O$_3$–B$_2$O$_3$ based glasses doped with Yb^{3+} and Nd^{3+} that emits at around 1000 nm.[6–8] The spectral width was from 950 nm to 1100 nm (with a central wavelength of 1014 nm and a full width at half maximum (FWHM) of 98 nm). Although this spectral width is larger than that of SLDs and LEDs, it is not enough to substitute this light source for the halogen lamp. Therefore, we next tried to increase the spectral width by stacking the Sm^{3+} doped glass phosphor between LED and the Yb^{3+}, Nd^{3+} co-doped glass phosphor.[9] In this technique, the Sm^{3+}-doped glass phosphor is excited by the LED, and the phosphor emits in the visible (550–730 nm) and NIR regions (860–970 nm, 1100–1200 nm). The Yb^{3+}, Nd^{3+} co-doped glass phosphor is excited by the visible light of the Sm^{3+}-doped glass phosphor and emits in the NIR region (970–1100 nm). As a result, we obtained NIR wideband light (860–1200 nm). In the study reported here, we optimise the Sm^{3+} concentration and the thickness of the Sm^{3+}-doped glass phosphor to increase the luminescence efficiency.

2. Experimental

The samples used in this study were synthesised by a melt quenching method. Powders of Sm$_2$O$_3$, Bi$_2$O$_3$, H$_3$BO$_3$, and Sb$_2$O$_3$ were mixed in nominal molar compositions of xSm$_2$O$_3$.(100–x)(45B$_2$O$_3$.45Sb$_2$O$_3$.10Bi$_2$O$_3$). The mixed powders were melted in an Al$_2$O$_3$ crucible at 1000°C in an electric furnace. After 10 min at 1000°C, the molten liquid was poured between two stainless steel mould plates kept at room temperature

Corresponding author. Email ohshima@mercury.numse.nagoya-u.ac.jp
Original version presented at VII Int. Conf. on Borate Glasses, Crystals and Melts, Dalhousie University, Halifax, Nova Scotia, Canada, 21–25 August 2011

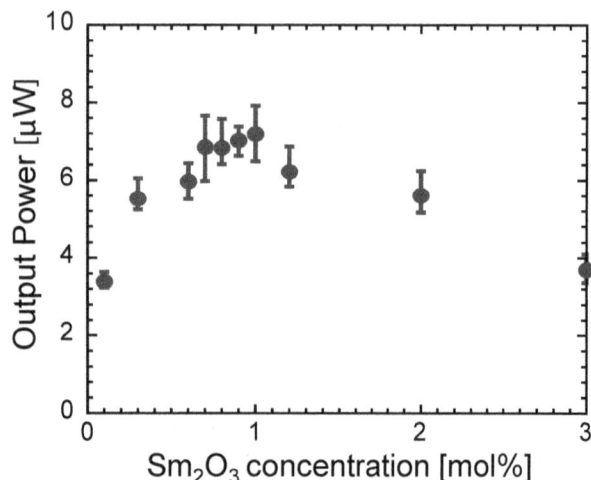

Figure 1. *Dependence of the glass phosphor output power on the Sm$_2$O$_3$ concentration [Colour available online]*

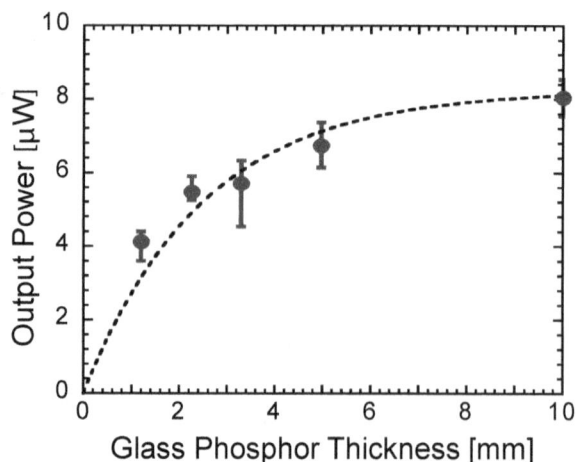

Figure 2. *Dependence of the output power on the glass phosphor thickness. The dotted line shows the theoretical output power considering only the absorption of the excitation light [Colour available online]*

for formation of the glass.

Output power measurements, photoluminescence (PL) measurements, absorption measurements, and photoluminescence excitation (PLE) measurements were carried out at room temperature. The photoexcitation light source for the PL measurement was a LED (with central wavelength 405 nm). A multichannel fibre optic spectrometer was used for detection in the visible region. An InGaAs photomultiplier mounted on a 0·20 m grating monochromator was adopted for lock-in detection in the NIR region. The light source for the PLE measurement was the emission of a halogen lamp dispersed by a monochromator.

3. Results

3.1 Dependence of output power on Sm$_2$O$_3$ concentration

We varied the Sm$_2$O$_3$ concentration from 0·1 to 3·0 mol% (the glass phosphor thickness was kept at 1 mm in this experiment). Figure 1 shows the dependence of the output power on the Sm$_2$O$_3$ concentration. The output power increases with increasing Sm$_2$O$_3$ concentration up to 1 mol%, then decreases for higher Sm$_2$O$_3$ concentration above 1 mol%. The decrease is due to concentration quenching. The highest output power is achieved at a Sm$_2$O$_3$ concentration of 1 mol%.

3.2 Dependence of output power on glass phosphor thickness

We varied the glass phosphor thickness from 1 mm to 10 mm (the Sm$_2$O$_3$ concentration was kept at 1 mol%). Figure 2 shows the dependence of the output power on the glass phosphor thickness. The output power increases with increasing glass phosphor thickness. The output power, I_{out} is given by

$$I_{out} = \frac{\eta I_{ex} \alpha_{ex} \exp\left(-\alpha_{lumi}d\right)\left(\exp\left((\alpha_{lumi} - \alpha_{ex})d\right) - 1\right)}{\left(\alpha_{lumi} - \alpha_{ex}\right)} \quad (1)$$

where η is the luminescence efficiency, I_{ex} is the excitation light intensity, α_{ex} is the absorption coefficient of the excitation light, α_{lumi} is the absorption coefficient of the luminescence light, and d is the glass phosphor thickness.[10] Assuming luminescence light absorption is zero (α_{lumi}=0), I_{out} is described as

$$I_{out} \propto (1 - \exp(-\alpha d)) \quad (2)$$

This equation is shown as a dotted line in Figure 2. The experimental results coincide with the dotted line. Therefore, the output power tends to saturate with increasing glass phosphor thickness. Moreover, the maximum glass phosphor thickness was 10 mm in our experiments. From these results, we concluded that the highest output power is achieved at the glass phosphor thickness of 10 mm.

3.3 Luminescence efficiency of Sm^{3+}-doped glass phosphor

We measured the luminescence efficiency of the Sm^{3+}-doped glass phosphor (1 mol%, 10 mm in thickness). Figure 3 shows the spectrum (A) of the LED that irradiates the glass phosphor, the transmitted spectrum of the spectrum (A) through the Sm^{3+}-doped glass phosphor, and the luminescence spectrum from the Sm^{3+}-doped glass phosphor. By dividing the irradiating intensity at each wavelength in the spectrum (A) of the LED with the photon energy, the number of photons at each wavelength can be obtained. Then, by integrating the number of photons, E_{LED} (the number of photons in the spectrum (A)) can be evaluated. In the same way, $T_{phosphor}$ (the number of photons in the transmitted spectrum of the spectrum (A) through the Sm^{3+}-doped glass phosphor), and $L_{phosphor}$ (the number of photons in the luminescence spectrum from the Sm^{3+}-doped glass phosphor) can be evaluated. In Fig. 3, the hatched region in blue shows (E_{LED}–$T_{phosphor}$),

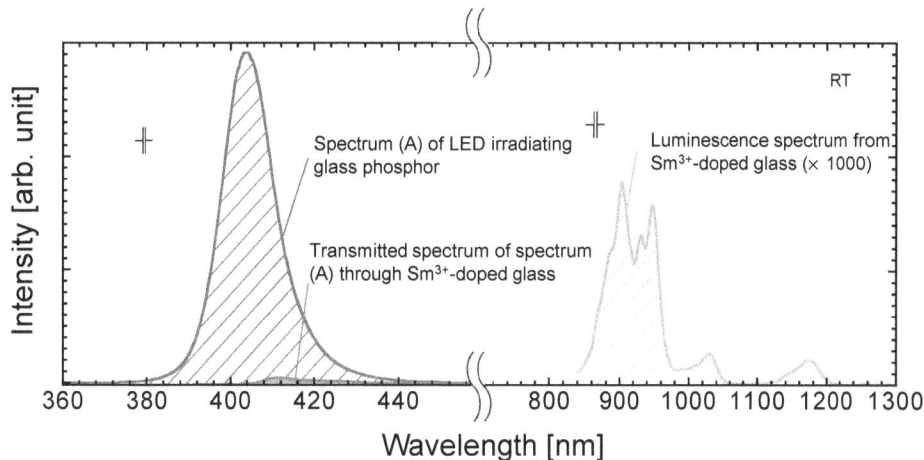

Figure 3. The spectrum (A) of the LED that irradiated the glass phosphor, the transmitted spectrum of the spectrum (A) through the Sm³⁺-doped glass phosphor (1 mol%, 10 mm in thickness), and the luminescence spectrum from the Sm³⁺-doped glass phosphor (hatched in orange) [Colour available online]

that is the loss of the photons irradiating the glass phosphor (which includes reflections at the surfaces and the absorption by the glass phosphor), and the violet area shows $L_{phosphor}$. We define the luminescence efficiency (LE) of the glass phosphor as

$$LE_{phosphor} = L_{phosphor}(E_{LED} - T_{phosphor}) \qquad (3)$$

By using Equation (3), the LE of the NIR light of the Sm³⁺-doped glass phosphor (1 mol%, 10 mm in thickness) was calculated to be 0·5%. The value of the LE of the Sm³⁺-doped glass phosphor used in this study is smaller than that of the Yb³⁺, Nd³⁺ co-doped glass (18%).[9]

4. Discussion

Figure 4 shows the absorption spectrum of the Sm³⁺-doped glass phosphor (blue line) and the PLE

spectrum (red line) of the Sm³⁺-doped glass phosphor monitored at 950 nm. The absorption spectrum indicates that the light at around 405 nm is absorbed by both Sm³⁺ (⁶H₅/₂→⁶P₃/₂) and the host glass. However, the PLE spectrum shows that the direct excitation of Sm³⁺ contributes to the NIR luminescence. Figure 5 shows the spectrum (A) of the LED that irradiates the glass phosphor, the transmitted spectrum of the spectrum (A) through the Sm³⁺-doped glass phosphor (1 mol%, 10 mm in thickness), and the transmitted spectrum of the spectrum (A) through the host glass (10 mm). By using the same method as in subsection 3.3, T_{host} (the number of photons in the transmitted spectrum of the spectrum (A) through the host glass) can be evaluated. In Figure 5 the hatched region in red shows $T_{host} - T_{phosphor}$, the hatched region in green

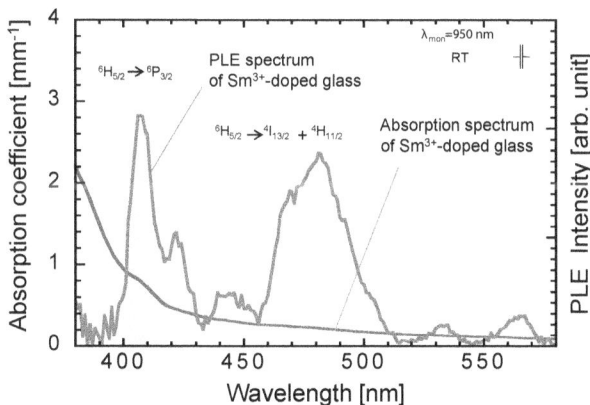

Figure 4. The absorption spectrum of the Sm³⁺-doped glass phosphor and the PLE spectrum of Sm³⁺-doped glass phosphor monitored at 950 nm. The absorption spectrum indicates that the 405 nm light is absorbed by both Sm³⁺ (⁶H₅/₂→⁶P₃/₂) and the host glass. The PLE spectrum suggests that the direct excitation of Sm³⁺ contributes to the NIR luminescence [Colour available online]

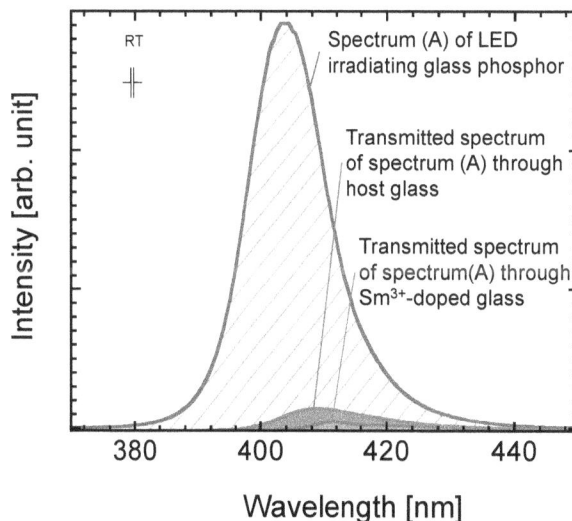

Figure 5. The spectrum (A) of the LED that irradiates the glass phosphor, the transmitted spectrum of the spectrum (A) through the Sm³⁺-doped glass phosphor (1 mol%, 10 mm in thickness), and the transmitted spectrum of the spectrum (A) through the host glass (10 mm in thickness) [Colour available online]

shows $E_{LED}-T_{phosphor}$. We define the luminescence efficiency of Sm^{3+} as

$$LE_{Sm^{3+}}=L_{phosphor}/(T_{host}-T_{phosphor}) \qquad (4)$$

$LE_{Sm^{3+}}$ is connected with the intrinsic quantum efficiency (*IQE*) for the excited state of Ln^{3+} ions (calculated by *IQE*=measured lifetime/radiative lifetime) by following equation

$$LE_{Sm^{3+}}=\eta_{eff}IQE \qquad (5)$$

where η_{eff} is the efficiency with which the energy is transferred from the excited level to the emitting level.[11] By using Equation (4), the *LE* of Sm^{3+} (1 mol%, 10 mm in thickness) was estimated to be 12%. Therefore, absorption by the host glass causes a lower *LE* of the Sm^{3+}-doped glass phosphor. The *LE* of Sm^{3+} (12%) is nearly the same as the *LE* of the Yb^{3+}, Nd^{3+} co-doped glass phosphor (18%). In order to increase the *LE* of the Sm^{3+}-doped glass phosphor, we need to design a host glass, with the high *LE* of Sm^{3+}, that has a lower absorption of the excitation light.

5. Conclusions

For the Sm^{3+}-doped $10Bi_2O_3.45Sb_2O_3.45B_2O_3$ glass phosphor, the highest output power was achieved for a Sm_2O_3 concentration of 1 mol% and a thickness of 10 mm. The luminescence efficiency of the Sm^{3+}-doped glass phosphor (1 mol%, 10 mm) was 0·5%. On the other hand, the luminescence efficiency of Sm^{3+} was estimated to be 12%. The excitation light absorption of the host glass causes a lower luminescence efficiency of the Sm^{3+}-doped glass phosphor. Therefore, we need to design a host glass, with the high luminescence efficiency of Sm^{3+}, that has a lower absorption of the excitation light.

Acknowledgements

This work was supported in part by the Technology Development Program for Advanced Measurement and Analysis of the Japan Science and Technology Agency (JST).

References

1. Siesler, H. W., Ozaki, Y., Kawano, S. & Heise, H. M. *Near-Infrared Spectroscopy-Principles, Instruments, Applications*, Wiley-VCH, Weinheim, Germany, 2002.
2. Cutler, C. C., Newton, S. A. & Shaw, H. J. *Opt. Lett.*, 1980, **5**, 488.
3. Blin, S., Kim, H. K., Digonnet, M. J. F. & Kino, G. S. *J. Lightwave Technol,* 2007, **25**, 861.
4. Fercher, A. F., Drexler, W., Hitzenberger, C. K. & Lasser, T. *Rep. Prog. Phys.*, 2003, **66**, 239.
5. Zhang, Y., Sato, M. & Tanno, N. *Opt. Lett.*, 2001, **26**, 205.
6. Fuchi, S., Sakano, A. & Takeda, Y. *Jpn. J. Appl. Phys.*, 2008, **47**, 7932.
7. Fuchi, S., Sakano, A., Mizutani R. & Takeda, Y. *Glass Technol.: Eur. J. Glass Sci. Technol. A*, 2009, **50**, 319.
8. Fuchi, S., Sakano, A., Mizutani, R. & Takeda, Y. *Appl. Phys. Express*, 2009, **2**, 32102.
9. Fuchi, S. & Takeda, Y. *Phys. Status Solidi C*, 2011, 8, 2653.
10. Fuchi, S., Sakano, A., Mizutani, R. & Takeda, Y. *Appl. Phys. B*, 2011, 105, 877.
11. Tan, M. C., Connolly, J. & Riman, R. E. *J. Phys. Chem. C*, 2011, 115, 17952.

Phys. Chem. Glasses: Eur. J. Glass Sci. Technol. B, October 2013, **54** (5), 199–205

Glass formation and crystalline phases in the ternary systems CaO–Bi$_2$O$_3$–B$_2$O$_3$ and SrO–Bi$_2$O$_3$–B$_2$O$_3$

Anna H. Barseghyan,* Rafael M. Hovhannisyan, Berta V. Petrosyan, Hovakim A. Aleksanyan

Institute of Electronic Materials,119 Arshakunyats Ave., 0007 Yerevan, Armenia

Vahan P. Toroyan

Institute of General and Inorganic Chemistry of NAS RA, Argutyan St., District 2,10, 0051 Yerevan, Armenia

Manuscript received 11 November 2011
Revised version received 8 April 2013
Accepted 8 April 2013

Large glass forming areas in the ternary systems CaO–Bi$_2$O$_3$–B$_2$O$_3$ and SrO–Bi$_2$O$_3$–B$_2$O$_3$ have been revealed by super-cooling of melts. Transparent glass samples have been obtained in a narrow bismuth-rich area of composition in the SrO–Bi$_2$O$_3$ system. Four pseudo-binary eutectic and four peritectic compositions have been revealed in the melting diagrams of the pseudo-binary systems Bi$_2$O$_3$–CaB$_2$O$_4$, BiBO$_3$–CaB$_2$O$_4$, Bi$_2$O$_3$–SrB$_2$O$_4$, BiBO$_3$–SrB$_2$O$_4$. The existence of ternary compounds CaBi$_2$B$_2$O$_7$ and CaBi$_2$B$_4$O$_{10}$ in the CaO–Bi$_2$O$_3$–B$_2$O$_3$ system, and SrBi$_2$B$_2$O$_7$ and SrBi$_2$B$_4$O$_{10}$ in the SrO–Bi$_2$O$_3$–B$_2$O$_3$ system has been confirmed. These compounds have good glass forming ability and are in the area of stable glasses. The melting behaviour of ternary compounds CaBi$_2$B$_2$O$_7$, CaBi$_2$B$_4$O$_{10}$, SrBi$_2$B$_2$O$_7$, and SrBi$_2$B$_4$O$_{10}$ has been studied. They melt incongruently at 770, 730, 755 and 790°C, respectively.

Introduction

A study of the CaO–Bi$_2$O$_3$–B$_2$O$_3$ and SrO–Bi$_2$O$_3$–B$_2$O$_3$ systems is a part of a long term program of developing new lead-free glassy and glass-ceramic solders, having wide range of rheological, dilatographic properties. The binary Bi$_2$O$_3$–B$_2$O$_3$ system, as a basis for many ternary systems, is most interesting for the development of new low melting compositions, because of both the presence of several low melting eutectics[1] and the propensity for glass formation.[2] Moreover, bismuth borate single crystals and glass ceramics have nonlinear optical (NLO) and other attractive properties.[3–6] Both these factors have led to the intensive study of binary and ternary bismuth borate systems and the glasses that they form.[7–15]

At present, the crystalline compounds Bi$_{24}$B$_2$O$_{39}$, Bi$_4$B$_2$O$_9$, Bi$_3$B$_5$O$_{12}$, BiB$_3$O$_6$, Bi$_2$B$_8$O$_{15}$ and BiBO$_3$, with melting points (m.p.s) 735, 675, 722, 708, 715 and 685°C, respectively, are known in the binary Bi$_2$O$_3$–B$_2$O$_3$ system.[1,10,16–18] Amongst the binary compounds, six eutectic compositions with m.p.s 622, 646, 665, 698, 695 and 709°C have been determined in this system.[1,18]

The Bi$_2$O$_3$–CaO and Bi$_2$O$_3$–SrO systems are complex ones and have been studied by various authors. Information about the following compounds has been reported: Ca$_2$Bi$_2$O$_5$, Ca$_3$Bi$_4$O$_9$, CaBi$_2$O$_4$, Ca$_2$Bi$_6$O$_{11}$, Ca$_7$Bi$_6$O$_{16}$, Ca$_7$Bi$_{10}$O$_{22}$, Ca$_5$Bi$_{14}$O$_{26}$, Sr$_3$Bi$_2$O$_6$, Sr$_2$Bi$_2$O$_5$, SrBi$_2$O$_4$, Sr$_2$Bi$_6$O$_{11}$.[19–22]

Calcium and boron oxides form congruently melt-

ing compounds Ca$_3$B$_2$O$_6$, Ca$_2$B$_2$O$_5$, CaB$_2$O$_4$ and CaB$_4$O$_7$ with m.p.s 1479, 1298, 1154 and 986°C, respectively.[19] Hovhannisyan[23] reinvestigated the CaO–B$_2$O$_3$ binary system and corrected the high B$_2$O$_3$ region. New eutectic compositions, 27·5CaO.72·5 B$_2$O$_3$ (975°C) and 36·5CaO.63·5B$_2$O$_3$ (980°C), were determined around the compound CaB$_4$O$_7$.

In the binary system SrO–B$_2$O$_3$, it is known that there are congruently melting compounds, Sr$_3$B$_2$O$_6$ (1285°C), Sr$_2$B$_2$O$_5$(1185°C), SrB$_2$O$_4$ (1170±5°C), SrB$_4$O$_7$ (1015±5°C), and also the incongruently melting compound SrB$_6$O$_{10}$ (920°C).[23–26] The new incongruently melting compound Sr$_4$B$_{14}$O$_{25}$ (at 1012±5°C) was recently revealed in a narrow range of composition (64–65 mol% B$_2$O$_3$).[27] The most recently investigated compound is Sr$_2$B$_{16}$O$_{26}$, but its melting character has not been studied yet.[28] Four eutectic compositions are known in this system with m.p.s 940, 971, 1062 and 1165°C.[23]

In 2006 Barbier & Cranswick[29] were the first to synthesise two novel non-centrosymmetric compounds, MBi$_2$B$_2$O$_7$ or MBi$_2$O(BO$_3$)$_2$ (M=Ca, Sr), by solid state reactions in air at temperatures in the 600–700°C range. The crystal structures of these compounds have been determined and refined using powder neutron diffraction data. CaBi$_2$B$_2$O$_7$ has SHG (second harmonic generation) response twice as high as the well known NLO, KDP (potassium dihydrophosphate) single crystal.[29] However, the authors did not study the melting behaviour of either compound. Egorisheva *et al*[30] have studied the phase relation in the CaO–Bi$_2$O$_3$–B$_2$O$_3$ system and constructed the subsolidus section of the phase diagram

Corresponding author. Email annbars@gmail.com
Original version presented at VII Int. Conf. on Borate Glasses, Crystals and Melts, Dalhousie University, Halifax, Nova Scotia, Canada, 21–25 August 2011

at 600°C. A new ternary compound $CaBi_2B_4O_{10}$ was identified and the existence of the ternary compound $CaBi_2B_2O_7$ was confirmed. Both compounds melted incongruently at 700 and 783°C, respectively, and had liquidus temperature about 900–930°C.

Kargin et al[31] have studied the phase relations in the $SrO–Bi_2O_3–B_2O_3$ system in the subsolidus section at 600°C and have found two new ternary compounds $Sr_7Bi_8B_{18}O_{46}$ and $SrBiBO_4$. Both compounds melt incongruently at 760 and 820°C, respectively, without indication of liquidus temperature. However, later Barbier & Cranswick[32] have described the novel centrosymmetric borate $SrBi_2OB_4O_9$ ($SrBi_2B_4O_{10}$) forming in the $SrO–Bi_2O_3–B_2O_3$ system, thereby having substituted the doubtful existence of the $Sr_7Bi_8B_{18}O_{46}$ compound previously reported by Kargin et al.[31] However, the available data are insufficient and additional investigations of the $CaO–Bi_2O_3–B_2O_3$ and $SrO–Bi_2O_3–B_2O_3$ systems are essential for the construction of phase and glass forming diagrams, and for revealing and characterising new eutectic and stoichiometric compositions.

Experimental

Hundreds of samples of various binary and ternary compositions in the $CaO–Bi_2O_3–B_2O_3$ and $SrO–Bi_2O_3–B_2O_3$ systems were synthesised and tested. Compositions were prepared from chemically pure grade $CaCO_3$, $SrCO_3$, Bi_2O_3 and H_3BO_3 at 2·5–5·0 mol% intervals. Most of the samples were obtained as glasses by various cooling methods, depending on the propensity for crystallisation: casting on metallic plates (bulk samples), and super cooling techniques (twin roller quenching) constructed by our group (ribbon samples with thickness 30–400 μm). Glass formation was determined visually or by x-ray analysis. The glass melting was performed at

800–1200°C for 15–20 min with a 20–50 g batch in a 20–50 ml uncovered quartz glass or corundum crucible, using an air atmosphere and the "Supertherm 17/08" electric furnace. The chemical composition of some glasses was determined by traditional chemical analysis, and the results indicate a good compatibility between the calculated and analytical amounts of B_2O_3, CaO, SrO and Bi_2O_3. SiO_2 contamination from quartz glass crucibles did not exceed 2 wt%, and alumina contamination did not exceed 0·5–1·0 wt%, according to the chemical analysis data. Some compositions were studied by solid state synthesis. 15–20 g of material was carefully mixed in an agate mortar, pressed as tablets, located on platinum plates and underwent a first thermal treatment at 450–600°C for 24 h in electrical muffles (supplied by Naber). After regrinding, the powders were tested by differential thermal analysis (DTA) and x-ray methods. Tablets of the pressed samples were subjected to repeated thermal treatment.

DTA of the glasses was conducted using a Q-1500 type DTA in platinum or pure Al_2O_3 crucibles, with powder glass samples of 0·5–0·6 g weight, a heating rate of 10 K/min, and sensitivity of 250 μV. The accuracy of temperature measurement is ±5 K.

Powder x-ray patterns were obtained on a DRON-3 type diffractometer with Cu K_α radiation and a Ni filter. Samples for glass crystallisation were prepared with glass powder pressed in the form of rods or tablets. The crystallisation was performed in electrical muffles (supplied by Naber) with one stage heat treatment. This was done for 24–48 h at around the temperature at which the maximum exothermal effects were observed by DTA on the glasses. Crystalline phases of binary and ternary compounds formed by both glass crystallisation and solid state sintering were identified by using JCPDS-ICDD PDF-2 release 2008 database.[33]

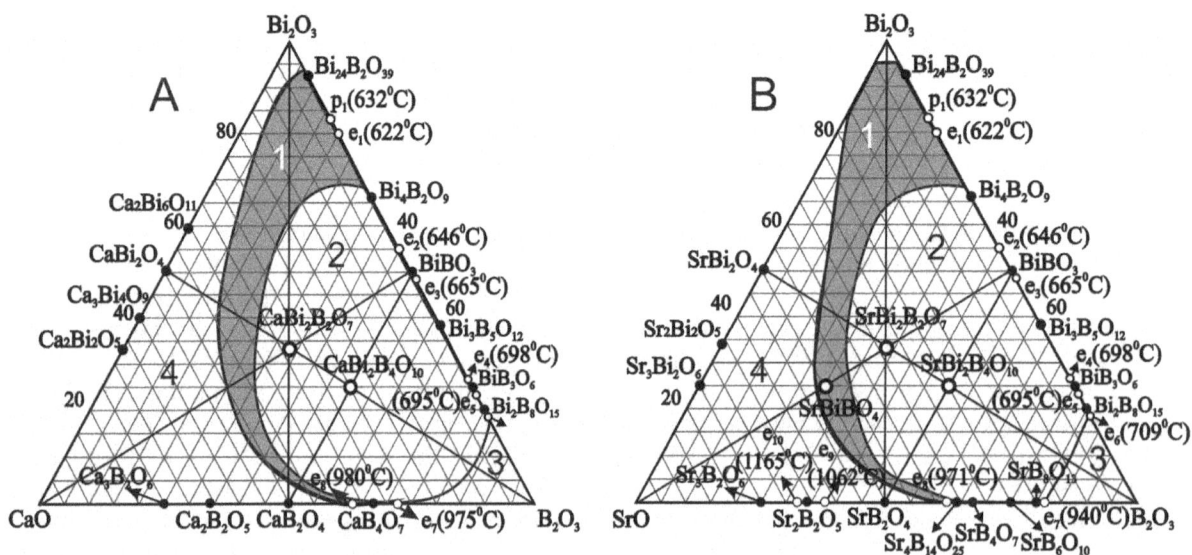

Figure 1. Glass forming diagrams of the $CaO–Bi_2O_3–B_2O_3$ (A) and $SrO–Bi_2O_3–B_2O_3$ (B) systems. 1 – area of glasses formed by the supercooling method, 2 – area of stable glasses, 3 – stable phase separation region, 4 – non-glass formation region

Results and discussion

Glass formation in the CaO–Bi$_2$O$_3$–B$_2$O$_3$ and SrO–Bi$_2$O$_3$–B$_2$O$_3$ systems

Glass formation in the CaO–Bi$_2$O$_3$–B$_2$O$_3$ and SrO–Bi$_2$O$_3$–B$_2$O$_3$ systems has previously been studied by Janakirama-Rao,[34] Imaoka & Yamazaki,[35] and Milyukov et al[36] All authors have shown that the studied compositions form glasses by the traditional melt casting method.

The glass forming diagrams of the CaO–Bi$_2$O$_3$–B$_2$O$_3$ and SrO–Bi$_2$O$_3$–B$_2$O$_3$ systems (Figure 1(A) and (B)) were constructed using different melt cooling rates. Large glass-forming areas have been revealed, which include all eutectic areas in the binary Bi$_2$O$_3$–B$_2$O$_3$ system, the low temperature eutectic areas in the binary CaO–B$_2$O$_3$ and SrO–B$_2$O$_3$ systems, and fields with low melting points in the ternary systems spreading up to the binary CaO–Bi$_2$O$_3$ and SrO–Bi$_2$O$_3$ systems. For the first time transparent glassy ribbons in a narrow bismuth-rich area of composition in the SrO–Bi$_2$O$_3$ system (5–17·5 mol% SrO) have been obtained. The compounds CaBi$_2$B$_2$O$_7$, CaBi$_2$B$_4$O$_{10}$, SrBi$_2$B$_2$O$_7$ and SrBi$_2$B$_4$O$_{10}$ have good glass forming ability and are in the area of stable glasses in the studied ternary systems (2 in Figure 1(A) and (B)). Stable phase separation regions were also observed in both systems for B$_2$O$_3$ content more than 81 mol% (3 in Figure 1(A) and (B)).

DTA and x-ray studies of the ternary compounds CaBi$_2$B$_2$O$_7$, CaBi$_2$B$_4$O$_{10}$, SrBi$_2$B$_2$O$_7$ and SrBi$_2$B$_4$O$_{10}$

Firstly we studied the DTA and x-ray characteristics of the known ternary compounds CaBi$_2$B$_2$O$_7$, CaBi$_2$B$_4$O$_{10}$, SrBi$_2$B$_2$O$_7$ and SrBi$_2$B$_4$O$_{10}$. Powders of glasses corresponding to stoichiometric compounds were used as initial samples for DTA studies and glass crystallisations. These glasses were melted at 900–1000°C in quartz glass crucibles (20–50 ml).

All characteristic points, T_g (glass transition), T_s (glass softening), T_c (peak of exothermal effects), and T_m (melting temperature), were observed in the DTA curves of the studied glasses (Figure 2, curves 1–4) and summarized in Table 1.

One exothermic effect of glass crystallisation (T_c) at 560°C and two endothermic effects of crystalline phases melting at 770 and 825°C are seen in the DTA curve of the 33·33CaO.33·33Bi$_2$O$_3$.33·33B$_2$O$_3$ (Ca-

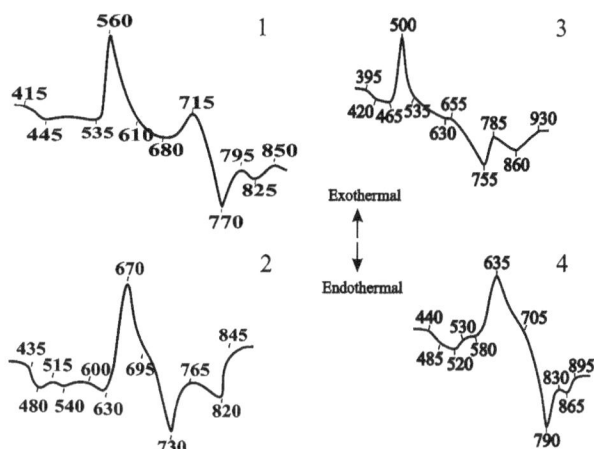

Figure 2. DTA curves of glasses with the composition of the ternary compounds in the CaO–Bi$_2$O$_3$–B$_2$O$_3$ and SrO–Bi$_2$O$_3$–B$_2$O$_3$ systems: 1 – CaBi$_2$B$_2$O$_7$, 2 – CaBi$_2$B$_4$O$_{10}$, 3 – SrBi$_2$B$_2$O$_7$, 4 – SrBi$_2$B$_4$O$_{10}$

Bi$_2$B$_2$O$_7$) glass composition (Figure 2, curve 1). X-ray diffraction patterns of products of glass crystallisation (at 560°C, for 24 h) show the formation of one CaBi$_2$B$_2$O$_7$ crystalline phase (Figure 3, curve 1) and fully correspond to the data of Barbier & Cranswick.[29] The melting temperature (T_{m1}) of this phase is 770°C and is different from the data of Egorisheva et al.[30] We have revealed that the presence of a second endothermic effect within the interval 795–850°C, with a minimum at 825°C, is associated with incongruent melting of CaBi$_2$B$_2$O$_7$ (Figure 2, curve 1). The crystalline compound CaBi$_2$B$_2$O$_7$ melted incongruently at 770°C with formation of the melt and crystalline CaB$_2$O$_4$ (Figure 3, curve 2). This latter phase was clearly observed and identified according to an x-ray database[33] in XRD patterns of products that have undergone thermal treatment at 770°C for 24 h and then been fast frozen in cold water. Dissolution of the CaB$_2$O$_4$ phase in a

Table 1. Characteristic temperatures for glasses with compositions corresponding to stoichiometric ternary compounds in the CaO–Bi$_2$O$_3$–B$_2$O$_3$ and SrO–Bi$_2$O$_3$–B$_2$O$_3$ systems

Glass Composition	T_g ±5°C	T_s ±5°C	T_c ±5°C	T_{m1} ±5°C	T_{m2} ±5°C
CaBi$_2$B$_2$O$_7$	415	445	560	770	825
CaBi$_2$B$_4$O$_{10}$	435	480	670	730	820
SrBi$_2$B$_2$O$_7$	395	420	500	735	860
SrBi$_2$B$_4$O$_{10}$	440	485	635	790	865

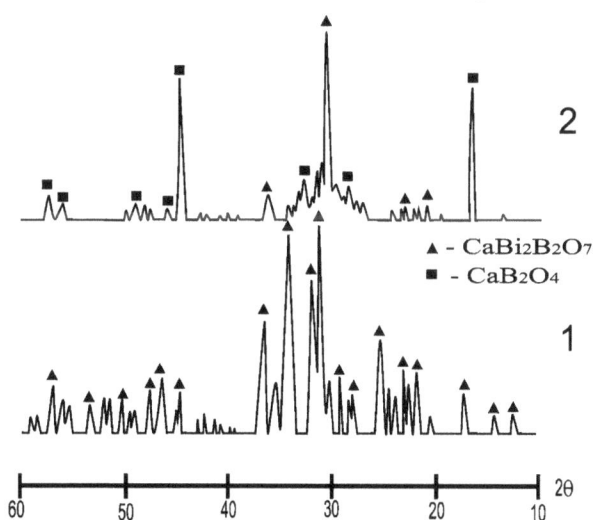

Figure 3. XRD patterns of the crystallised glasses with the composition of the ternary compound CaBi$_2$B$_2$O$_7$: 1 – 560° 24 h, cooling in the furnace; 2 – 770° 24 h, casting in cold water

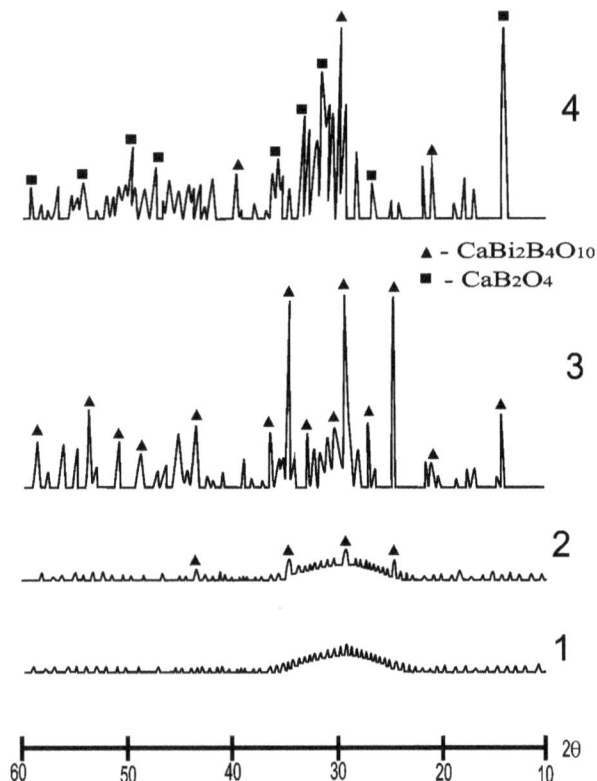

Figure 4. XRD patterns of the crystallised glasses with the composition of the ternary compound $CaBi_2B_4O_{10}$: 1 – 515°C 18 h, cooling in the furnace; 2 – 600°C 18 h, cooling in the furnace;3 – 670°C 24 h, cooling in the furnace; 4 – 730°C 24 h, casting in cold water

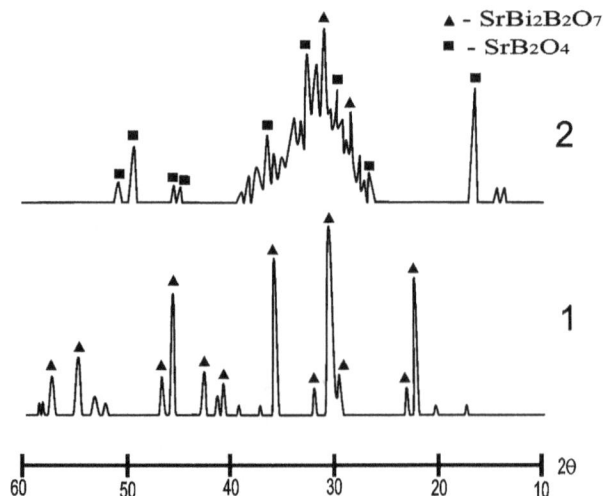

Figure 5. XRD patterns of the crystallised glasses with the composition of the ternary compound $SrBi_2B_2O_7$: 1 – 500°C 24 h, cooling in the furnace; 2 – 755°C 24 h, casting in cold water

melt leads to the appearance in the DTA curve of a second endothermic effect in the interval 795–850°C, with a minimum at 825°C (T_{m2}) (Figure 2, curve 1). Above 850°C we have glass forming $CaBi_2B_2O_7$ melt without the presence of any crystalline phase.

Two weak exothermic effects at 515 and 600°C, one strong exothermic effect of glass crystallisation (T_c) at 670°C and two endothermic effects of crystalline phases melting at 730 and 820°C are seen in the DTA curve of the $25CaO.25Bi_2O_3.50B_2O_3$ ($CaBi_2B_4O_{10}$) glass composition (Figure 2, curve 2). It is reported[37,38] that the presence of two exothermic effects in DTA curves of $Li_2O.4Ge_2O_3$ and $LaBSiO_5$ glass compositions is connected with the nucleation (weak exothermic effect) and nucleus growth at the temperatures of second exothermic effect. It is possible to assume that weak exothermic effects at 515 and 600°C observed in our DTA curve (Figure 2, curve2) are connected with precrystallisation fluctuations taking place in the glass matrix.[37] In our opinion precrystallisation fluctuation includes the nucleation and nucleus growth which is accompanied by a heat release (exothermic effect) and observed in the DTA curve. An amorphous matrix has been observed in the XRD patterns of $CaBi_2B_4O_{10}$ samples (Figure 4, curve 1) crystallised at the first weak exothermic effect (at 515°C for 18 h). Crystalline phase formation has been observed

in the XRD patterns of $CaBi_2B_4O_{10}$ samples (Figure 4, curve 2) that are thermally treated at the second weak exothermic effect at 600°C for 18 h. The weak reflections then become strong in the XRD patterns of $CaBi_2B_4O_{10}$ samples (Figure 4, curve 3) that are crystallised at the exothermic effect at 670°C. The x-ray diffraction patterns of crystallisation products of this glass (produced at 670°C for 24 h) show the formation of one $CaBi_2B_4O_{10}$ crystalline phase, fully corresponding to the data of Kargin *et al*, who first synthesised and described the compound $CaBi_2B_4O_{10}$.[39] The melting temperature (T_{m1}) of this phase is 730°C and is different from the data of Egorisheva *et al*[30] The presence of a second endothermic effect within the interval 765–845°C, with a minimum at 820°C (T_{m2}), is associated with incongruent melting of $CaBi_2B_4O_{10}$ (Figure 2, curve 2). The crystalline compound $CaBi_2B_4O_{10}$ melts incongruently at 770°C with the formation of melt and crystalline CaB_2O_4 (Figure 4, curve 4). This was clearly observed in the XRD patterns of products that have undergone thermal treatment at 820°C for 24 h, and then rapid quenching in cold water, with identification according to an x-ray database.[33] Dissolution of this CaB_2O_4 phase in a melt leads to the appearance in the DTA curve of the second endothermic effect in the interval 765–845°C, with a minimum at 820°C (T_{m2}) (Figure 2, curve 2). Above 850°C we have a glass forming $CaBi_2B_4O_{10}$ melt without the presence of any crystalline phase.

One exothermic effect of glass crystallisation (T_c) at 500°C and two endothermic effects of crystalline phases melting at 750 and 860°C are observed in the DTA curve of the $33.33SrO.33.33Bi_2O_3.33.33B_2O_3$ ($SrBi_2B_2O_7$) glass (Figure 2, curve 3). The x-ray diffraction patterns of the crystallisation products for this glass (at 500°C for 24 h) show the formation of one $SrBi_2B_2O_7$ crystalline phase (Figure 5, curve 1), cor-

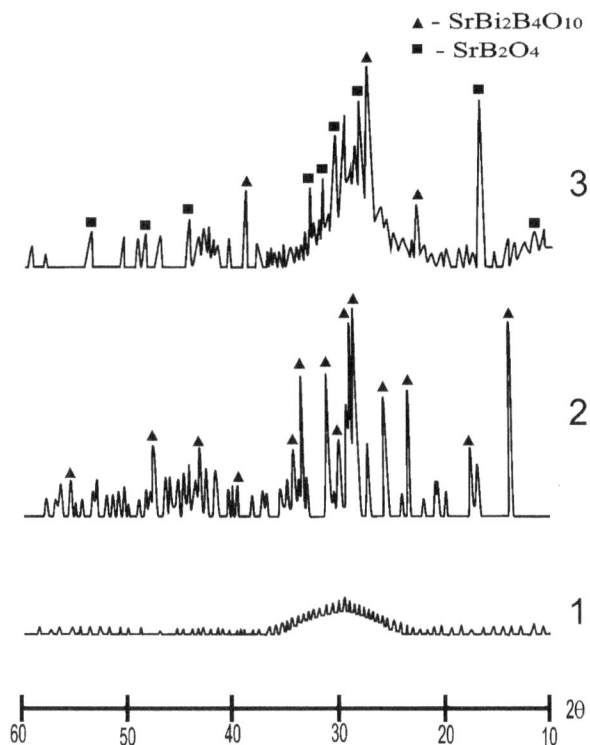

Figure 6. XRD patterns of the crystallised glasses with the composition of the ternary compound $SrBi_2B_4O_{10}$: 1 – 530°C 18 h, cooling in the furnace; 2 – 635°C 24 h, cooling in the furnace; 3 – 790°C 24 h, casting in cold water

responding fully to the data of Barbier & Cranswick, who first synthesised and described the compound $SrBi_2B_2O_7$.[29] The melting temperature (T_{m1}) of this phase is 750°C. The presence of a second endothermic effect within the interval 785–930°C, with a minimum at 860°C, is associated with incongruent melting of $SrBi_2B_2O_7$ (Figure 2, curve 3). The crystalline compound $SrBi_2B_2O_7$ melts incongruently at 750°C with the formation of melt and crystalline SrB_2O_4 (Figure 5, curve 2). This was clearly observed in the XRD patterns of products that have undergone thermal treatment at 750°C for 24 h, and then rapid quenching in cold water, with identification according to an x-ray database.[33] Dissolution of this SrB_2O_4 phase in a melt leads to the appearance in the DTA curve of the second endothermic effect in the interval 785–930°C, with a minimum at 860°C (T_{m2}) (Figure 2, curve 3). Above 950°C we have a glass forming $SrBi_2B_2O_7$ melt without the presence of any crystalline phase.

One weak exothermic effect at 530°C, one strong exothermal effect of glass crystallisation (T_c) at 635°C and two endothermic effects of crystalline phase melting at 790 and 865°C are seen in the DTA curve of the $25SrO.25Bi_2O_3.50B_2O_3$ ($SrBi_2B_4O_{10}$) glass (Figure 2, curve 4). The weak exothermic effect at 530°C observed in the DTA curve (Figure 2, curve 4) is apparently connected with precrystallisation

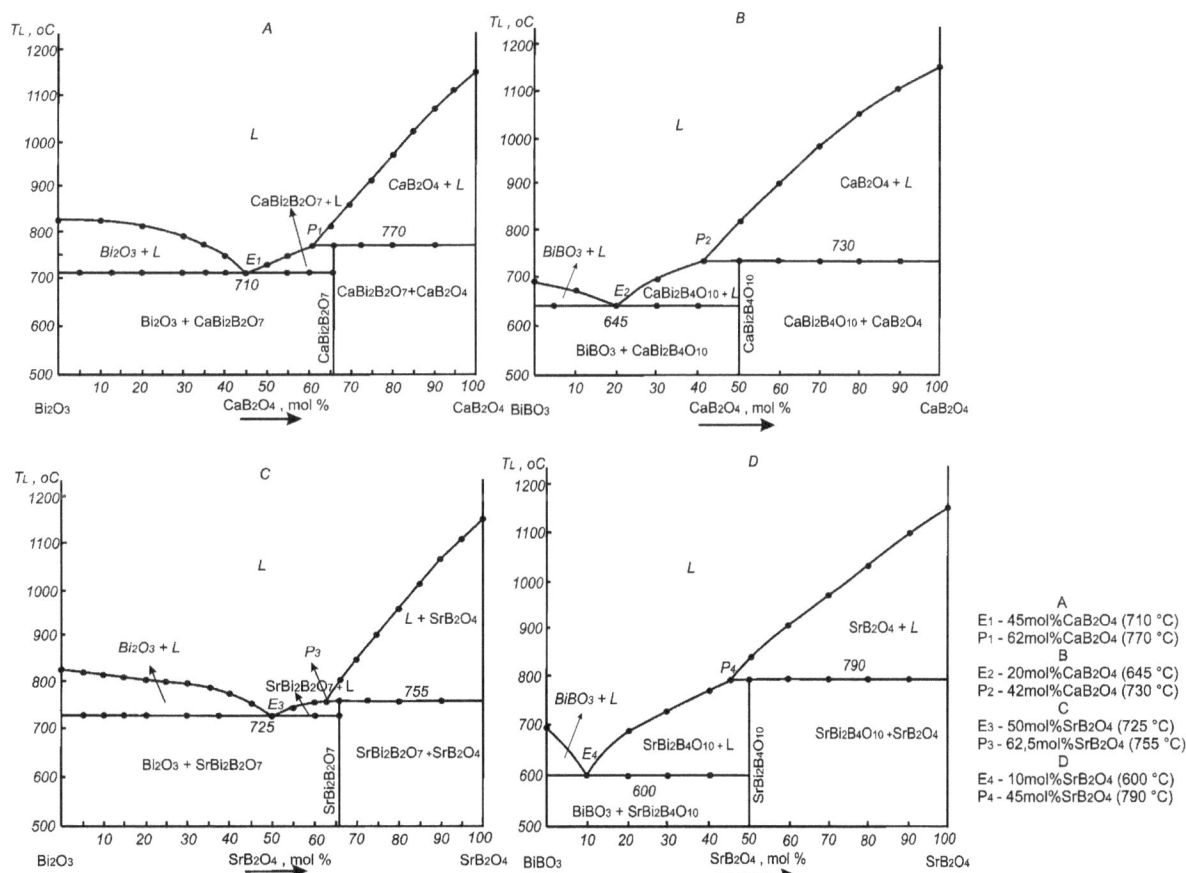

Figure 7. Phase diagrams of the pseudo-binary systems: $A – Bi_2O_3–CaB_2O_4$; $B – BiBO_3–CaB_2O_4$; $C – Bi_2O_3–SrB_2O_4$; $D – BiBO_3–SrB_2O_4$

fluctuations taking place in the glass matrix.[37] The amorphous matrix has been observed in the XRD patterns (Figure 6, curve 1) of $SrBi_2B_4O_{10}$ samples crystallised at the weak exothermal effect at 530°C for 18 h and nearby to this temperature. At this temperature the nucleation process started and the crystalline phase is observed in the x-ray diffraction patterns of the crystallisation (at 635 °C for 24 h) products of this glass (Figure 6, curve 2). The x-ray diffraction patterns of these glass crystallisation (at 635°C for 24 h) products show the formation of one $SrBi_2B_4O_{10}$ crystalline phase, corresponding fully to the data of Barbier & Cranswick, who first synthesised and described the compound $SrBi_2B_4O_{10}$.[32] The melting temperature (T_{m1}) of this phase is 790°C. The presence of a second endothermic effect within the interval 830–895°C, with a minimum at 865°C (T_{m2}), is associated with incongruent melting of $SrBi_2B_4O_{10}$ (Figure 2, curve 4). The crystalline compound $SrBi_2B_4O_{10}$ melts incongruently at 790°C with the formation of melt and crystalline SrB_2O_4 (Figure 6, curve 3). This was clearly observed in the XRD patterns of products that have undergone thermal treatment at 865°C 24 h, and then rapid quenching in cold water, with identification according to an x-ray database.[33] Dissolution of this SrB_2O_4 phase in a melt leads to the appearance in the DTA curve of the second endothermic effect in the interval 830–895°C, with a minimum at 865°C (T_{m2}) (Figure 2, curve 4). Above 900°C we have a glass forming $SrBi_2B_4O_{10}$ melt without the presence of any crystalline phase.

Phase diagrams of pseudo-binary systems Bi_2O_3–CaB_2O_4, $BiBO_3$–CaB_2O_4, Bi_2O_3–SrB_2O_4 and $BiBO_3$–SrB_2O_4

The process of phase diagram construction is time consuming. The traditional method based on the investigation of solid phase sintered samples takes a long time and is not effective. The glass samples investigation technique is progressive, because the DTA curves of glass samples register all processes, including the processes of crystallisation and melting of formed crystalline phases. The large glass forming areas which we have determined in the systems CaO–Bi_2O_3–B_2O_3 and SrO–Bi_2O_3–B_2O_3 made it possible to have enough glass samples for DTA investigations and phase diagram construction. Four pseudo-binary systems, Bi_2O_3–CaB_2O_4, $BiBO_3$–CaB_2O_4, Bi_2O_3–SrB_2O_4 and $BiBO_3$–SrB_2O_4, were studied and their phase diagrams were constructed (Figure 7).

Bi_2O_3–CaB_2O_4 pseudo-binary system

The system between the congruently melting compounds Bi_2O_3 and CaB_2O_4 is very important. The introduction of 45 mol% CaB_2O_4 reduced the melting point of Bi_2O_3 from 825°C and resulted in

the formation of eutectic E_1, with m.p. 710°C (Figure 7(A)). Further increase of CaB_2O_4 (62 mol%) leads to the formation of peritectic P_1. The presence of the peritectic in the liquidus curve is connected with the formation of the incongruently melting ternary compound $CaBiB_2O_7$ with m.p. 770°C (Figure 7(A)).

$BiBO_3$–CaB_2O_4 pseudo-binary system

Congruently melting $BiBO_3$ and CaB_2O_4 have m.p. 690°C and 1150°C, respectively. The introduction of 20 mol% CaB_2O_4 reduced the melting point of $BiBO_3$ from 690°C and resulted in the formation of eutectic E_2, with m.p. 645°C (Figure 7(B)). Further increase of CaB_2O_4 (42 mol%) leads to the formation of peritectic P_2. The constructed phase diagram confirms the presence of the stoichiometric ternary compound $CaBi_2B_4O_{10}$ that melts incongruently at 730°C (Figure 7(B)).

Bi_2O_3–SrB_2O_4 pseudo-binary system

The system between the congruently melting compounds Bi_2O_3 and SrB_2O_4 is very important. The introduction of 50 mol% SrB_2O_4 reduced the melting point of Bi_2O_3 from 825°C and resulted in the formation of eutectic E_3, with m.p. 725°C (Figure 7(C)). Further increase of SrB_2O_4 (62·5 mol%) leads to the formation of peritectic P_3. The presence of the peritectic in the liquidus curve is connected with the formation of the incongruently melting ternary compound $SrBiB_2O_7$ with m.p. 755°C (Figure 7(C)).

$BiBO_3$–SrB_2O_4 pseudo-binary system

The introduction of 10 mol% SrB_2O_4 reduced the melting point of $BiBO_3$ from 690°C and resulted in the formation of eutectic E_4, with m.p. 600°C (Figure 7(D)). Further increase of SrB_2O_4 (45 mol%) leads to the formation of peritectic P_4. The constructed phase diagram confirms the presence of the stoichiometric ternary compound $SrBi_2B_4O_{10}$ that melts incongruently at 790°C (Figure 7(D)).

Conclusions

Large glass forming areas have been determined in the systems CaO–Bi_2O_3–B_2O_3 and SrO–Bi_2O_3–B_2O_3, which include all the eutectic areas in the binary system Bi_2O_3–B_2O_3, the low temperature eutectic areas in the binary systems CaO–B_2O_3 and SrO–B_2O_3, and fields with low melting points in the ternary systems spreading up to the binary CaO–Bi_2O_3 and SrO–Bi_2O_3 systems. Transparent glassy ribbons in the narrow bismuth-rich area of composition in the SrO–Bi_2O_3 system have been obtained for the first time. The ternary compounds $CaBi_2B_2O_7$, $CaBi_2B_4O_{10}$, $SrBi_2B_2O_7$ and $SrBi_2B_4O_{10}$ have good glass forming ability and are in the area of stable glasses. The melting character of the ternary compounds $CaBi_2B_2O_7$, $CaBi_2B_4O_{10}$, $SrBi_2B_2O_7$ and $SrBi_2B_4O_{10}$ have been studied. All compounds melt incongruently, at 770, 730, 755 and 790°C, respectively. They formed melts and CaB_2O_4, SrB_2O_4 crystalline phases, respectively, as a result of incongruent melting. The phase diagrams were constructed for four pseudo-binary systems, Bi_2O_3–CaB_2O_4, $BiBO_3$–CaB_2O_4, Bi_2O_3–SrB_2O_4, and $BiBO_3$–SrB_2O_4. Each of these systems has one eutectic (E) and one peritectic (P) composition, with the following melting points: Bi_2O_3–CaB_2O_4, E_1=710°C and P_1=770°C; $BiBO_3$–CaB_2O_4, E_2=645°C and P_2=730°C; Bi_2O_3–SrB_2O_4, E_3=725°C and P_3=755°C; $BiBO_3$–SrB_2O_4, E_4=600°C and P_4=790°C.

Acknowledgments

This work was supported by the International Science and Technology Center (ISTC), Project #A-1591.

References

1. Levin, E. M. & McDaniel, C. L. *J. Am. Ceram. Soc.*, 1962, **45**, 355.
2. Mazurin, O. V., Streltsina, M. V & Shvaiko-Shvaikovskaya, T. P. *The properties of glasses and glass-forming melts*. Vol.2, Nauka, Leningrad,1975.
3. Becker, P. & Bohaty, L. *Phys. Chem. Glasses*, 2003, **44**, 91.
4. Muehlberg, M, Burianek, M., Edongue, H. & Poetsch, C. *J. Cryst. Growth*, 2002, **237**, 740.
5. Becker, P. *Cryst. Res. Technol.*, 2004, **38**, 74.
6. Oprea, I.-I. *Optical Properties of Borate Glass-Ceramic*. Dissertation zur Erlangung des Grades Doktor der Naturwissenschaften,Osnabruck University, Germany, 2005.
7. Hwang, C. & Fujino, S. *XX International Congress on Glass Proceedings*, Sept 27–Oct 1, 2004, Kyoto. PDF file No O-07-52.
8. Honma, T., Benino, Y., Fujiwara, T., Komatsu, T. & Sato, R. *Appl. Phys. Lett.*, 2003, **82**, 892.
9. Elwell, D., Morris, A. W. & Neate, B. W. *J. Cryst. Growth*, 1972, **16**, 67.
10. Zargarova, M. I. & Kasumova, M. F. *Zh. Neorg. Mater.*, 1990, **26**, 1678.
11. Egorisheva, A., Skorikov, V., Volodin, V., Myslitskii, O & Kargin, Yu. *Russ. J. Inorg. Chem.*, 2006, **51**, 1956. (In Russian.)
12. Barbier, J., Penin, N., Denoyer, A. & Cranswick, L. M. D. *Solid State Sci.*, 2005, **7**, 1055.
13. Kim, J.-M. & Jung, B.-H. *Mater. Sci. Forum*, 2003, **439**, 18.
14. Kim, J.-M. & Kim, H.-S. *Mater. Sci. Forum*, 2006, **510–511**, 574.
15. Kim, Y.-J., Hwang, S.-J. & Kim, H.-S. *Mater. Sci. Forum*, 2006, **510–511**, 578.
16. Pottier M. J. *Bull. Soc. Chim. Belg.*, 1974, **83**, 235.
17. Becker, P. & Froehlich, R. *Z. Naturforsch. B*, 2004, **59**, 256.
18. Hovhannisyan, M. R., Hovhannisyan, R. M., Grigoryan, B. V., Alexanyan, H. A. & Knyazyan, N. B. *Glass Technol.: Eur. J. Glass Sci. Technol. A*, 2009, **50** (6), 323.
19. Phase EquilibriaDiagrams, CD-ROM Database, Version 3.0, ACerS-NIST, 2004.
20. Vstavskaya, E. Yu., Zuev, A. Yu. & Cherepova, V. A. *Mater. Res. Bull.*, 1994, **29**, 12, 1233
21. Conflant, P., Boivin, J.-C. & Thomas, D., *J. Solid State Chem.*,1976, **18** (2), 133.
22. Roth, R. S., Rawn, C. J., Burton, B. P. & Beech, F. *J. Res. Natl. Inst. Stand. Technol.*, 1990, **95**, 291.
23. Hovhannisyan, R. M. *Phys. Chem. Glasses: Eur. J. Glass Sci. Technol. B*, 2006, **47**, 460.
24. Witzmann, H. & Beulich, W. *Z. Phys. Chem.*, 1964, **225**, (5-6), 336.
25. Witzmann, H. & Beulich, W. *Z. Phys. Chem.*, 1964, **225**, (3-4), 197.
26. Chenot, C., F. *J. Am. Ceram. Soc.*, 1967, **50**, (2), 117.
27. Kudrjavtcev, D. P., Oseledchik, Yu. S., Prosvirin, A. L. & Svitanko, N. V. *J. Cryst. Growth*, 2003, **254**, 456.
28. Tang, Z-H., Chen, X., & Li, M. *Solid State Sci.* 2008, **10**, 894
29. Barbier, J. & Cranswick, L. M. D. *J. Solid State Chem.*, 2006, **179**, 3958.
30. Egorisheva, A., Volodin, V. & Skorikov, V. *Inorg. Mater.*, 2008, **44**, 70.
31. Kargin, Y. F., Ivicheva, S. N., Komova, M. G. & Krut'ko, V. A. *Russ. J. Inorg. Chem.*, 2008, **53**, 425.
32. Barbier, J. & Cranswick, L. M. D. *Powder Diffraction*, 2009, **24** (1), 35.
33. Powder Diffraction File. PDF2 Release 2008. International Center forDiffraction Data, 2008, USA.
34. Janakirama-Rao, B. V. *Compt. Rend. VII Congr. Int. Du Verre, Bruxelles*, 1965, **1**, (104-104.6)
35. Imaoka, M. & Yamazaki, T. *Rep. Inst. Ind. Sci. Univ. Tokyo*, 1972, **22**, (3), 173.
36. Milyukov, E. M., Lunkin, S. P. & Maltseva, Z. S. *Fiz. I Khim. St.*, 1979, **5**, (5), 612.
37. Sarkisov, P. D. *The Directed Glass Crystallization - Basis for Development of Multipurpose Glass Ceramics*, Mendeleev Russian Chemical Technological University Press, Moscow, 1997. (In Russian.)
38. Pernice P., Aronne A. & Marotta A. *Mater. Chem. Phys.*, 1992, **30**, 195.
39. Kargin, Y., Ivicheva, S., Shvornieva. L., Komova, M. & Krutko, V., *Zh. Neorg. Mater*, 2008, **53**, 1614.

Phys. Chem. Glasses: Eur. J. Glass Sci. Technol. B, April 2013, **54** (2), 65–75

Zinc and manganese borate glasses – phase separation, crystallisation, photoluminescence and structure

Doris Ehrt

Otto-Schott-Institut, Friedrich-Schiller-Universität Jena, Fraunhoferstr.6 D-07743 Jena, Germany

Manuscript received 24 November 2011
Revised version received 3 April 2012
Accepted 10 September 2012

Zinc and manganese borates and borosilicates were prepared and investigated. Binary $ZnO–B_2O_3$ and $MnO–B_2O_3$ melts show similar stable phase separation below 50 mol% ZnO or MnO. Two layers were separated in the melts. The upper layer was nearly pure B_2O_3 and contained crystals. The lower layer was zinc or manganese borate with around 50 mol% ZnO or MnO. Melts of ternary systems with SiO_2 were separated into SiO_2 rich and zinc or manganese rich phases. The addition of Na_2O decreases phase separation. Compositions with more than 50 mol% ZnO provided clear homogeneous glass samples. Mn^{2+} was substituted for Zn^{2+} in glasses and crystals, with a coordination number from 6 to 4, and a variation of Mn^{2+} photoluminescence from orange-red to yellow-green depending on local structure. A drastic change of luminescence was detected when the glass was transformed to a glass ceramic by thermal treatment. Adding Eu_2O_3 or Tb_2O_3 to zinc borates provides strong orange-red (Eu^{3+}) or green (Tb^{3+}) luminescence. In phase separated samples, Mn^{2+}, Eu^{3+} and Tb^{3+} were only accumulated in the zinc borate and not in the B_2O_3 phase. The large Eu^{3+} and Tb^{3+} ions could not substitute for Zn^{2+} in the crystal phases.

Introduction

Zn^{2+} and Mn^{2+} are very interesting for glass structures and for many applications. They are so-called intermediates and can act as network modifiers in six-fold coordination or as network formers in four-fold coordination with oxygen, depending on the other components. However, ZnO and MnO cannot form a glass *per se*, unlike SiO_2 or B_2O_3. The binary systems $ZnO–B_2O_3$ and $MnO–B_2O_3$ are also very interesting for glass forming because B_2O_3 is able to build different species, mainly $BØ_3$, $BØ_4^-$, and $BØ_2O^-$ groups, with various connections. Krogh-Moe[1] found by infrared spectroscopy that (in contrast to a number of other borates) some vitreous and crystalline anhydrous zinc borates (such as $4ZnO.3B_2O_3$ and $ZnO.B_2O_3$) have structures which are not similar. He found that all boron atoms are four-fold coordinated by oxygen in the cubic crystalline zinc borate modification and that a coordination change occurs during the melting process to three-fold coordination. He concluded from his experimental results that the coordination number of boron 'depends on a precarious balance' of the B–Ô–B angle. The phase diagram for the system $ZnO–B_2O_3$[2] indicates also an extensive structural rearrangement by fusion with a very flat field of crystallisation up to 50 mol% ZnO and a very broad liquid–liquid phase separation for less than 50 mol% ZnO. There appear to be no reports of the phase diagram for the system $MnO–B_2O_3$, and

hence the existence of $MnO.2B_2O_3$ and $MnO.3B_2O_3$ was assumed.[1,2] The investigation of the $MnO–B_2O_3$ system is experimentally difficult, because Mn^{2+} ions in melts are normally easily partially oxidized to Mn^{3+}, which leads to a more or less pink to brown colouring of the glass[3]. Zn^{2+} and Mn^{2+} ions have similar ionic radii, approximately 0·6 Å, and often show similar behaviour in the structure, e.g. green luminescent willemite mineral due to four-fold coordinated Mn^{2+} ions on four-fold coordinated Zn^{2+} sites. It was shown in our previous papers[4–7] that the visible Mn^{2+} and Zn^{2+} photoluminescence is very sensitive to their local structure and a drastic change of luminescence could be detected when the glass was transformed to a glass ceramic by thermal treatment. Photoluminescence of solid state materials is an important optical property for many classical and new applications and an increasingly large numbers of papers have been published in recent years. This paper reports the detection of similar stable phase separation of zinc and manganese borate melts, the use of Mn^{2+} photoluminescence as a sensor for the local structure of Mn^{2+} and Zn^{2+} in glasses and glass ceramics, and the effect of adding Tb_2O_3 and Eu_2O_3 on the properties of homogeneous and phase separated zinc borate glasses.

Experimental

The batch compositions and measured characteristic properties of the zinc and manganese borate glasses prepared here are given in Tables 1 and 2, and Figures

Email doris.ehrt@uni-jena.de
Original version presented at VII Int. Conf. on Borate Glasses, Crystals and Melts, Dalhousie University, Halifax, Nova Scotia, Canada, 21–25 August 2011

Table 1. Composition and measured properties of the undoped zinc borate glasses

	~4ZnO.3B₂O₃	ZnO.B₂O₃	~3ZnO.4B₂O₃ (ZnB2)		ZnO.B₂O₃.3SiO₂
Composition (mol%/wt%)	56/59·8 ZnO	50/53·9 ZnO	44/47·9 ZnO		20/24·6 ZnO
	44/40·2 B₂O₃	50/46·1 B₂O₃	56/52·1 B₂O₃		20/21·0 B₂O₃
					60/54·4 SiO₂
Melting temperature (°C)	1100	1100	1100		1650
Comments	homogeneous	homogeneous	phase separation, 2 layers:		phase separation, opaque
			B₂O₃	ZnO–B₂O₃	
Density (g/cm³) ±0·01	3·55	3·37	1·99	3·38	
T_g (°C) (5 K/min) ±3°C	560	570	230	570	560
Thermal expansion coefficient (ppm/K) ±0·1 (100–300°C)	4·4	4·4	n.m.	4·5	3·7
Refractive index n_e (546 nm) ±0·0001	1·6594	1·6460	n.m.	1·6430	n.m.
Abbe number v_e ±1	51	52	n.m.	53	n.m.

(n.m.=not measured)

1–17. Raw materials were high purity (Fe<1 ppm, other metals <0·5 ppm) ZnO, H_3BO_3, $MnCO_3$, SiO_2, Tb_3O_5, and Eu_2O_3. Doping concentrations of zinc borate glasses were 1×10^{20} and 2×10^{20} Mn^{2+} per cm^3, and 1×10^{19}, 1×10^{20} and 1×10^{21} Eu^{3+} or Tb^{3+} per cm^3 glass. 50–200 g batches were melted in covered Pt crucibles. Manganese containing samples were prepared under reducing conditions by adding of 1% sugar as a reducing agent in silica glass crucibles. Glasses were obtained by pouring the melts into a mould. After annealing from 50 K above T_g to room temperature at 3–5 K/min, the glasses were cut, ground and polished to produce samples for different measurements. Optical absorption spectra in the range 190–3200 nm were obtained with an error <1% using a commercial double beam spectrometer (UV-3102PC, Shimadzu). Photoluminescence excitation and emission spectra were recorded in the range 200–900 nm in reflection mode using a commercial spectrometer (RF-5301PC, Shimadzu). The error of the measured intensity in arbitrary units under the same conditions was <5%. The excitation source was a 150 W Xe lamp. Polished plates (~20×20 mm², thickness 2–10 mm) were used. The experimental setup for the decay time measurements was published in our previous papers.[8,9] Excitation sources were a N_2 laser (337 nm, <500 ps pulse) for Mn^{2+} and Tb^{3+}, and a 395 nm LED for Eu^{3+}.

The refractive indices in the visible range were measured with a Pulfrich refractometer (Carl Zeiss Jena) with an error Δn of ±2×10⁻⁵. The density of the glass samples was determined using Archimedes'

principle with an error ±0·001 $g\,cm^{-3}$. Differential thermal analysis (DSC and DTA; powdered sample heated at 10 $K\,min^{-1}$) and dilatometer (heated at 5 $K\,min^{-1}$) measurements were carried out to obtain values for thermal properties. X-ray diffraction (XRD), electron microscopy (SEM with EDX and WDX), and optical microscopy measurements (commercial equipment) were used to characterise the phase separation and crystallisation behaviour.

Results

The compositions selected are given in Tables 1 and 2. The homogeneous undoped binary zinc borate glass with high ZnO content ($56ZnO.44B_2O_3$) was already prepared and investigated in our previous paper.[10] It has interesting physical and chemical properties and the glass composition is near to the crystalline compound $4ZnO.3B_2O_3$ described by Krogh-Moe[1]. Previously a ratio ZnO to B_2O_3 of 1:1 was also assumed for this compound.[2] This is why we melted these two compositions which gave stable homogenous optical glasses with only small difference in properties (Table 1). The phase diagram[2] also shows a broad immiscibility gap in the melt (T>982°C) at ZnO<50 mol%, splitting into nearly pure B_2O_3 and a zinc borate phase with composition near 1:1. The composition $44ZnO.56B_2O_3$ was melted to study phase separation. Surprisingly, the melt was separated into two layers, an upper opalescent B_2O_3 and a lower clear ZnO–B_2O_3 phase.

Table 2. Composition and measured properties of manganese borate glasses

	MnO.5B₂O₃		MnO.4B₂O₃		MnO.B₂O₃	MnO–Na₂O–B₂O₃	MnO–Na₂O–B₂O₃–SiO₂
Composition (mol% / wt%)	16·7/17·0 MnO		20/20·3 MnO		50/50·5 MnO	4·0/4·13 MnO	4/4·6 MnO
	83·3/83·0 B₂O₃		80/79·7 B₂O₃		50/49·5 B₂O₃	12·7/11·5 Na₂O	5/5·0 Na₂O
						83·3/84·4 B₂O₃	15/16·8 B₂O₃
							76/73·6 SiO₂
Melting temperature (°C)	1300		1300		1100	1130	1600
Comments	phase separation, 2 layers 2 layers		phase separation, black glass		homogeneous, red-brown glass	homogeneous, light brown glass	phase separation,
	B₂O₃	MnO–B₂O₃	B₂O₃	MnO–B₂O₃			
Density (g/cm³) ±0·001	1·821	2·974	1·853	2·973	3·337	2·190	2·292
T_g (°C) (5 K/min) ±3°C	290	588	292	580	545	455	520+660
Thermal expansion coefficient (ppm/K) ±0·1 (100–300°C)	n.m.	6·2			7·6	8·8	13·9+3·7
Refractive index n_e (546 nm) ±0·0001	n.m.	1·6540					
Abbe number v_e ±1	n.m.	55					

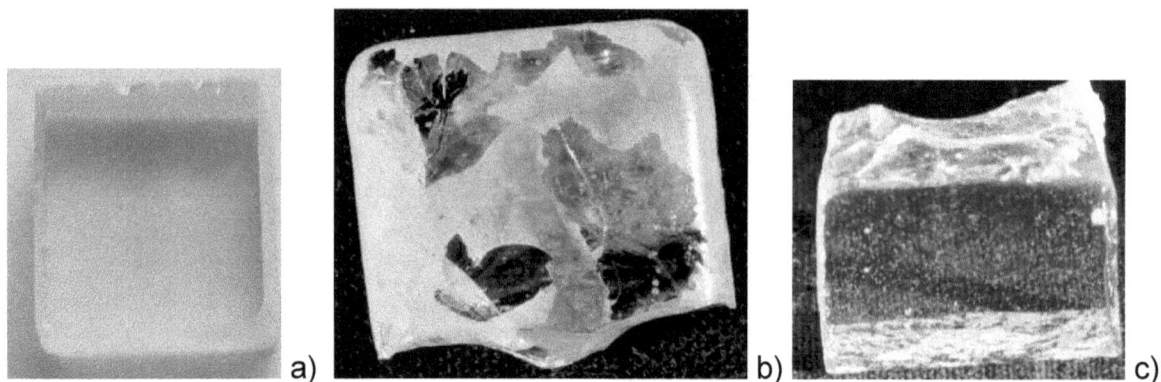

Figure 1. Photographs of different colourless ZnO–B$_2$O$_3$ glass samples: (a) polished plate (20×20×10 mm^3) of homogeneous 56ZnO.44B$_2$O$_3$ glass sample; (b) cast glass (50×20×15 mm^3) of phase separated 44ZnO.56B$_2$O$_3$ sample with an upper opalescent B$_2$O$_3$-rich phase containing crystals and a lower clear ZnO–B$_2$O$_3$ phase; and (c) polished plate (thickness 10 mm) of the clear ZnO–B$_2$O$_3$ phase of the phase separated 44ZnO.56B$_2$O$_3$ glass sample [Colour available online]

The phases could be separated very easily (Figure 1(c)) and their properties were determined (Table 1). Typical glass samples are shown in Figures 1(a)–(c). The composition 50ZnO.50B$_2$O$_3$ yielded a similar

Figure 2. (a) Absorption and (b) photoluminescence excitation (dashed lines) and emission (solid lines) spectra of the clear ZnO–B$_2$O$_3$ phases from the phase separated 44ZnO.56B$_2$O$_3$ glasses doped with 1×10^{20} and 2×10^{20} Mn^{2+} per cm^3 (sample thickness=10 mm)

homogeneous glass to the composition 56ZnO.44B$_2$O$_3$ (Figure 1(a)). Absorption and photoluminescence spectra of the clear ZnO–B$_2$O$_3$ phases of the separated 44ZnO.56B$_2$O$_3$ glasses doped with 1×10^{20} and 2×10^{20} Mn^{2+} per cm^3 are shown in Figures 2(a) and (b). Mn^{2+} ions were accumulated in the lower ZnO–B$_2$O$_3$ phase, and the upper B$_2$O$_3$ phase has protected Mn^{2+} from oxidation to Mn^{3+} by atmospheric oxygen. The glass samples were colourless and only Mn^{2+} was detected. The intensities of the absorption and excitation bands at 411 nm correlate with the Mn^{2+} concentration. The spectra are typical for Mn^{2+} mainly in six-fold coordination with strong yellow-orange emission (Figure 2(b)). Lifetimes τ_e~7 ms at ~600 nm were measured without concentration quenching of the photoluminescence. The upper B$_2$O$_3$ layer was opalescent due to the presence of crystallites detected by XRD measurements (Figure 7(c)). The B$_2$O$_3$ layer was not mechanically stable. Cracks were formed (Figure 1(b)) and it was not possible to obtain bulk samples for optical and dilatometer measurements. However, DSC, XRD and density measurements were made with powder samples from both separated layers.

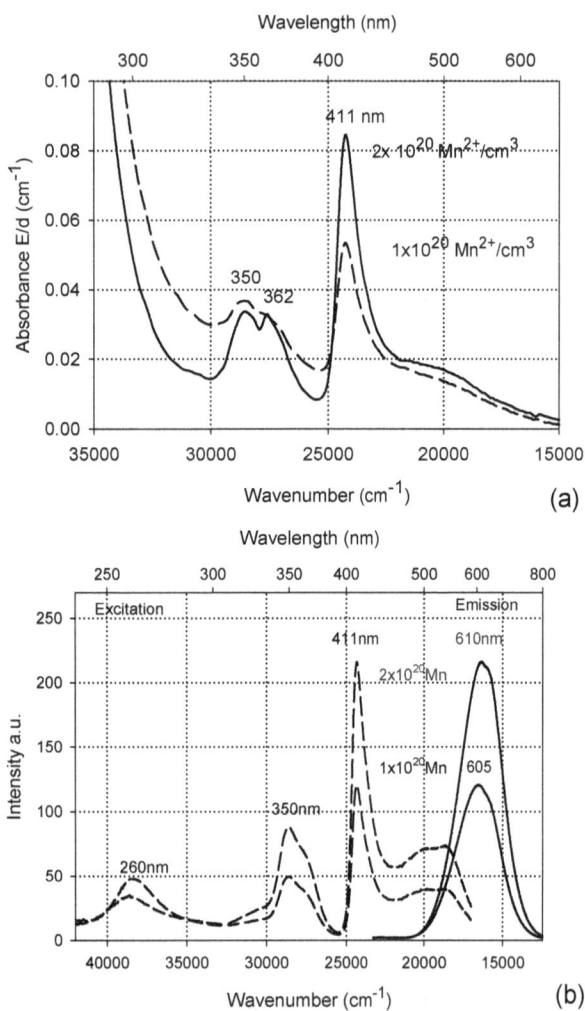

A ternary zinc borosilicate glass, 20ZnO.20B$_2$O$_3$. 60SiO$_2$, in the immiscibility gap of the phase diagram,[11] was melted in a Pt crucible at 1650°C. The melt was phase separated into a ZnO–SiO$_2$ phase and a SiO$_2$(B$_2$O$_3$) phase. The viscosity of the melt was very high, so that casting of the melt was not possible. The melt was cooled to room temperature in the Pt crucible. The melt was phase separated without any additional heat treatment. Opaque samples with very high mechanical stability were drilled out for measurements; T_g was 560°C and the thermal expansion coefficient (TEC) was 3·7 ppm/K (Table 1).

Table 2 shows the composition and properties of the manganese borate glass samples. Similar to zinc borates, the melts of compositions MnO.5B$_2$O$_3$ and MnO.4B$_2$O$_3$ separated into two layers (Figures 3(a) and (b)), an upper opalescent and colourless B$_2$O$_3$ phase with small H$_3$BO$_3$ crystals, and a lower

Figure 3. Photographs of the separated phases from MnO.4B_2O_3 glass (MnO.5B_2O_3 glass shows the same behaviour): (a) upper opalescent colourless B_2O_3-rich phase with crystals; and (b) lower clear yellow-orange MnO–B_2O_3 phase [Colour available online]

orange coloured, clear MnO–B_2O_3 phase, which was very stable, and plane parallel polished plates, with thickness ~6 mm, could be prepared for absorption, photoluminescence, refractive index and dilatometer measurements (Figures 4(a) and (b)). The MnO–B_2O_3

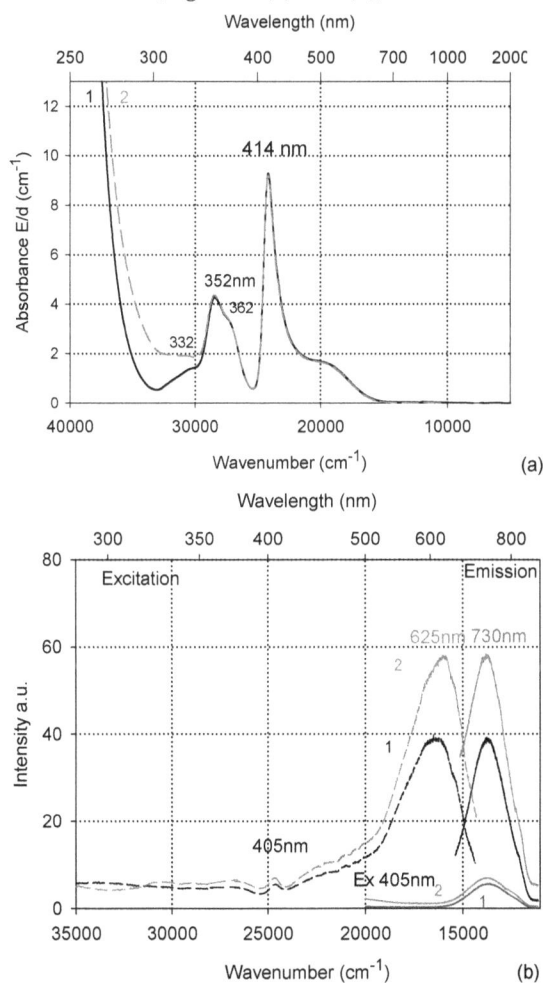

Figure 4. (a) Absorption and (b) photoluminescence excitation (dashed lines) and emission (solid lines) spectra of the clear yellow MnO–B_2O_3 phases from the MnO.5B_2O_3 (1) and MnO.4B_2O_3 (2) glasses [Colour available online]

phases of both compositions have identical Mn^{2+} absorbance, Figure 4(a), with maxima at 352 and 414 nm, and similar deep red luminescence emission with a maximum at 730 nm and with two very short decay times, τ_{e1}~10 μs and τ_{e2}~100 μs. Mn^{3+} could not be detected. The excitation spectra (Figure 4(b)) are very different from the absorption spectra (Figure 4(a)) or from the spectra from glasses with lower Mn^{2+} content (Figure 2). The excitation maxima are shifted to much longer wavelength, ~625 nm. The excitation band at 405 nm was much lower and provided much lower emission intensity at 730 nm (Figure 4(b)). Assuming a phase composition of 50MnO.50B_2O_3, similar to the zinc borate system, then the Mn^{2+} concentration would be very high, ~10^{22} Mn^{2+} per cm^3. Unfortunately it was not possible to prepare a glass with 50MnO.50B_2O_3 under comparable reducing conditions containing only Mn^{2+}. The glass obtained was homogenous but with a nearly black colour due to Mn^{3+} and Mn^{2+}. The absorption spectrum (not shown) of a thin glass sample (~250 μm) had two very broad bands with maxima at 300 and 450 nm due to direct charge transfer transitions of electrons between Mn^{2+} and Mn^{3+} ions. It was not possible to measure Mn^{2+} luminescence, because the strong absorption of Mn^{3+} prevents the excitation of Mn^{2+}. The density and TEC values were higher than these of the separated MnO–B_2O_3 phases (Table 2).

Substitution of MnO by Na_2O in binary borate and ternary borosilicate glasses decreases the immiscibility. Homogeneous glass samples, light brown coloured by only small amounts of Mn^{3+}, were obtained with the composition 4MnO.12·7Na_2O.83·3B_2O_3 with 0·8×10^{21} Mn^{2+} per cm^3. A typical Mn^{2+} photoluminescence spectrum with red emission, maximum at 650 nm, is shown in Figure 5(a). Multi-exponential decay behaviour with lifetime around 5 ms was measured. Much shorter lifetime, <0·3 ms, and photoluminescence spectra (Figure 5(b)) more similar to the separated MnO–B_2O_3 phase with assumed Mn^{2+}

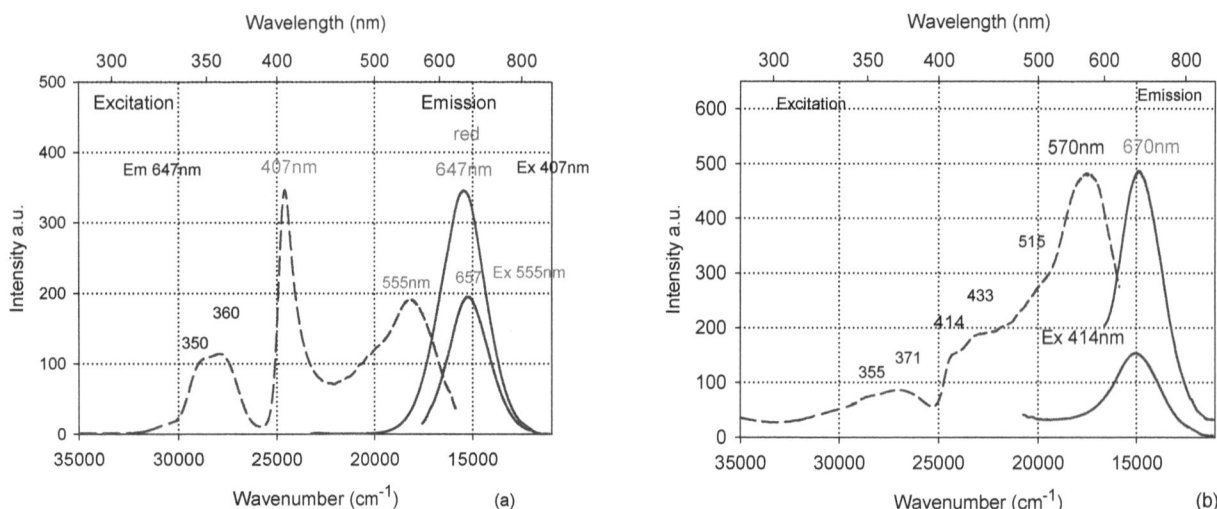

Figure 5. Mn^{2+} photoluminescence excitation (dashed lines) and emission (solid lines) spectra of (a) a clear homogeneous glass sample (in mol%) $12.7Na_2O.4MnO.83.3B_2O_3$; (b) a phase separated borosilicate glass sample $5Na_2O.4MnO.15B_2O_3.76SiO_2$ [Colour available online]

~10^{22} per cm^3 (Figure 4(b)), were measured for the borosilicate glass, $4MnO.5Na_2O.15B_2O_3.76SiO_2$, with a calculated Mn^{2+} concentration, 0.9×10^{21} per cm^3, which is nearly equal to the Mn^{2+} content in the homogeneous $4MnO.12.7Na_2O.83.3B_2O_3$ glass sample. However, the borosilicate glass sample was opalescent due to sub-liquidus phase separation, as shown in the SEM image (Figure 6(a)). A dark SiO_2 rich matrix and small light Na_2O–MnO–B_2O_3 rich droplets can be recognised. The accumulation of the Mn^{2+} ions in the sodium borate droplets, and not in the SiO_2 rich matrix, leads to much higher Mn^{2+} concentration in the droplets and the shift of the photoluminescence excitation and emission maxima to longer wavelengths (Figure 5(b)). Two different values for T_g and for the TEC (Table 2) were found with dilatometer measurements for $4MnO.5Na_2O.15B_2O_3.76SiO_2$ glass samples due to the phase separation.

The SEM images of the opaque $20ZnO.20B_2O_3.60SiO_2$ glass samples (Figures 6(b) and (c)) with liquid–liquid phase separation show a light ZnO–SiO_2 rich matrix (higher electron density) and large dark $SiO_2(B_2O_3?)$[11] rich droplets with secondary ZnO–SiO_2 rich droplets inside and vice versa. Usually the SiO_2–B_2O_3 system also separates into two phases.

The crystallisation behaviour of different undoped and doped zinc borate glasses was studied. The glass samples were transformed to glass ceramics by thermal annealing at 750°C and 850°C for 2 h. Figures 7(a)–(c) show typical XRD patterns of the measured samples. Very interesting effects were found in the crystallised zinc borate glass samples doped with manganese, which is demonstrated in Figures 8–11. Different crystalline zinc borate phases with manganese are formed. They were recognised by various Mn^{2+} photoluminescence which could be detected by UV illumination in Figures 8–10, and with the change of the photoluminescence excitation and emission spectra during the crystallisation process (Figure 11(a)–(c)).

Photoluminescence and properties of zinc borate glasses and glass ceramics doped with Tb^{3+} and Eu^{3+} in a concentration range 1×10^{19}, 1×10^{20}, and 1×10^{21} ions per cm^3, which is equivalent with approximately 0.1, 1.0 and 10 wt% RE_2O_3, were investigated (Figures 12–17). Strong green emission was found with Tb^{3+} (Figure 12(b)), and strong red emission with Eu^{3+} (Figure 15(b)) under UV illumination. Doped

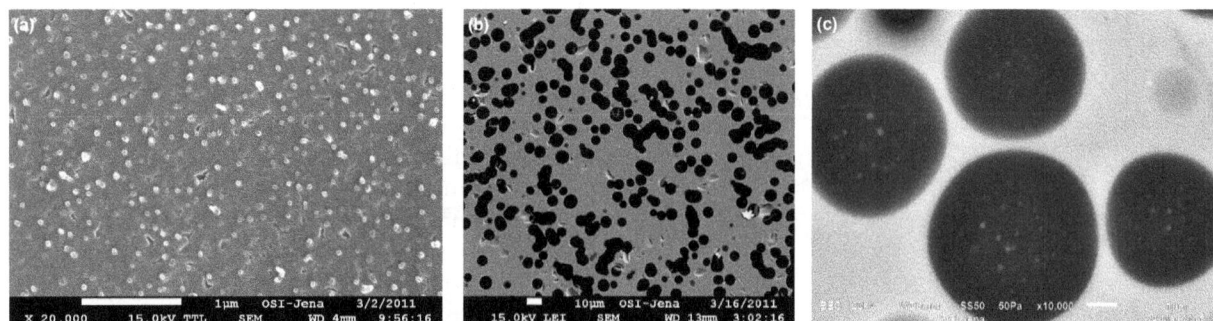

Figure 6. SEM images of opaque glass samples with phase separation: (a) $5Na_2O.4MnO.15B_2O_3.76SiO_2$ with SiO_2 rich matrix (dark) and Na_2O–MnO–B_2O_3 rich droplets (light); (b) and (c) $20ZnO.20B_2O_3.60SiO_2$ with SiO_2 rich droplets (with B_2O_3) (dark) and ZnO–SiO_2 rich matrix (light)

Figure 8. Photographs of a crystallized zinc borate phases from a $44ZnO.56B_2O_3$ sample doped with 1×10^{20} Mn^{2+} per cm^3 after thermal treatment, 2 h 750°C, (size 20×20×2 mm^3): (a) under normal white light illumination white opaque, (b) under UV light (Hg lamp, 254 nm) crystals with green, yellow and orange-red emission colours [Colour available online]

lower clear $ZnO–B_2O_3$ phase with an accumulation of the RE^{3+} ions (Figures 12(a) and (b)). The lifetimes were nearly constant for Tb^{3+} at 486 nm (blue) very short, τ_e~20 μs, and at 543 nm (green) longer, τ_e~2·4

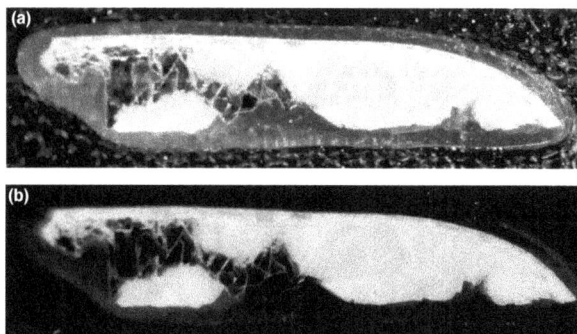

Figure 9. Photographs of the cross-sections of the samples of Figure 8 (microscopic section, thickness ~100 μm): (a) under white light various white and colourless crystals, (b) under UV light (Hg lamp, 254 nm) crystals with green (on the surface), yellow and orange-red emission colours (in the volume) [Colour available online]

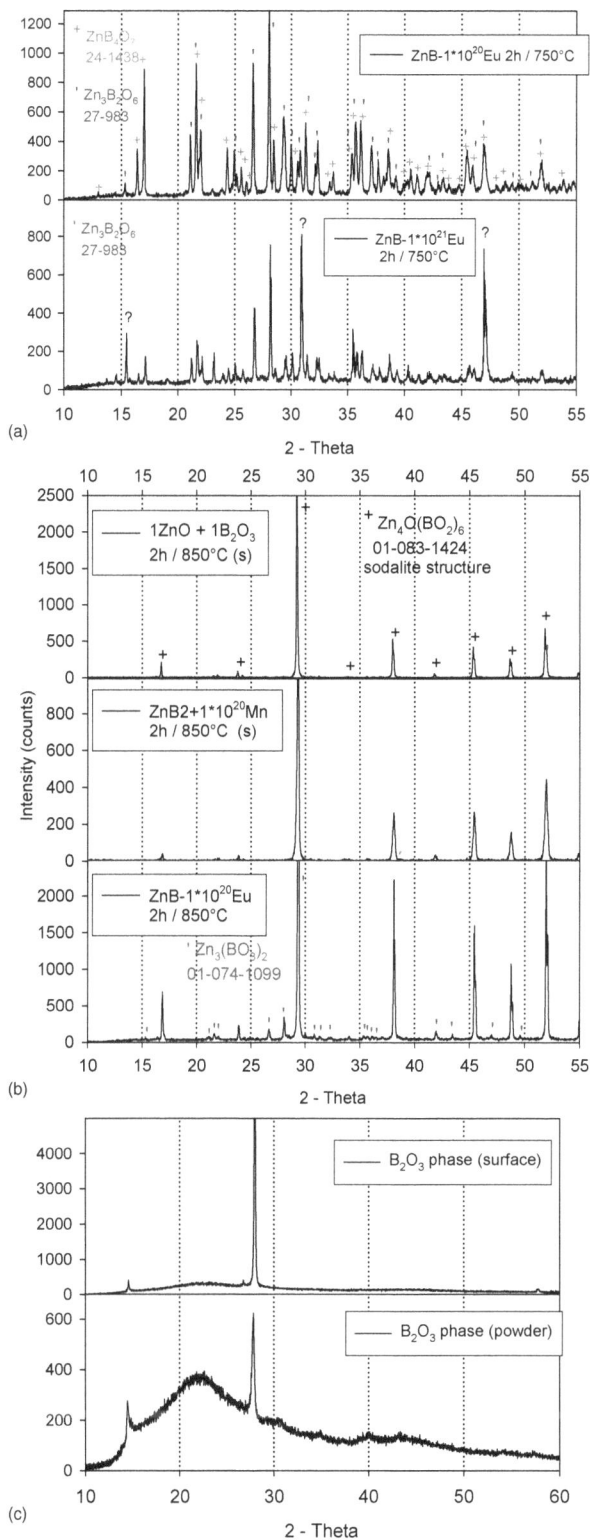

Figure 7. XRD patterns of various $ZnO–B_2O_3$ glass ceramic samples (compositions: $ZnB=56ZnO.44B_2O_3$; $ZnB2=ZnO–B_2O_3$ phase from phase separated $44ZnO.56B_2O_3$ glass sample): (a) 2 h 750°C thermal treatment, (b) 2 h 850°C thermal treatment, (c) opalescent B_2O_3 rich phase of phase separated $44ZnO.56B_2O_3$ glass sample [Colour available online]

$44ZnO.56B_2O_3$ shows similar liquid–liquid phase separation, split into an upper opalescent B_2O_3 phase with a small amount of crystals (Figure7(c)), and a

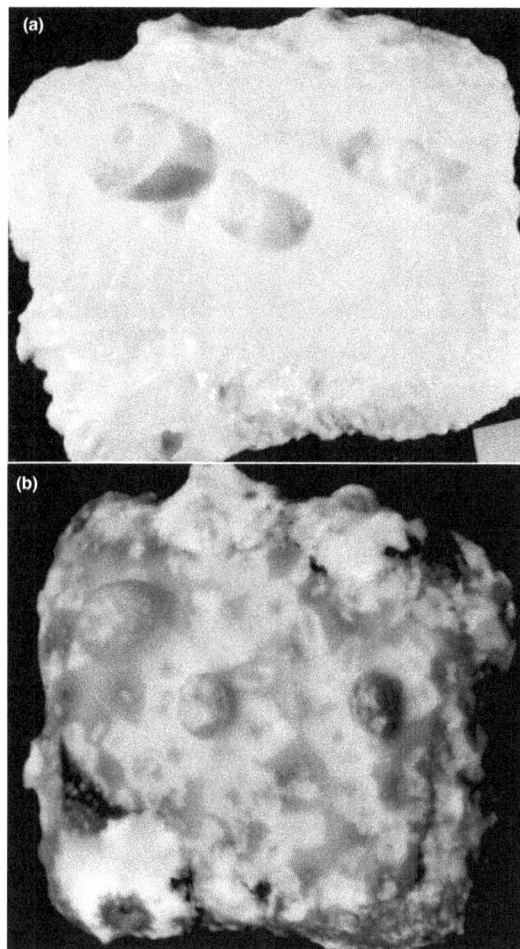

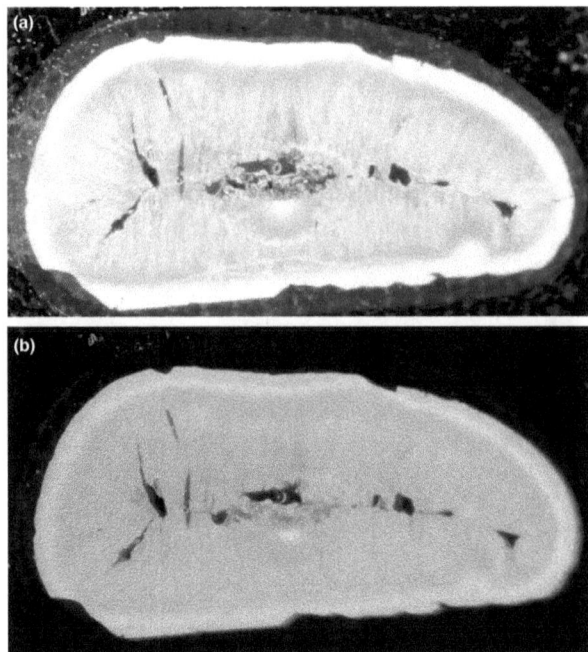

Figure 10. Photographs of a crystallised zinc borate phase from $44ZnO.56B_2O_3$ doped with $1\times10^{20}\ Mn^{2+}$ per cm^3 after thermal treatment, 2 h 850°C, (microscopic section, thickness ~100 μm): (a) under normal white light strong white surface crystallisation growing in the volume, (b) under UV light (Hg lamp, 254 nm) crystals with green emission colour only [Colour available online]

ms. Lifetimes τ_e~2 ms were measured for Eu^{3+} at 612 nm in all glass and glass ceramic samples.

Typical DSC curves of a clear zinc borate and an opaque B_2O_3 phase are shown in Figure 13(a). The endothermic peaks at 117°C and at 158°C (with shoulder) could be caused by the loss of water due to OH groups at the surface of the powdered B_2O_3 rich phase, which can easily form H_3BO_3. Typical DTA curves of homogeneous samples are shown in Figure 13(b). The DTA curves of $50ZnO.50B_2O_3$ and the zinc borate phases of $44ZnO.56B_2O_3$ are very similar. The values for T_g were 550–570°C. Two significant crystallisation maxima around 750°C and 850°C and a melting temperature at ~1000°C were found. The crystallisation maxima decreased with higher RE_2O_3 doping. Significant changes in the photoluminescence spectra of the glass ceramic samples were detected (Figures 16(a) and (b)). The opaque glass ceramic samples had much higher emission intensity than the glass samples and the peaks are split, which is typical for crystals. An additional europium borate crystal phase was detected by SEM measurements with EDX analysis (Figures 17(a)–(d)).

Discussion

Materials with high ZnO content are interesting for many applications. Crystalline zinc borate hydrates are commercially produced. They are non-toxic (eco-

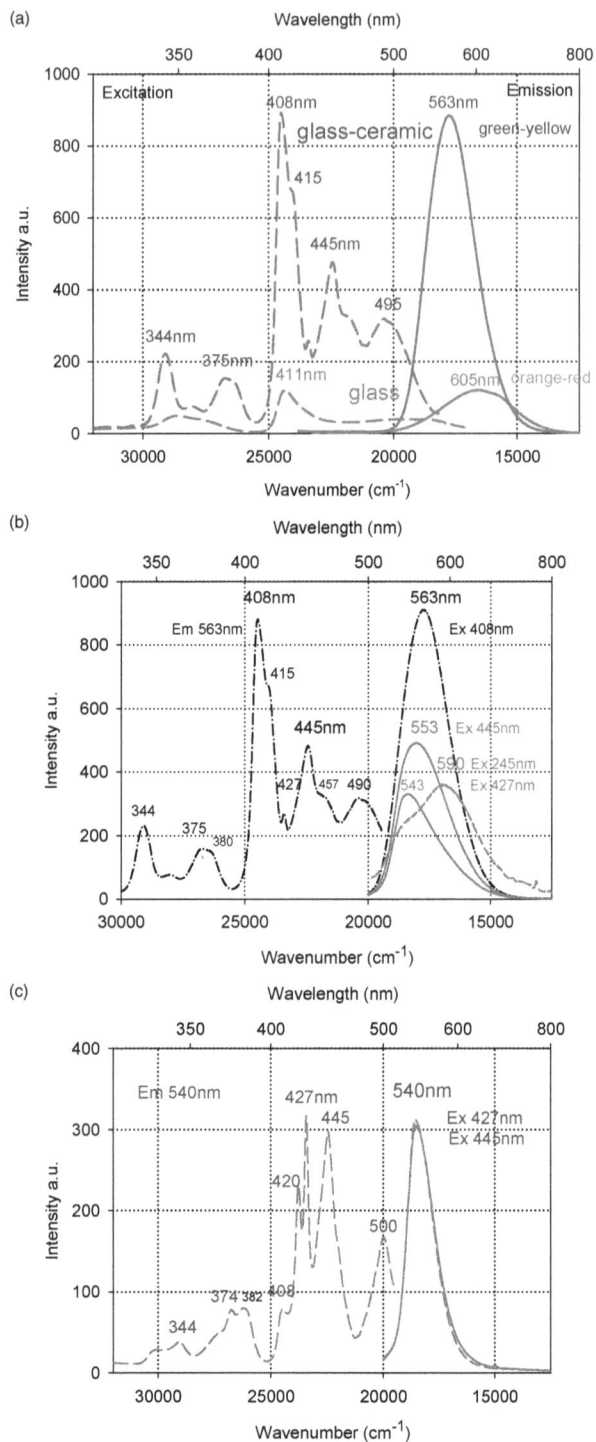

Figure 11. Photoluminescence excitation and emission spectra of glass and glass ceramic samples of the ZnO–B_2O_3 phase from $44ZnO.56B_2O_3$ doped with $1\times10^{20}\ Mn^{2+}$ per cm^3. (a) Low intensity of the broad orange-red emission (maximum at 605 nm) of the glass sample after excitation at the maximum at 411 nm, and high intensity of the broad green-yellow emission (maximum at 563 nm) of the glass ceramic sample (treated at 750°C for 2 h) after excitation at the maximum at 408 nm. (b) Emission spectra with four different excitation wavelengths (245, 408, 427 and 445 nm) for the glass ceramic sample (750°C, 2 h) shown in (a), also showing the excitation spectrum. (c) Spectra for the glass ceramic sample treated at 850°C for 2 h [Colour available online]

Figure 12. Photographs of the separated phases from 44ZnO.56B₂O₃ glass doped with 1×10²¹, 1×10²⁰ and 1×10¹⁹ Tb³⁺ per cm³ (from left to right) with opaque B₂O₃ rich and clear colourless ZnO–B₂O₃–Tb₂O₃ rich phase which could be polished as a plane parallel sample with thickness ~6 mm: (a) under daylight, (b) under UV light (370 nm) with only strong green emission colour of the clear zinc borate phases [Colour available online]

friendly), have good thermal stability and low water solubility. They are extensively used in plastic, rubber or coating industry[18], in medicine, in cosmetics, as fungicides, in ceramic glazes, paints, as flame retardants, as fertilisers, or as raw materials for the ceramic and glass industry. In recent years, water free zinc borate glasses and crystals began to obtain interest as a host for luminescence applications. Binary borate glasses with more than 50 mol% ZnO with good glass forming ability were described in various papers. Their structure was compared with the structure of crystals related to the borate anions.[1,10,12–15] In contrast to the metaborate anion of calcium, potassium or sodium with a chain structure of triangles, a new structural type of anhydrous metaborate anion in cubic $Zn_4O(BO_2)_6$, with a three-dimensional framework of $BØ_4$ tetraedra, similar to the $(Al_3Si_3O_{22})^{3-}$ anion of sodalite, was described by Smith et al.[14] in 1961. A re-determination of the cubic $Zn_4O(BO_2)_6$ structure with ZnO_4 and BO_4 groups was reported in 1980.[15] Krogh-Moe investigated the infrared spectra of cubic zinc borate and zinc borate glass of the same chemical composition. He found that all boron atoms are fourfold coordinated by oxygen in the crystalline modification, but in the glass boron is only threefold coordinated, and the spectrum of $4ZnO.3B_2O_3$ glass was almost identical with the spectrum of $ZnO.B_2O_3$

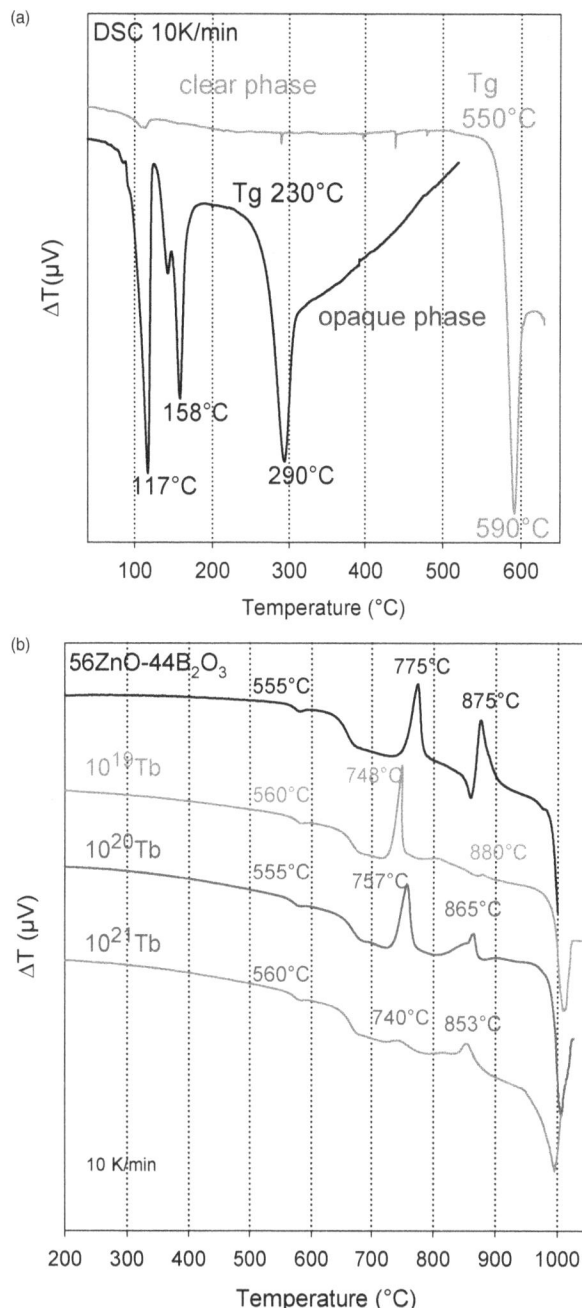

Figure 13. DSC curves: (a) of a clear zinc borate and an opaque B₂O₃ rich phase of the phase separated 44ZnO.56B₂O₃ glass sample, (b) of homogeneous 56ZnO.44B₂O₃ glass samples with increasing Tb₂O₃ doping [Colour available online]

glass.[1] A theoretical investigation of the geometrical and electronic structure of cubic $Zn_4O(BO_2)_6$ was reported by Casarin et al.[16] They concluded that Zn_4O_{13} clusters exist in the borate sodalite framework with $BØ_4$ groups. The Zn_4O_{13} cluster in cubic $Zn_4O(BO_2)_6$ crystals should be responsible for intrinsic photoluminescence, thermoluminescence and other interesting optical properties. However, glasses of the same chemical composition have a completely different structure with mainly three-fold coordinated boron groups.[12] We prepared B_2O_3 and $56ZnO.44B_2O_3$ glasses and found very different properties.[10] The

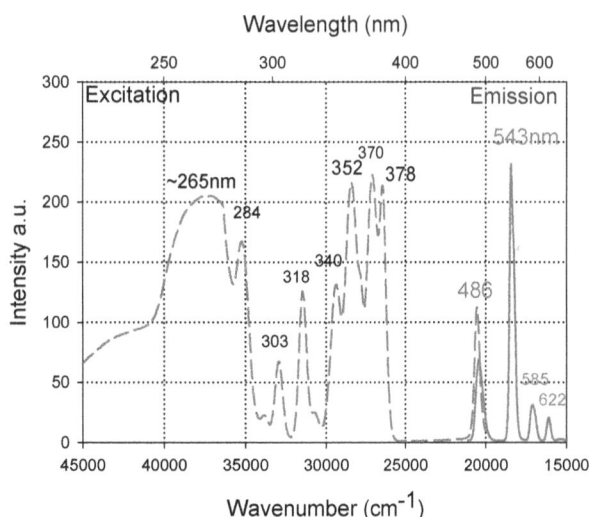

Figure 14. Photoluminescence excitation (dashed line) and emission (solid line) spectra of Tb³⁺ (1×10²⁰ per cm³) in a homogeneous 56ZnO.44B₂O₃ glass sample [Colour available online]

zinc borate glass had very high thermal and chemical stability. Blue luminescence emission of the zinc borate glass was found by UV excitation, ~250 nm. Small amounts of Mn^{2+}, which can substitute for Zn^{2+} in zinc borate glasses, provided the typical absorption and photoluminescence spectra of Mn^{2+} ions six-fold coordinated with oxygen (Figure 2). The narrow absorption transition with maximum at 411 nm ($^6A_1{\to}^4A_1$ and ^4E) is very sensitive to the

Figure 15. Photographs of polished clear 56ZnO.44B₂O₃ glass samples doped with 1×10²¹ and 1×10²⁰ Eu³⁺ per cm³ (thickness 10 mm and 2 mm): (a) colourless under daylight, (b) orange-red emission colour under UV light (370 nm) [Colour available online]

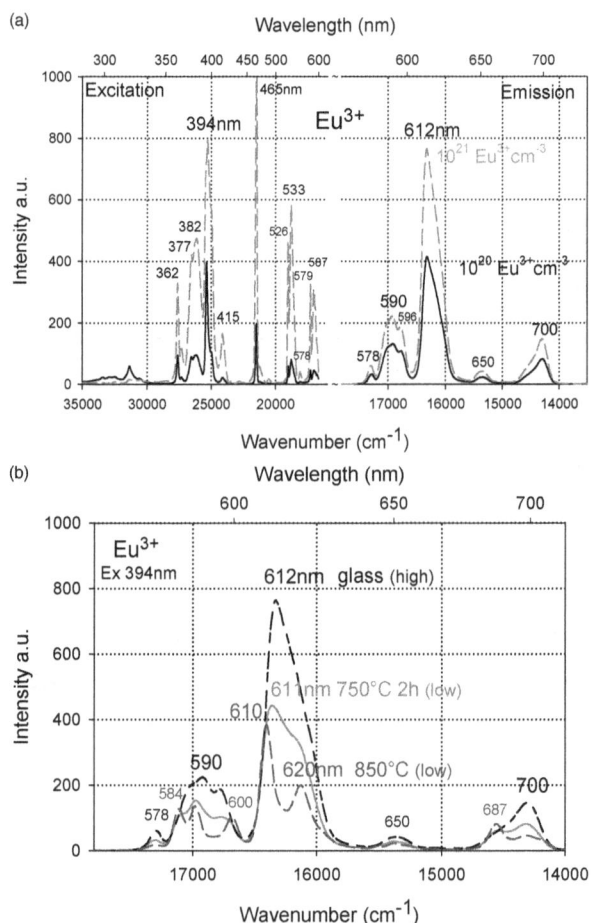

Figure 16. Photoluminescence of Eu³⁺ in 56ZnO.44B₂O₃: (a) excitation and emission spectra of glass samples with different Eu³⁺doping, (b) emission spectra of glass and glass ceramic samples with 1×10²¹ Eu³⁺ per cm³ measured with different sensitivities, high for glass and low for glass ceramic samples [Colour available online]

local structure of Mn^{2+} ions and to the local optical basicity of the surrounding matrix.[3,4,17,19] The maximum for the manganese borate glass containing 12·7 mol% Na₂O is shifted to 407 nm (Figure 5(a)) and for the separated MnO.B₂O₃ phases it is shifted to 414 nm (Figure 4(a)). In MnO.P₂O₅ glass, an absorption maximum at 409 nm and an emission maximum at 685 nm with lifetime <0·1 ms have been measured.[4,8] The emission bands of all glasses are very broad and the maxima are shifted from ~600 nm to 730 nm as the Mn^{2+} concentration increases from ~1×10²⁰ to ~1×10²² ions per cm³. A strong shift of the excitation maximum to longer wavelengths was detected at high Mn^{2+} content (Figure 4(b) and 5(b)), which should be caused by clustering and by ion–ion interaction due to very small Mn^{2+}–Mn^{2+} distances.[3,17,19] These effects also decrease the luminescence intensity and multi-exponential lifetimes, from ~7 ms to ~10 μs.

Phase separation of glass melts into separate liquid and glass layers is seldom if ever observed. Nevertheless, it was found in both binary borate systems, with ZnO and MnO. A nearly pure B₂O₃

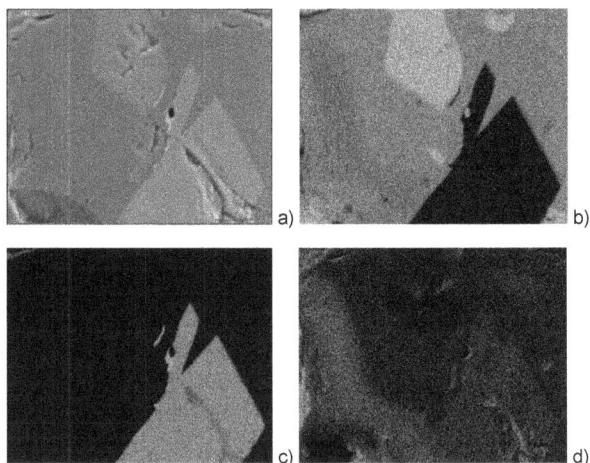

Figure 17. SEM and EDX images of a 56ZnO.44B$_2$O$_3$/1×10^{21} Eu^{3+} per cm^3 glass ceramic sample (2 h 850°C): (a) SEM with two crystals, (b) Zn distribution with one Zn rich crystal, (c) Eu distribution with one Eu rich and Zn free crystal, (d) B distribution with Eu-borate crystal [Colour available online]

phase as upper layer has low density, ~1·8–2·0 g/cm^3, low T_g~250–290°C, low surface tension*, 88 mN/m, low viscosity*, 1·1 dPa s, (*calculated for a composition in mol% 98B$_2$O$_3$.2ZnO or MnO for 1000–1200°C with SciGlass6.6 with Priven2000) and high thermal expansion coefficient, ~17 ppm/K,[10] which can cause stresses and cracks. The weak surface crystallisation (Figure 7(c)) could be initiated by contact with water from the atmosphere and formation of H$_3$BO$_3$, or also by small amounts of ZnO or MnO dissolved in the B$_2$O$_3$ phase and the slow cooling rate of the melt from 600°C. XRD measurements of all opalescent B$_2$O$_3$ phases have given the same typical pattern, shown in Figure 7(c). It provides stronger crystallisation at the surface (measured in reflection mode) than in the volume (measured with a powdered sample). According to the XRD database (card 6-0297) this pattern should be a cubic B$_2$O$_3$ phase with water as impurity and with a melting point 294°C. However, a calculation of the XRD pattern from hexagonal H$_3$BO$_3$ corresponds also to the measured data shown in Figure 7(c). The card 6-0297 was added to the database in 1953. It seems that the authors made a mistake by taking H$_3$BO$_3$ to be B$_2$O$_3$ which is extremely hard to crystallise. However, in the case of liquid–liquid phase separation, the B$_2$O$_3$ phase contains 1–2% ZnO or MnO which could favour the crystallisation of B$_2$O$_3$ besides H$_3$BO$_3$. It was amazing that the formed B$_2$O$_3$ phases had relatively high chemical stability in comparison to pure B$_2$O$_3$ glass samples.

The lower layer, with an approximate composition ZnO.B$_2$O$_3$ or MnO.B$_2$O$_3$, has high density, ~3·3 or 3·0 g/cm^3, high T_g~550°C, low TEC, ~4 and 6 ppm/K, low viscosity too, but much higher surface tension >200 mN/m. Phase separation occurs by decrease of the free energy. That means in this case, that Zn–O–Zn

and Mn–O–Mn bonding should be preferred to Zn–O–B or Mn–O–B bonding when the ZnO or MnO content is less than 50 mol%. A broad distribution of the local sites in melts and glasses is possible. The dominating coordination number to oxygen for Zn^{2+} and Mn^{2+} is six, and for boron three. The change of the structure during the crystallisation process was investigated for zinc borate glasses with Mn^{2+} as a sensor ion. The glass samples were annealed for 2 h at 750°C and 850°C and studied by XRD and luminescence measurements (Figure 7–11). Surface crystallisation was found at both temperatures. Different zinc borate phases were detected in samples, annealed at 750°C. The assignment of the XRD reflections to definite zinc borate crystal phases with JCPDS data was not possible. Many XRD reflections were measured after heat treatment at 750°C (Figure 7(a)). An assignment to ZnB$_4$O$_7$ (24-1438) and Zn$_3$B$_2$O$_6$ (27-983) was made, but other very similar (and confusing) JCPDS data, with many reflections were found. It should be assumed that different crystal phases of zinc borates can form by heat treatment at lower temperature.

Nevertheless, interesting crystallisation effects were observed by various luminescence colours, orange-yellow-green, under UV light (Figure 8(b) and 9(b)), and spectra (Figure 11(a) and (b)). The cross-section of the sample (Figure 9) shows green luminescent crystals on the surface and yellow, orange and red (transparent) crystals in the volume due to Mn^{2+} ions in various zinc borate crystals. The samples annealed at 850°C yield only green luminescence (Figure 10(b)), due to needle-like Zn$_4$O(BO$_2$)$_6$ crystals (XRD in Figure7(b)). The change of the structure by crystallisation is demonstrated with typical photoluminescence spectra of glass and glass ceramic samples in Figures 11(a) to (c). The glass sample provides much broader bands with lower luminescence intensity (Figure 11(a)). The glass ceramic samples show much higher luminescence intensity and splitting into narrow bands, especially in the excitation spectra (Figure 11(b) and (c)). The shape of the one broad emission band is also dependent on the excitation wavelength. The broadest emission band was obtained by excitation at 245 nm, in the maximum of the charge transfer transition (Figure 11(b)). The intensities of the excitation bands change drastically and the emission bands become smaller and shift to shorter wavelengths with increasing annealing temperature from 750 to 850°C.

Zinc borate glasses with Tb^{3+} (4f^8) are efficient luminescent materials for green and with Eu^{3+} (4f^6) for orange-red emission with typical sharp f–f transition bands. The RE^{3+} ions are accumulated in the zinc borate phase and not in the B$_2$O$_3$ phase. The crystallisation tendency of zinc borates decreases with increasing RE^{3+} content. The large Eu^{3+} and Tb^{3+} ions cannot substitute for Zn^{2+} in the crystal phases. However, the glass ceramic samples showed much higher luminescence intensity, and the emission bands for Eu^{3+} split into more very narrow bands (Figure 16(b)). Europium borate and zinc borate phases were detected by SEM with EDX measurements (Figure 17(a)–(d)). This is in good agreement with the results from Bettinelli et al.[12] They found by spectroscopic investigations of $4ZnO.3B_2O_3$ glasses doped with Eu^{3+} an Eu^{3+}–O stretching/BO_3^{3-} mode.

Conclusions

$ZnO–B_2O_3$ and $MnO–B_2O_3$ melts show similar stable phase separation. They split into two layers which can be separated. The upper layer is nearly pure B_2O_3 with low density, low T_g, and opalescence due to a few crystallites. The lower layer has much higher density and a composition near 50:50 $ZnO–B_2O_3$ or $MnO–B_2O_3$. Their thermal and optical properties were determined. Ternary systems with SiO_2 also show stable phase separation with a SiO_2 rich phase. Substitution of Na_2O for ZnO or MnO decreases phase separation. Mn^{2+} can substitute for Zn^{2+} in glasses and crystals with coordination numbers 6 to 4. Mn^{2+} photoluminescence is a sensor for the local structure of glasses and crystals. It was shown that the local structure of Mn^{2+} and Zn^{2+} changes when the glass is transformed into a glass ceramic. In the glassy state, Mn^{2+} and Zn^{2+} ions prefer six-fold coordination with broad distribution to four-fold coordination in combination with various borate groups. In glass ceramics, formed at ~750°C, different crystalline zinc borate phases with different Mn^{2+} luminescence spectra and colours from orange-red (six-coordinated) to yellow-green (four-coordinated) were detected. In glass ceramics, formed at 850°C, only the cubic crystalline phase with ZnO_4, MnO_4, and BO_4 groups (sodalite structure), and green Mn^{2+} emission were detected. Zinc borate samples doped with Eu^{3+} or Tb^{3+} yield very strong orange-red or green photoluminescence by UV excitation. In phase separated samples the Eu^{3+} and Tb^{3+} ions were accumulated only in the $ZnO.B_2O_3$ phase and not in the B_2O_3 phase. The crystallisation tendency of zinc borate glasses was decreased by adding rare earth oxides. Crystallisation of europium or terbium borates was observed at high doping content.

Acknowledgements

The author wishes to thank her co-workers R. Atzrodt, S. Ebbinghaus, A. Herrmann, G. Möller, M. Müller, T. Kittel, G. Völksch, B. Müller and the students A. Gräfe, M. Kracker, C. Segel, F. Lindner and C. Funke for different measurements and discussions.

References

1. Krogh-Moe, J. Z. Kristallogr., 1962, 117, 166.
2. Harrison, D. E., Hummel, F. A. J. Electrochem. Soc. 1956, 103, 491.
3. Möncke, D., Kamitsos, E., Herrmann, A., Ehrt, D., Friedrich, M. J. Non-Cryst. Solids, 2011, 357, 2542.
4. Ehrt, D. J. Non-Cryst. Solids, 2004, 348, 22.
5. Ehrt, D. IOP Conf. Ser.: Mater. Sci. Eng., 2009, 2, 012001.
6. Ehrt, D. IOP Conf. Ser.: Mater. Sci. Eng., 2011, 21, 012001.
7. Ehrt, D., Herrmann, A., Tiegel, M. Phys. Chem. Glasses: Eur. J. Glass Sci. Technol. B, 2011, 52, 68.
8. Herrmann, A., Ehrt, D. Phys. Chem. Glasses: Eur. J. Glass Sci. Technol. B, 2005, 78, 99.
9. Herrmann, A., Fibikar, S., Ehrt, D. J. Non-Cryst. Solids, 2009, 355, 2093.
10. Ehrt, D. Phys. Chem. Glasses: Eur. J. Glass Sci. Technol. B, 2006, 47, 669.
11. Ingerson, E., Morey, G. W., Tuttle, O. F. Am. J. Sci., 1948, 246, 39.
12. Bettinelli, M., Speghini, A., Ferrari, M., Montagna, M. J. Non-Cryst. Solids, 1996, 201, 211.
13. Masai, H., Ueno, T., Takahashi, Y., Fujiwara, T. J. Non-Cryst. Solids, 2010, 356, 2689.
14. Smith, P., Garcia-Blanco, S., Rivoir, L. Z. Kristallogr., 1961, 115, 460.
15. Smith-Verdier, P., Garcia-Blanco, S. Z. Kristallogr., 1980, 151, 175.
16. Casarin, M., Maccato, C., Vittadini, A. J. Phys. Chem. B, 2002, 106, 2569.
17. Konidakis, I., Varsamis, C. E., Kamitsos, E. I., Möncke, D., Ehrt, D. J. Phys. Chem. C, 2010, 114, 9125.
18. Mu, Z., Hu, Y., Chen, L., Wang, X. Opt. Mater. 2011, 34, 89.
19. Möncke, D., Ehrt, D., Kamitsos, E. Phys. Chem. Glasses: Eur. J. Glass Sci. Technol. B, 2013, 54 (1), 42–51.

Phys. Chem. Glasses: Eur. J. Glass Sci. Technol. B, December 2013, **54** (6), 241–246

Crystallisation of a lead borate glass and its influence on the thermoluminescence response

M. Rodríguez Chialanza,[1,]* *E. Castiglioni,*[2] *J. Castiglioni*[3] *& L. Fornaro*[4]

[1] *Grupo de Semiconductores Compuestos, Cátedra de Radioquímica, Facultad de Química, Universidad de la República, Montevideo, Uruguay*

[2] *Laboratorio de Datación por Luminiscencia, Facultad de Ciencias, Universidad de la República, Montevideo, Uruguay*

[3] *Departamento de Experimentación y Teoría de la Estructura de la Materia y sus Aplicaciones, Facultad de Química, Universidad de la República, Montevideo, Uruguay*

[4] *Grupo de Semiconductores Compuestos, Centro Universitario de la Región Este (CURE), Universidad de la República, Rocha, Uruguay*

Manuscript received 31 October 2011
Revision received 7 February 2013
Manuscript accepted 25 June 2013

Lead borate glasses of composition 33PbO.67B$_2$O$_3$ were prepared by the conventional melting/quenching method. Glass-ceramics were prepared from powdered glasses by heat treatment at 590°C for 60, 120, 180 and 240 min. The crystallisation process was followed by scanning electron microscopy. The thermoluminescence of the glasses and glass-ceramics was measured and the response was correlated with the structural changes that occur during crystallisation. A great improvement in the thermoluminescence response was obtained after 60 min of isothermal crystallisation, which corresponds to a degree of crystallisation of more than 50%. As the crystallisation proceeds, the glass structure changes and a localised state which improves the response of the materials is generated. These results indicate that the studied lead borate glass-ceramics are promising materials for thermoluminescence dosimetric applications.

1. Introduction

Borates have previously been studied for their glass forming ability and their interesting structure, which presents some particularities such as the well known borate anomaly. The accepted model for the structure of vitreous boric oxide consists of equal numbers of boroxol groups and independent BO$_3$ triangles.[1] Furthermore, the initial addition of a network modifier to B$_2$O$_3$ leads to an increase in the coordination number of some of the boron atoms from three to four.[1]

According to the literature, borates (vitreous or crystalline) are being used for detection applications in medical dosimetry; and their performance strongly depends on the way the material was obtained. For this reason, the development of these materials is focused on obtaining better knowledge and control of their microstructure and their optical properties.[2,3] On the other hand, thermoluminescence has been induced in certain borate glasses by the development of ceramic-glass materials, as is the case for calcium borate glasses,[3] and also luminescence was induced on heat treatment of doped lead borate glasses.[4]

Glass-ceramic materials can be produced in a controlled manner by the process of surface induced nucleation of glass particles. Granular glass samples crystallise more easily than bulk glass samples in a process of simultaneous amorphous state sintering and crystallisation. Generally, in the case of glass particles, crystallisation starts on athermal surface sites, such us scratches, solids impurities, etc.[5–7]

Following on from these previous studies, the aim of our work is to study the influence of the structural glass-ceramic transformation of a lead borate glass, with nominal composition of 33PbO.67B$_2$O$_3$, on its thermoluminescence response.

2. Experimental

In order to rigorously study the glass crystallisation we selected the congruent composition 33PbO.67B$_2$O$_3$ (in mol%) from the phase diagram[8] for our work. Lead borate glasses of that nominal composition were prepared from H$_3$BO$_3$ (99·5% grade purity) and PbO (99·9% grade purity) by the conventional melting/quenching method. Complete fusion was achieved at 950°C and maintained for 1 h. Once the sample was homogeneously melted, the fused glass was poured into a steel mould at room temperature. Glass samples were re-melted to ensure homogeneity. The as-prepared samples were stored in a desiccator to avoid moisture absorption and therefore the possibility of undesired crystallisation during further thermal treatment. Glass samples were crushed, and then fractions with particles 23–65, 65–190, 190–250, 250–315, 315–468 µm in size were separated and stored in a desiccator. Fully crystallised samples were obtained by cooling down the fused glass at the

Corresponding author. Email mrodrig@fq.edu.uy
Original version presented at VII Int. Conf. on Borate Glasses, Crystals and Melts, Dalhousie University, Halifax, Nova Scotia, Canada, 21–25 August 2011

crystallisation temperature for 24 h. An inductively coupled plasma atomic emission spectrometer – ICP-AES Simultaneous CCD-VISTA-MPX (Varian, Mulgrave, Australia) with radial configuration was used for chemical analyses of 25 samples of glass.

Differential thermal analysis (DTA) was carried out using a Shimazdu DTA-50. Air was used as atmosphere during the DTA experiments; 16 mg of representative samples were used in all the experiments. This amount was chosen to ensure good thermal contact between sample and crucible, in order to avoid temperature gradients into the sample. α-Al_2O_3 was employed as a reference in DTA, and the heating rate was 10°C/min.

Granular glass particles in the range 65–190 mm were pressed into pellets and heat treated at the temperature corresponding to the maximum of the DTA crystallisation peak for different period of times (0, 60, 120, 180 and 240 min) in order to obtain glass-ceramic samples.

The sintered discs were sectioned along a diameter. One of the sections was polished for microscopy, whilst the other was used for x-ray diffraction characterisation. The identity of the crystalline phase in the glass-ceramic samples obtained after heat treatment was investigated by x-ray powder diffraction at room temperature, using a θ–2θ Rigaku Ultima IV diffractometer and Cu K_α radiation (λ=1·5418 Å).

The morphology of the crystallised samples was observed by scanning electron microscopy (SEM) with JEOL 5900 Low Vacuum equipment. Polished sections of crystallised samples were used, in order to determine the crystalline volume fraction. Eyepiece reticles and objective powers were calibrated using a stage micrometer. Ten images of each section, randomly taken, were analysed. The experimental crystallised volume fraction was determined as is described elsewhere[7] by measuring the porosity and the crystallised fractions in the micrographs. For each image, by using an area selector (and dividing the results by the total area of the image), the porosity and the crystallised area were determined.

The infrared absorption spectra of the glasses, in the 400–1600 cm^{-1} range, were measured (Shimazdu FTIR spectrophotometer, Prestige-21) at room temperature. The spectra were measured with a standard resolution of 2 cm^{-1}. A normalisation procedure was applied before the deconvolution analysis.

Thermoluminescence (TL) measurements were performed using conventional Daybreak TL equipment with a Corning 7–59 and a Schott BG-39 infrared rejecting filter to reduce thermal noise, and a 9635 QB photomultiplier tube for photon counting. Prior to any measurements, the zero-dose TL signal was measured and it was found to be in the order of the background. All TL glow curves were recorded with a linear heating rate of 10°C/s under constant nitrogen flow. The glass samples were irradiated on plate us-

ing a calibrated ^{90}Sr source with a nominal dose rate of 2 Gy/min. The black-body radiation was digitally subtracted from the TL glow curves.

3. Results

ICP analysis showed that the average content of lead, x, (expressed as $Pb_xB_4O_7$ instead of $33PbO.67B_2O_3$) in glasses corresponds to 0·99 mol for the entire samples, and indicated that the glass has the desired composition.

X-ray powder diffraction of the glass samples showed typical amorphous behaviour, whereas diffractograms of the crystalline samples showed PbB_4O_7 as the only identified crystalline phase (Figure 1).

Figure 2(a) shows how the peak maximum height, δT_p, from DTA diagrams, decreases as particle size increases, and the same behaviour is observed in Figure 2(b) for the quantity $T_p^2/\Delta T_p$ relation as a function of particle size (where T_p is temperature of the peak maximum, and ΔT_p is the full width at half maximum of the peak).

The evolution of the degree of transformation from glass to crystallised sample, estimated by optical microscopy, as a function of crystallisation time, is given in Table 1. The transformation was also followed by scanning electron microscopy; the morphology of the crystals grown for 60 and 180 min is shown in Figure 3.

The infrared transmission spectrum of a glass sample $33PbO.67B_2O_3$ recorded at room temperature (Figure 4) exhibits three groups of bands: (I) in the region 1200–1400 cm^{-1}, (II) in the region 850–1100 cm^{-1}, and (III) another band at about 650–710 cm^{-1}. It is accepted that the broad absorption region 1200–1450 cm^{-1} can be attributed to B–O bond stretching of trigonal BO_3 units, while that in the region 850–1100 cm^{-1} originates from B–O bond stretching of tetrahedral, BO_4, units, and the third band is identified as being

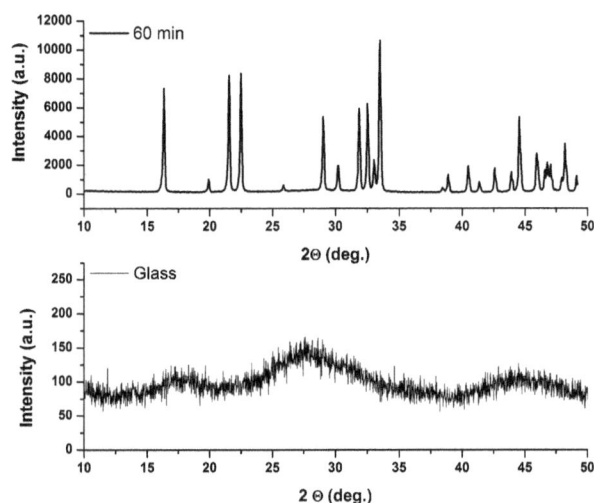

Figure 1. XRD diffractogram for glass and for glass-ceramic heat treated at 590°C for 60 min

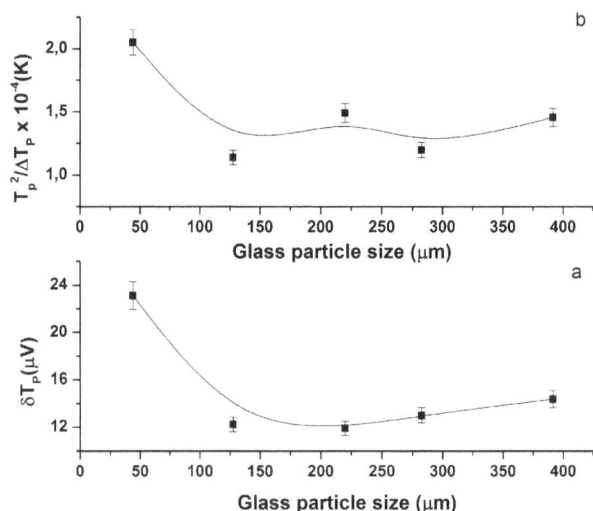

Figure 2. (a) DTA crystallisation peak maximum height, δT_p, as a function of particle size for lead borate glass. (b) The quantity $T_p^2/\Delta T_p$ (see text for details) as a function of particle size for lead borate glass. DTA heating rate, 10°C/min; sample weight, 16 mg

due to the bending vibration of B–O–B linkages in the borate network.[9,10]

The FTIR spectra for the heat treated samples exhibit changes, namely those illustrated by the narrowing of the previous bands and the appearance of some new bands in the spectra. New bands appear in the FTIR spectra of the glass-ceramic samples, for example at 1150 and 941 cm^{-1}.

Figure 5 shows the dose dependence of the thermoluminescence response (evaluated as the area under the glow curve) for samples crystallised for 60, 180 min and crystallised directly from the melt (Figure 5(a)) and for glassy samples (Figure 5(b)) at doses in the range 4·2–37·8 Gy. The thermoluminescence response greatly increases for crystallised samples.

4. Discussion

As we used a congruent composition of $33PbO.67B_2O_3$ for the glass, the same composition was expected in the parent glass and the crystalline phase. This fact was verified by the x-ray powder diffraction results, which gave the congruent composition PbB_4O_7 (JCPDS file 15-278) as the only crystalline material in the heat treated samples.

A temperature of 591±2°C for the peak maximum, T_p, and a glass transition temperature, T_g, of 440±2°C were estimated from thermal analysis. Therefore, a temperature of 590°C was selected for performing

Table 1. Evolution of the volume crystallised fraction during heat treatment at 590°C

Crystallisation time (min)	Degree of transformation (%)
0	0
60	58±6
120	68±8
180	83±8

Figure 3. SEM micrographs for the granular glass sample heat treated for (a) 60 min, (b) 180 min

the heat treatment for crystallisation.

It has been demonstrated that differential thermal analysis identifies and distinguishes between surface and internal crystallisation of glasses.[11] If we consider that crystal growth proceeds at a constant rate,[12] then the maximum height of the DTA peak,

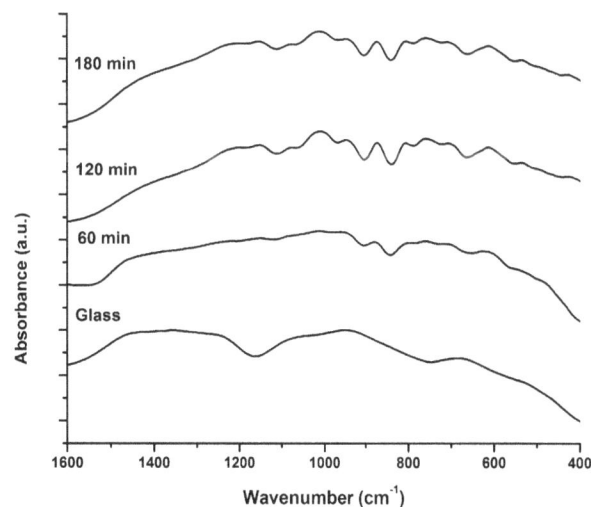

Figure 4. FTIR spectra for glass sample and samples heat treated for 60, 120 and 240 min, showing the changes in borate structural groups

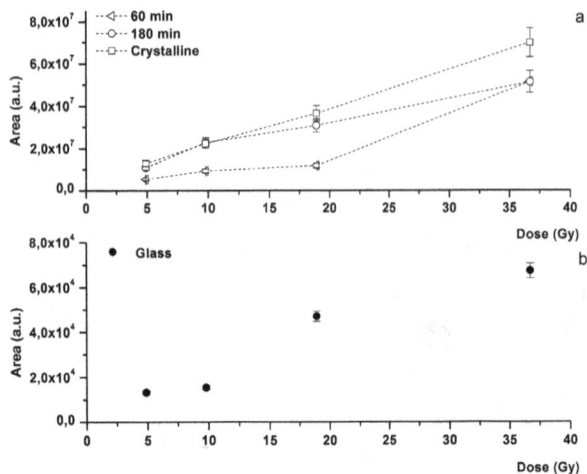

Figure 5. Thermoluminescence response as a function of the irradiated dose for (a) samples heat treated for 60 and 180 minutes, and fully crystallised, and (b) glass samples

δT_p, is proportional to the total number of nuclei (internal and surface nuclei combined) present in the sample, and the quantity $T_p^2/\Delta T_p$ is related to the crystal growth dimension, which in turn depends on the specific mechanism for crystallisation.[11] For dominant surface crystallisation a decrease in δT_p should be observed when the particle size increases, because the total effective surface area of the glass particles (the number of surface nuclei available for crystallisation) decreases as well. This is what Figure 2(a) shows for the studied borate glass, for which a decrease in the peak maximum height with increasing particle size was observed. In Figure 2(b) it is shown that the quantity $T_p^2/\Delta T_p$ also decreases as the particle size increases. The decrease of both parameters as the partice size increases indicates a surface crystallisation mechanism.

As shown by the data in Table 1, crystallisation of our lead borate glass proceeds rapidly, and after 60 min of heat treatment more than 50% of the glass is transformed. The results in Table 1 agree with the description of the evolution of crystallisation of the JMAK (Johnson, Mehl, Avrami, Kolmogorov) theory,[13–15] which predicts that, after a certain period of time, all of the sample is crystalline.

From FTIR spectrometry and based on literature data, we attempted to assign the infrared bands to the presence of specific borate groups in our glass and glass-ceramics. The bands around 1240 and 685 cm^{-1} for glass samples, are assigned to the B–O stretching vibration in BO$_3$ units and to the B–O–B bending vibration between two BO$_3$ units respectively. The band at 961 cm^{-1} can be assigned to B–O stretching in tetrahedral units (BO$_4$). These findings suggest that the structural groups present in our glasses contains BO$_4$ and BO$_3$ units. Furthermore, the bands around 1080 and 870 cm^{-1} can be assigned to B–O stretching in tetrahedral units (BO$_4$). Both bands become thinner, reducing their area as the crystallisation proceeds,

indicating the tranformation of these groups as crystallisation time increases. The assigment of the previous bands was based on results for other compounds. For example, in the case of silver lead borate glass, a band near 1030 cm^{-1} has been assigned to B–O bond stretching vibrations of BO$_4$ tetrahedra in tri-, tetra- and pentaborate groups.[16] Moreover, a band at 1050 cm^{-1} has been asigned to BO$_4$ stretching in tri-, tetra- and pentaborate groups, and a band in the region between 900 and 1000 cm^{-1} has been assigned to diborate groups in magnesium sodium borate glass.[10] According to these assignments, we suggest the presence of pentaborate and diborate groups in our glass samples. These results are in agreement with previous reports for lead borate glasses, in which diborate and pentaborate groups were found.[17] Additionally, using Raman spectroscopy, pentaborate and diborate groups were found in glasses with compositions in the range 30PbO.70B$_2$O$_3$ to 35PbO.65B$_2$O$_3$.[18]

For the glass-ceramic samples, we observed a small shift in the above-mentioned bands (e.g. 961 to 941 cm^{-1}) and the appearence of a new band near 1150 cm^{-1} for samples crystallised for 240 min. In the Raman spectrum of lithium borate glasses and melts, a band at 950 cm^{-1} has been interpreted as indicating the formation of BO$_4$ tetrahedra (both included and not included in superstructural groupings), and a weak band near 1125 cm^{-1} has been assigned to the vibrational mode of diborate groups.[19] On the other hand, a band at 1180 cm^{-1} for lead fluoroborate glasses[20] has been assigned to vibrations of BO$_4$ units. Also, in the infrared spectra of crystalline SrO.2B$_2$O$_3$ (an iso-structural compound), the presence of a band at 1100 cm^{-1} has been associated with BO$_4$ units.[21,22]

Considering that the structure of crystalline PbO–B$_2$O$_3$ consists of a three-dimensional borate network where all boron are tetrahedrally coordinated with corner-linking between the tetrahedra,[23,24] we assign the mentioned bands to BO$_4$ units. In addition to this, our results (bands near 1230, 1150, 1077, 1006, 941, 870, 813, 620 cm^{-1} for the case of 240 min of crystallisation) present considerable similarities with those previously reported for crystalline PbO.2B$_2$O$_3$.[25] In the aforementioned work it is concluded from infrared measurements that BO$_4$ units are presents in the crystalline studied samples. The structural difference between crystalline and glassy samples was reported some time ago for lead borate glass.[26,27]

As shown in Figure 5(a), the thermoluminescence response (the area under the glow curve) increases when crystallisation time increases. There is no shift in the temperature for the peak maximum, indicating first order kinetics for the thermoluminescence process.[28] The analysis of the TL glow curve in order to obtain kinetic parameters such as trap depth, E_{TL}, and frequency factor, s, has recently been reported by our group.[29] The thermoluminescence of the studied lead borate glass-ceramic could be consistent with traps involving localised transitions through excited states below the conduction band.[30] It is important to remark that glass samples (Figure 5(b)) show thermoluminescence which increases with crystallisation time, and is similar to that obtained for calcium borate glass.[3]

Following on from these observations, we measured the thermoluminescence of samples crystallised directly from the melt. The response (again evaluated as the area under the curve) is shown in Figure 5(a). We consider the relation $A/A_{60\,min}$ (where A refers to the area under the curve for a fully crystalline sample) which gives an idea of the proportion of the response with the crystallisation fraction. For fully crystalline samples, this relation gives $2\cdot3\pm0\cdot5$. For 180 min crystallised samples the ratio is $1\cdot2$ as reported elsewhere,[29] and for samples heat treated for 24 h a value of $2\cdot2\pm0\cdot5$ was obtained. These results are consistent with the results on the degree of crystallisation, which shows that at 180 min we have more than 80% crystallisation.

The same linear behaviour of response as a function of irradiating dose was observed for glass and glass-ceramic samples. Therefore, the species that are responsible for the thermoluminescence should be present in the glass and increase in their concentration as the crystallisation proceeds. In the case of quartz or silica, broken silicon and oxygen bonds are important defects in the SiO_2 network that contribute to the thermoluminescence response. Nonbridging oxygen is a potential hole trap, and empty Si orbitals can trap electrons.[31] In the case of our samples, the conversion of BO_3 units into BO_4 units during the crystallisation, or the presence of BO_4, may be generating localised states between the valence band and the conduction band, which may be responsible for the thermoluminescence phenomena. The effects that radiation produces in lead borate glasses lead to the formation of hole centres, such as OHCs (oxygen hole centres) or BOHCs (boron oxygen hole centres). Furthermore, Pb^{2+} can also act as a hole trap.[31] Future work will be focused on the study of electronic states in our samples, and on deepening the study of structural changes during the crystallisation.

6. Conclusions

Surface crystallisation and structural changes with crystallisation time have been shown for glass-ceramics of composition $33PbO.67B_2O_3$. The increase in the thermoluminescence response of these glass-ceramics, when the crystallisation time increases, was also correlated with the observed structural changes. The results agree with previous reports for similar systems, and indicate these glass-ceramics to be potential materials for thermoluminescent dosimetry.

Acknowledgements

The authors thank Agencia Nacional de Investigación e Innovación (ANII) and Comisión Sectorial de Investigación Científica (CSIC) for financial support, Prof. Dr. A. C. Hernandes and M. Basso from Grupo de Crescimento de Cristais y Materiais Ceramicos, Instituto de Fisica, Universidad de Sao Paulo, Sao Carlos, SP, Brasil, for the ICP analysis, and Prof. L. Suescun from Crystallography, Solid State and Materials Laboratory (Cryssmat-Lab), DETEMA, Facultad de Química, Universidad de la República, Montevideo, Uruguay, for the x-ray analysis.

References

1. Wright, A. C. Borate structures: crystalline and vitreous. *Phys. Chem. Glasses: Eur. J. Glass Sci. Technol. Part B*, 2010, **51** (1), 1–39.
2. Santiago, M., et al., Thermo- and radioluminescence of undoped and Dy-doped strontium borates prepared by sol-gel method. *Radiat. Meas.*, 2011, **46** (12), 1488.
3. Rojas, S. S., Yukimitu, K., de Camargo, A. S. S., Nunes, L. A. O. & Hernandes, A. C. Undoped and calcium doped borate glass system for thermoluminescent dosimeter. *J. Non-Cryst. Solids*, 2006, **352**, 3608–12.
4. Pisarska, J., Lisiecki, R., Ryba-Romanowski, W., Goryczkad, T. & Pisarskia, W. A. Unusual luminescence behavior of Dy^{3+}-doped lead borate glass after heat treatment. *Chem. Phys. Lett.*, 2010, **489**, (4–6), 198–201.
5. Gutzow, I., Pascova, R., Karamanov, A. & Schmelzer, J. The kinetics of surface induced sinter crystallisation and the formation of glass-ceramic materials. *J. Mater. Sci.*, 1998, **33**, 5265–73.
6. Prado, M. O., Ferreira, E. B. & Zanotto, E. D. Sintering kinetics of crystallising glass particles. A review. *Proc. 106th Ann. Meeting of the ACerS - Symp. #26 - Melt Chemistry, Relaxation, and Solidification Kinetics of Glasses*, USA, 2004.
7. Ferreira, E. B., Nascimento, M. L. F., Stoppa, H. & Zanotto, E. D. Methods to estimate the number of surface nucleation sites on glass particles. *Phys. Chem. Glasses: Eur. J. Glass Sci. Technol. B*, 2008, **49** (2), 81–9.
8. Kassem, Y. M., Mahdy, A. N. & Abidir, M. F. The $PbO_2–PbO–B_2O_3$ system in air. *Thermochim. Acta*, 1989, **149**, 71–86.
9. Pisarski, W. A., Pisarska, J. & Romanowski, W. R. Structural role of rare earth ions in lead borate glasses evidenced by infrared spectroscopy: $BO_3 \leftrightarrow BO_4$ conversion. *J. Molec. Struct.*, 2005, **744–747**, 515–20.
10. Kamitsos, E. I., Karakassides, M. A. & Chryssikos, G. D. Vibrational spectra of magnesium-sodium-borate glasses. 2. Raman and mid-infrared investigation of the network structure, *J. Phys. Chem.*, 1987, **91** (5), 1073–9.
11. Ray, C., Yang, Q., Huang, W. & Day, D. Surface and internal crystallisation in glasses as determined by differential thermal analysis. *J. Am. Ceram. Soc.*, 1996, **79** (12), 3155–60.
12. Fokin, V., Cabral, A. A., Reis, R., Nascimento, M. L. & Zanotto, E. D. Critical assessment of DTA–DSC methods for the study of nucleation kinetics in glasses. *J. Non-Cryst. Solids*, **356** (6–8), 358–67
13. Kolmogorov, A. N. Zur Statistik der Kristallisationsvorgange in Metallen. *Izv. Akad. Nauk SSSR*, 1937, **1** (3), 355. (In Russian.)
14. Johnson, W. A., & Mehl, R. F. Reaction kinetics in processes of nucleation and growth. *Trans. AIME*, 1939, **135**, 416.
15. Avrami, M. *J. Chem. Phys.*, 1939, **7**, 1103.
16. Ciceo-Lucacel, R. & Ardelean, I. FT-IR and Raman study of silver lead borate-based glasses. *J. Non-Cryst. Solids*, 2007, **353**, 2020.
17. Harder, U., Reich, P. & Willfahrt, M. Infrared reflection measurements of lead borate glasses. *J. Molec. Struct.*, **349**, 297–300.
18. Meera, B. N. & Ramakrishna, J. Raman study of borate glasses. *J. Non-Cryst. Solids*, 1993, **159**, 1.
19. Osipov, A. A. & Osipova, L. M., Structure of lithium borate glasses and melts: investigation by high temperature Raman spectroscopy. *Phys. Chem. Glasses: Eur. J. Glass Sci. Technol. B*, 2009, **50** (6) 343–54.
20. Filho, A. G., Souza, Guedes, I., Freire, P. T. C., Filho, J. Mendes, Custódio, M. C. C., Lebullenger, R. M. & Hernandes, A. C. Raman spectroscopy study of high B_2O_3 content lead fluoroborate glasses. *J. Raman Spectrosc.*, 1999, **30**, 525–9.
21. Stefani, R., Maia, A. D., Teotonio, E. E. S.,Monteiro, M. A. F.,Felinto, M.

C. F. C., Brito, H. F., Photoluminescent behavior of SrB_4O_7:RE^{2+} (RE=Sm and Eu) prepared by Pechini, combustion and ceramic methods. *J. Solid State Chem.*, 2006, **179**, 1086–92.

22. Lavat, A., Graselli, C., Santiago, M., Pomarico, J. & Caselli, E. Influence of the preparation route on the optical properties of dosimetric phosphors based on rare-earth doped polycrystalline strontium borates. *Crystal Res. Technol.*, 2004, **39** (10), 840–8.

23. Corker, D. L., Glazer, A. M. Structure and optical non-linearity of $PbO.2B_2O_3$. *Acta Crystall. B*, 1996, **52**, 260.

24. Wang, H., Wang, Y., Cao, X., Zhang, L., Feng, M. & Lan, G. Electronic density of states and optical properties of PbB_4O_7. *Phys. Status Solidi B*, 2009, **246**, 437.

25. Weir, C. E. & Schroeder, R. A. Infrared spectra of the crystalline inorganic borates, *J. Res. NBS, A: Phys. Chem.*, 1964, (5), 465–87.

26. Konijnendijk, W. L. & Verweij, H. Structural aspects of vitreous $PbO.2B_2O_3$ studied by Raman scattering. *J. Am. Ceram. Soc.*, 1976, **59**,

459.

27. Blair, S., Stentz, D., Feller, H., Leelasager, R., Goater, C., Feller, S. & Affatigato, M. A comparison of crystalline and vitreous lead borates using laser photoionisation mass spectroscopy. *J. Non-Cryst. Solids*, 2001, **293–295**, 490–5.

28. Bos, A. J. J. Theory of thermoluminescence. *Radiat. Meas.*, 2006, **41**, S45.

29. Rodriguez Chialanza, M., Castiglioni, J. & Fornaro, L. Crystallisation as a way for inducing thermoluminescence in a lead borate glass. *J. Mater. Sci.*, 2012, **47**, 2339.

30. McKeever, S. W. S. *Thermoluminescence of Solids.* Cambridge University Press, 1997, pp. 187–98.

31. Sharma, G., Thind, K. S., Monika, Singh, H., Manupriya & Gerward, L. Optical properties of heavy metal oxide glasses before and after γ-irradiation. *Phys. Status, Solidi A*, 2007, **204** (2), 591–601. DOI 10.1002/pssa.200622124

Phys. Chem. Glasses: Eur. J. Glass Sci. Technol. B, December 2013, **54** (6), 247–253

Increase of light extraction from Bi₂O₃–B₂O₃ based glass phosphor doped with Yb³⁺ and Nd³⁺

Shunichi Kobayashi, Shingo Fuchi,* Koji Oshima & Yoshikazu Takeda

Department of Crystalline Materials Science, Graduate School of Engineering, Nagoya University, Furo-cho, Chikusa-ku, Nagoya 464-8603, Japan

Manuscript received 12 March 2012
Revised version received 14 July 2013
Accepted 15 November 2013

We investigated the dependence of detected light power on the sidewall angle of Bi₂O₃–B₂O₃ based glass phosphors doped with Yb³⁺ and Nd³⁺. The detected light power of the light source with a sidewall angle of 30° was about 1·8 times higher than for an angle of 0°. Calculations show that the light extraction efficiency (LEE) increases with increasing sidewall angle when the detection area is infinite. However, in the case of restricted detection area, the dependence of the LEE on the sidewall angle is not the same. The LEE decreases when the sidewall angle increases above 30°. From these results, it is shown that the sidewall angle is the key factor to achieve the highest detected light power.

1. Introduction

Wideband light sources are used in the fibre gyroscope[1,2] and in optical coherence tomography (OCT), a novel cross-sectional imaging technique for biological tissues.[3] Since OCT is based on the Michelson interferometer, a low coherence, wideband light source is desirable for a high depth resolution. The depth resolution, Δz, which is equal to the coherence length in the Michelson interferometer, is calculated from the equation:

$$\Delta z = \frac{2\ln 2}{\pi} \frac{\lambda_c^2}{\Delta \lambda} \qquad (1)$$

where λ_c is the central wavelength and $\Delta \lambda$ is the full width at half maximum (FWHM) of a Gaussian shaped spectrum. It is obvious from this equation that a shorter central wavelength results in a higher depth resolution. However, the central wavelength is limited to the near infrared (NIR) region in present applications, leading to penetration depths in biological tissues larger than that at other wavelengths.[3] The water absorption profile has a local minimum at ~1060 nm. In addition to this, a recent study has shown that imaging biological tissue at a central wavelength around 1000 nm yields minimal degradation of the OCT axial resolution, which results from the water dispersion.[4] Therefore, the central wavelength λ_c should be ~1000 nm. Furthermore, in order to achieve a higher depth resolution, a wider spectral width is required, as is clear from the above equation.

Super luminescent diodes (SLDs) and light emitting diodes (LEDs) in the NIR region are usually used in OCT. The spectral width of SLDs and LEDs is approximately 50 nm, corresponding to a coherence length around 10 μm. These light sources are simple and not expensive. However, this coherence length of around 10 μm is insufficient for some medical applications. Recently, using a super continuum laser, a coherence length less than 10 μm has been achieved.[3] However, this light source is complex, quite expensive, and difficult to use in practical applications in a hospital setting. Zhang *et al* have reported a method for the synthesis of several LEDs to shorten the coherence length.[5] However, synthesis of several LEDs and control of the light intensity of individual LEDs are complicated for practical uses.

Therefore, we propose a new type of light source for OCT by combining an LED and a NIR light emitting phosphor in one package.[6,7] Since this light source is similar to a white LED that is composed of a blue LED and a yellow phosphor, it is simple and useful for practical applications. It has been reported that a high refractive index glass is effective for broadening a luminescence spectrum.[8] Therefore, in this study, from among various high refractive index glasses, we chose a Bi₂O₃–B₂O₃ based glass, because it is of great interest for optical device fabrication due to its low melting temperature, and because of its wide composition range for glass formation. It is well known that the luminescence of Yb³⁺ and Nd³⁺ is located near 1000 nm. Thus, Yb³⁺ and Nd³⁺ were used for the luminescence centre in this study. The detailed preparation conditions of the Yb³⁺, Nd³⁺ co-doped Bi₂O₃–B₂O₃ glasses are given in Ref. 7. We did not observe degradation of the Yb³⁺, Nd³⁺ co-doped Bi₂O₃–B₂O₃ glasses in air. The glass transition temperature was 453°C and the crystallisation temperature was 500–525°C.

We optimised the composition and thickness of the glass phosphor by considering the output power and

* Corresponding author. Email fuchi@nagoya-u.jp
Original version presented at VII Int. Conf. on Borate Glasses, Crystals and Melts, Dalhousie University, Halifax, Nova Scotia, Canada, 21–25 August 2011

the FWHM. To estimate the output power, a simple model including absorption of excitation light, absorption of the luminescent light, and luminescence efficiency was used.[9] The optimised glass phosphor has nominal molar composition $1.0Yb_2O_3.4\cdot0Nd_2O_3.47\cdot0Bi_2O_3.47\cdot0B_2O_3.1\cdot0Sb_2O_3$ with a thickness of 3 mm. The excitation bands of this glass phosphor are at ~584, ~748, and ~800 nm. We achieved a central wavelength of 1020 nm, a FWHM of 98 nm, and an output power of over 1 mW by using a high power amber LED.[7] However, this output power is smaller than that of a conventional OCT light source. Therefore, we need to increase the output power of the light source. The refractive index of this glass phosphor is around 2. A higher refractive index should cause a lower light extraction efficiency (LEE), due to the total internal reflection at the surface of the glass phosphor. Since glasses easily form many types of shape, we tried to change the incident angle at the surface of the glass phosphor by changing the sidewall angle of the glass phosphor. The major goal of this study is to achieve an output power of 1 mW at the end surface of the light guide combined with the proposed light source, applicable to instruments in practical use.

2. Experimental

We synthesised the glass phosphor by a melt quenching method. We mixed the powders of Yb_2O_3, Nd_2O_3, Bi_2O_3, H_3BO_3, and Sb_2O_3 in a nominal molar composition of $1.0Yb_2O_3.4\cdot0Nd_2O_3.47\cdot0Bi_2O_3.47\cdot0B_2O_3.1\cdot0Sb_2O_3$. We used Sb_2O_3 to suppress reduction of Bi^{3+}. We melted the mixed powders in an Al_2O_3 crucible at 1250°C in an electric furnace. After 10 min, we poured the molten liquid into stainless steel mould plates kept at room temperature for formation of the glass phosphor. The sidewall angles of the stainless steel mould plates were set at 0°, 15°, 30°, and 45°. The volume, composition, and thickness of the glass phosphor were the same in all samples. The distance between the glass phosphor and the detector was 5 mm. The diameter of the detector was 10 mm. The proposed light source consists of a high power amber LED, an aluminium block with a hole, the glass phos-

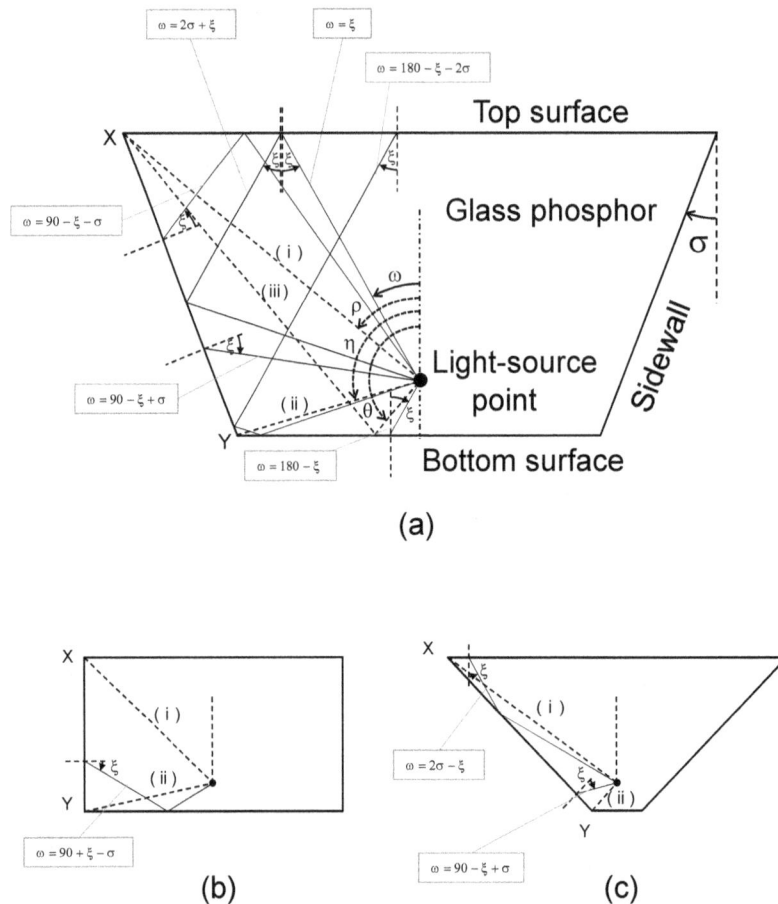

Figure 1. The geometry for the model calculation, showing the light paths at critical angles. ω is the angle between the line from the light source point and the vertical line (·····). ξ is the critical angle of the glass phosphor (refractive index n), i.e. $\sin^{-1}(1/n)=\xi$. ρ is the angle of the line (i) between the light source point and the corner X. η is the angle of the line (ii) between the light source point and the corner Y. θ is the angle of the line (iii) between the light source point and the point at the bottom surface where the light is totally reflected and hits the corner X. Finally, σ is the angle of the sidewall from the vertical line. (a) σ is between 0° and 45°, (b) $\sigma=0°$, and (c) $\sigma=45°$. ρ, η, θ, and σ are geometrically determined due to the shape of the phosphor and the position of the light source point [colour available online]

phor, and a visible cut filter.[7] The high power amber LED was the excitation light source. The aluminium block was set on the amber LED. The glass phosphor was set in the hole. The visible light cut filter was set on the aluminium block to cut the excitation light. However, in this study, no mirrors at the bottom or surrounding the sidewall were used so as to estimate the LEE correctly.

3. Model calculation

We calculated the LEE of the NIR light for each sidewall angle. The size of the glass phosphor and the distance from the glass phosphor to the detector were the same as the experimental setting. Many light source points are needed to estimate the LEE correctly. However, increasing the number of light source points increases the complexity of the model calculation. Therefore, we arranged 55 light source points in the glass phosphor. We assumed that the light source points radiated all angles with the same intensity. We ignored light that reflected more than two times in most cases.

Figure 1(a) shows the light paths at critical angles of the phosphor glass with a sidewall angle of σ (between 0° and 45°), Figure 1(b) is for $\sigma=0°$, and Figure 1(c) is for $\sigma=45°$, the maximum sidewall angle considered. Only the left half of the phosphor glass is shown for simplicity, and the vertical dash-dotted line is taken as $\omega=0°$. Several critical lines are shown in the figure. Solid lines are those at critical angle ξ at the top surface, the bottom surface, and the sidewall surface. The angles for those lines are shown in squares. Broken lines (i), (ii), and (iii) are light paths that go through the points X and Y. The angles are defined as ρ, ξ, and θ for lines (i), (ii), and (iii), respectively. Detailed geometrical explanations of the light paths are given in the Appendix.

5. Results

5. 1 Experimental results

Figure 2 shows the experimentally detected light power for light sources with sidewall angles of 0, 15, 30, and 45°. In the sidewall angle range 0°–30°, the detected light power increases with increasing sidewall angle. However, beyond a sidewall angle of 30°, the detected light power decreases with increasing sidewall angle. The highest detected light power is achieved at a sidewall angle of 30°. The detected light power with a sidewall angle of 30° is about 1·8 times higher than that for 0°. Thus we successfully increased the detected light power by changing the sidewall angle. By using the glass phosphor with a sidewall angle of 30°, we achieved an output power of 1 mW at the end surface of the light guide combined with the proposed light source. Since 1 mW is one of the milestones for practical uses, we are

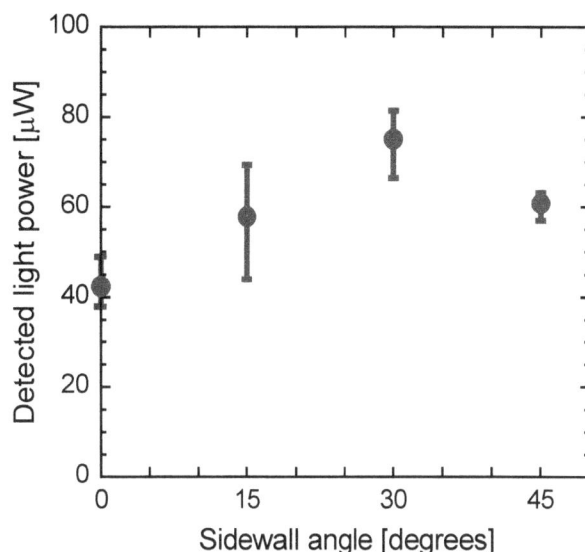

Figure 2. Experimental dependence of the detected light power on the sidewall angle [colour available online]

now trying to apply this light source for medical instruments.

5. 2 Calculated results

Figure 3 shows the calculated LEE. When the detection area is infinite, the LEE increases with increasing sidewall angle. However, for a detection area of 10 mm diameter (this is the same as the experimental setting), the dependence of the LEE on the sidewall angle is not the same as for infinite detection area. For the sidewall angle range 0°–30°, the LEE increases with increasing sidewall angle. In this sidewall angle range, the ratio of the LEE with infinite detection area to that with 10 mm (in diameter) detection area is almost constant. However, beyond a sidewall angle of 30°, the LEE decreases with increasing sidewall

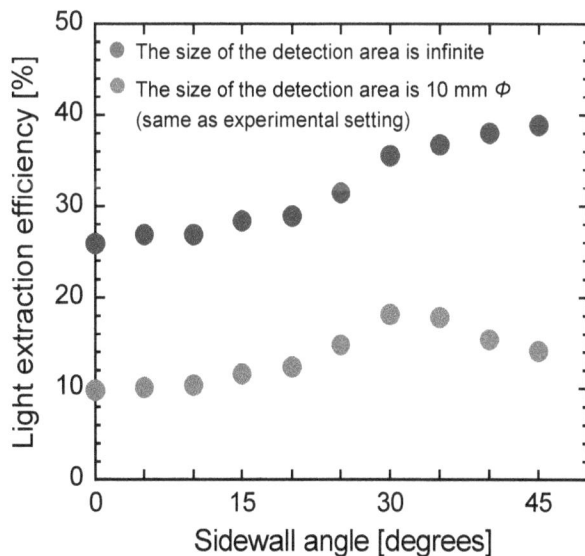

Figure 3. Dependence of the calculated light extraction efficiency (LEE) on the sidewall angle [colour available online]

angle; at a sidewall angle of 30°, the LEE reaches its maximum. The LEE with a sidewall angle of 30° is about 1·8 times higher than that for an angle of 0°. The dependence of the calculated LEE on the sidewall angle coincides with that of the detected light power. In the LEE calculation, we only changed the sidewall angle. Therefore, the increase of the detected light power should be due to the increase of the LEE. The sidewall angle of 30° is the best angle in $1\cdot0Yb_2O_3.4\cdot0Nd_2O_3.47\cdot0Bi_2O_3.47\cdot0B_2O_3.1\cdot0Sb_2O_3$ glass phosphor.

4. Discussion

In Figure 4, several typical light paths are shown. Line A, Line C, and Line F are extracted from the top surface (red line). Line B, Line D, and Line E are extracted from the sidewall (blue line). Line G is extracted from the bottom surface (green line). Line D and E are not detected by the 10 mm diameter detector, but they are detected by the infinite size detector if the angle of the arrow is upward from the parallel to the bottom surface.

Figure 5 shows the schematic drawing of the light extraction in the case of a sidewall angle of 0° (a) and 30° (b) with the same angle division (5°) of w. With a sidewall angle of 0°, ρ is around 45° and η is around 100°. Then, the NIR light is extracted as the Line A in Figure 4, in the angle range 0°–30°. Light paths in this range are drawn with red lines. In the other angle range, the NIR light is not extracted from the top surface, as shown by blue, green, and black lines. Black lines show the light that is not extracted from the glass phosphor. In the angle ranges of 65°–85° and

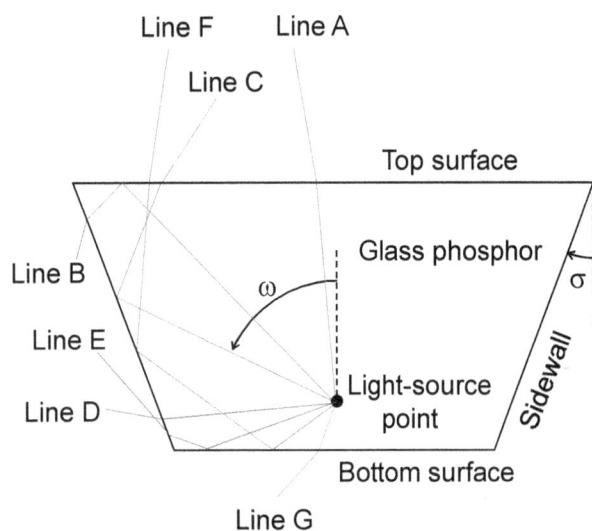

Glass phosphor

(a)   (b)

Figure 5. Schematic drawing of the extraction of the NIR light for sidewall angles of (a) 0°, and (b) 30° [colour available online]

110°–120°, the NIR light is detected by the infinite size detector, since the angle of the arrow is upward from the bottom surface.

In the case of a sidewall angle of 30°, ρ is around 50° and η is around 110°. Then, the NIR light is extracted as the Line A in Figure 4, in the angle range 0°–30°. In the angle range 55°–85°, the NIR light is extracted as the Line C in Figure 4. In the angle range 115°–135°, the NIR light is extracted as the Line F in Figure 4. All these light paths are drawn by red lines. In the other angle ranges, the NIR light is not extracted from the top surface, as shown by blue, green, and black lines. In the angle range of 90°–100°, light is detected by the infinite size detector, since the angle of the arrow is upward from the bottom surface.

It is shown in Figure 5 that the light paths that are extracted from the top surface are much more dense for a sidewall angle of 30° (Figure 5(b)).

Figure 6 shows a schematic drawing of the detection of the NIR light with a finite size detector. In the case of a sidewall angle of 30° (Fig. 6(a)), the NIR light is extracted from the top surface in angle ranges of 0°–30°, 55°–85°, and 115°–135° (solid red lines) and detected by the finite size detector. However, in the angle ranges of 20°–30°, 85°, and 135°, the NIR light

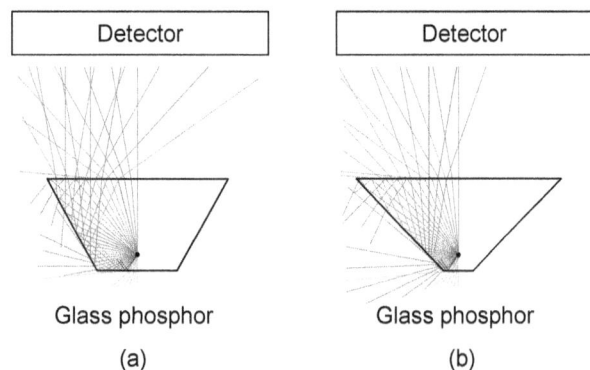

Line F   Line A

Line C

Top surface

Line B

Glass phosphor

ω

Line E

Light-source point

Line D

Bottom surface

Line G

Figure 4. Schematic drawing of the NIR light paths. Red lines A, C, and F are paths that go out through the top surface, and blue line B and green line G are paths that go out through the sidewall or bottom surface. Blue lines D and E are those lights that are not detected by the 10 mm diameter detector, but they are detected by the infinite size detector [colour available online]

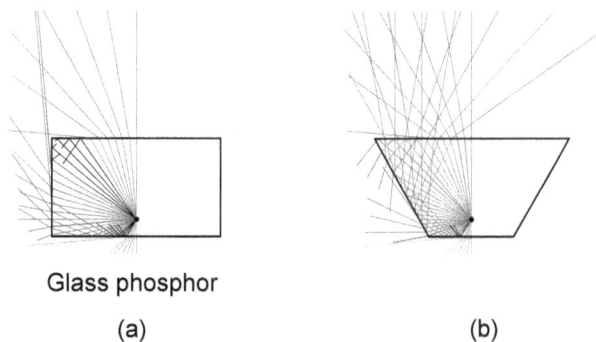

Detector   Detector

Glass phosphor   Glass phosphor

(a)   (b)

Figure 6. Schematic drawing of the detection of the NIR light for sidewall angles of (a) 30°, and (b) 45°. Red solid and dashed lines show the NIR light extracted from the top surface. Only red solid lines are detected by the 10 mm diameter detector [colour available online]

is not detected (dashed red lines). In the other angle ranges, NIR light is not detected (dashed blue, green, and black lines). On the other hand, in the case of a sidewall angle of 45° (Figure 6(b)), the NIR light is extracted from the top surface in angle ranges of 0°–30° and 60°–100° (solid red lines) and detected by the finite size detector. However, in the angle ranges of 20°–30° and 60°–75°, the NIR light is not detected (dashed red lines). In the other angle ranges, NIR light is not detected (dashed blue, green, and black lines). As was shown in Figure 3, the NIR light that is extracted from the top surface and is detected decreases with increasing sidewall angle.

From the above discussions, it is obvious that the extracted NIR light detection depends on the sidewall angle and the detection area. Therefore, the sidewall angle has an optimum angle when the detection area is limited.

5. Conclusion

We have investigated the dependence of the detected light power on the sidewall angle for a glass phosphor. We achieved the highest detected light power at a sidewall angle of 30° in $1\cdot0Yb_2O_3.4\cdot0Nd_2O_3.47\cdot0Bi_2O_3.47\cdot0B_2O_3.1\cdot0Sb_2O_3$ glass phosphor. The calculated LEE (light extraction efficiency) increased with increasing sidewall angle in the case of infinite

Appendix

When the light from the source-point has the angle w between 0° and the blue line in Figure A1 then the light (Line A in Figure 4) is extracted from the top surface, i.e. light is extracted if the angle ω satisfies the following inequality

$$0° \leq \omega < \xi \tag{A1}$$

When the light from the source-point has a value of w that satisfies inequalities (A2a) and (A2b) at the same time, the light (Line B in Figure 4) goes out through the left sidewall and is never extracted from the top surface

$$\xi \leq \omega < \varrho \tag{A2a}$$

$$90° - \xi - \sigma \leq \omega < \varrho \tag{A2b}$$

When ω is at (90°−ξ−σ), the light satisfies the critical angle ξ at the sidewall surface. The ranges for inequalities (A2a) and (A2b) are shown in Figure A2.

If ω satisfies the following inequality at the same time, the light (Line C in Figure 4) is extracted from the top surface after reflection at the sidewall, and is then detected

$$\varrho \leq \omega < 2\sigma + \xi \tag{A3a}$$

$$2\sigma - \xi < \omega \leq 90° - \xi + \sigma \tag{A3b}$$

detection area. On the other hand, the calculated LEE decreased with increasing sidewall angle above 30° in the case of restricted detection area. From these results, it was shown that the light extraction and the detection of the extracted light depend on the sidewall angle of the glass phosphor.

Acknowledgements

This work was supported in part by the Technology Development Program for Advanced Measurement and Analysis of the Japan Science and Technology Agency (JST).

References

1. Cutler, C. C., Newton, S. A. & Shaw, H. J. *Opt. Lett.*, 1980, **5**, 488.
2. Blin, S., Kim, H. K., Digonnet, M. J. F. & Kino, G. S. *J. Lightwave Technol.*, 2007, **25**, 861.
3. Fercher, A. F., Drexler, W., Hitzenberger, C. K. & Lasser, T. *Rep. Progr. Phys.*, 2003, **66**, 239.
4. Wang, Y., Nelson, J., Chen, Z., Reiser, B., Chuck, R. & Windeler, R. *Opt. Express*, 2003, **11**, 1411.
5. Zhang, Y., Sato, M. and Tanno, N. *Opt. Lett.*, 2001, **26**, 205
6. Fuchi, S., Sakano, A. & Takeda, Y. *Jpn. J. Appl. Phys.*, 2008, **47**, 7932.
7. Fuchi, S., Sakano, A., Mizutani, R. & Takeda, Y. *Appl. Phys. Express*, 2009, **2**, 32102.
8. Tanabe, S., Sugimoto, N., Ito, S. & Hanada, T. *J. Lumin.*, 2000, **87-89**, 670.
9. Fuchi, S., Sakano, A., Mizutani, R. & Takeda, Y. *Appl. Phys. B*, 2011, **105**, 877.

Inequality (A3a) is for a phosphor with a smaller value of s (Figure A3(a)), and inequality (A3b) is for a larger value of s (Figure A3(b)). When ω equals r, the light goes through the point X (see Figure 1(a)).

If ω satisfies the following inequality, the light path is shown by Line D (in Figure 4), and light is extracted from the sidewall directly.

$$90° - \xi + \sigma < \omega \leq \eta \tag{A4}$$

At $w = 90° - \xi + \sigma$, the light is totally reflected at the sidewall, and it goes along the sidewall. At $w = \eta$, the light goes out through the point Y (Figure 1(a)), as shown in Figure A4(a).

When s is very small, there is a light path, such as Line E in Figure 5, that is extracted from the sidewall after reflection at the bottom surface. This is when ω satisfies the following inequality;

$$\eta \leq \omega \leq 90° + \xi - \sigma \tag{A5}$$

At $\omega = 90° + \xi - \sigma$, the light is totally reflected at the sidewall, and it goes along the sidewall. The difference from Figure A4 is that the light is reflected once at the bottom surface.

Line F in Figure A6(a) and A6(b) shows the light that is extracted from the top surface after reflection at the bottom surface and sidewall. Light is reflected more than two times only in this case. If ω satisfies the following inequality at the same time, the light goes out from the top surface shown by Line F in Figure A6(a) and A6(b);

$$180°-\xi-2\sigma<\omega\leq q \qquad \text{(A6a)}$$

$$90°+\xi-\sigma\leq\omega<180°-\xi \text{ or } q \qquad \text{(A6b)}$$

Inequality (6a) is for a general case and (6b) is for a very small sidewall angle. At $\omega=q$, the light goes along

the broken line (iii). At $\omega=180°-\xi$, the light is totally reflected and goes along the bottom surface and goes out through the point Y (Figure 1(a)).

Line G (Figure A7) shows the light that is extracted from the bottom surface directly. This is when ω satisfies the following inequality.

$$180°-\xi\leq\omega\leq180° \qquad \text{(A7)}$$

At $\omega=180°-\xi$, the light is at the critical angle, and it goes out through the point Y (Figure 1(a)).

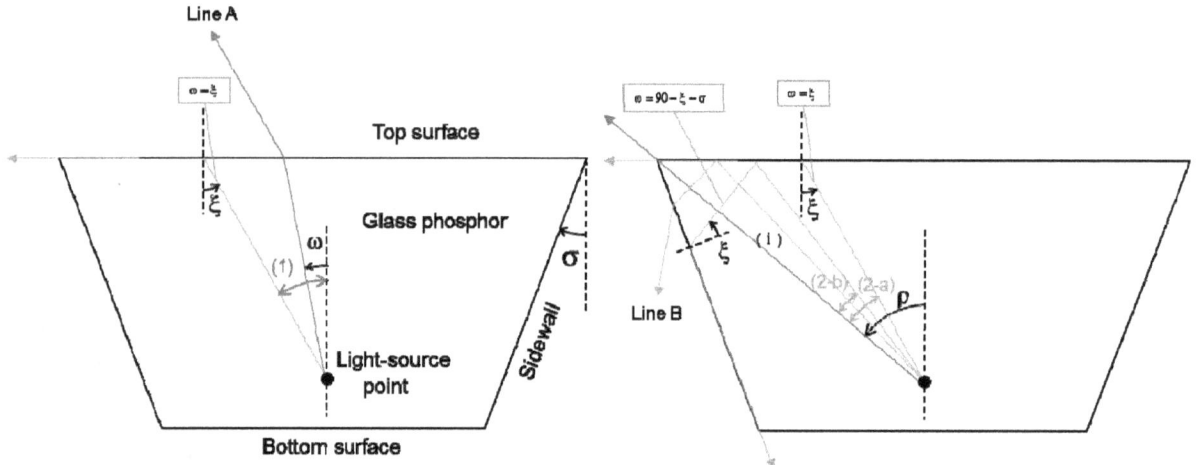

Figure A1　　　　　　　　　　　　Figure A2

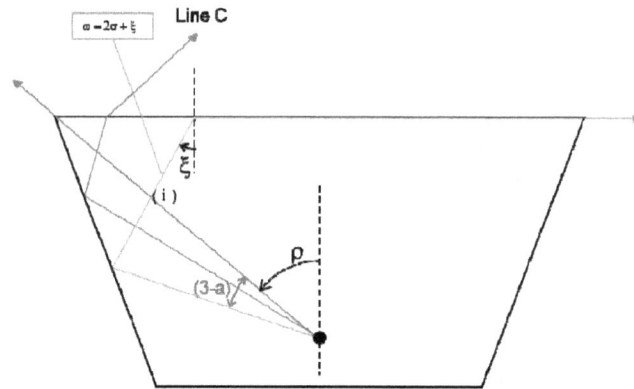

Figure A3(a)

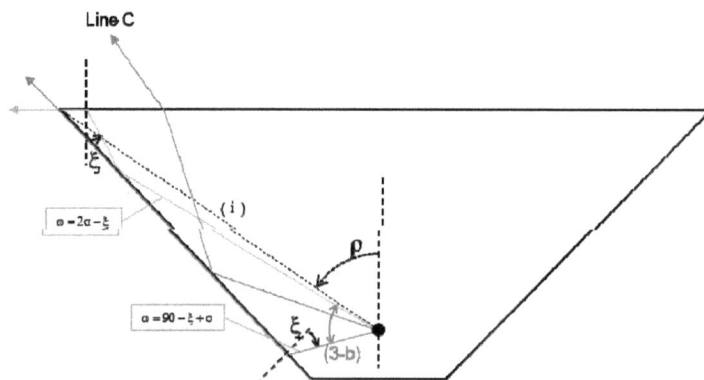

Figure A3(b)

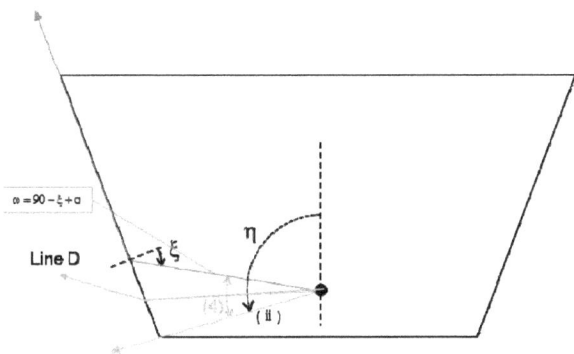

Figure A4

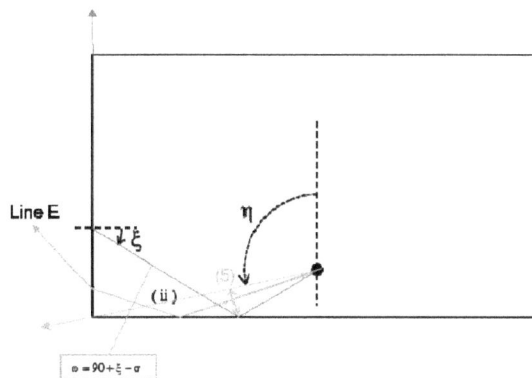

Figure A5

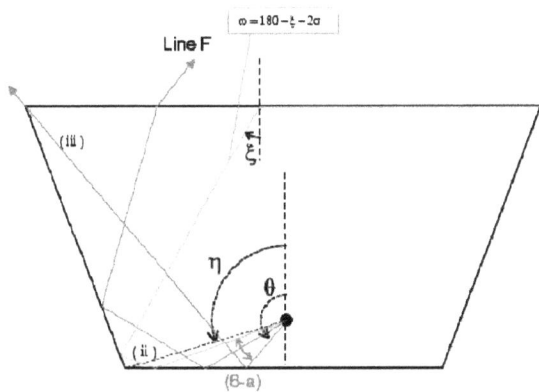

Figure A6(a)

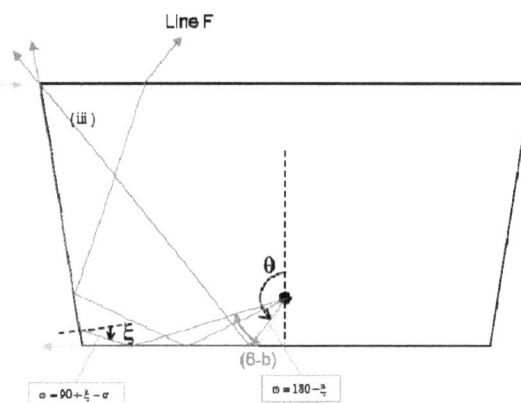

Figure A6(b)

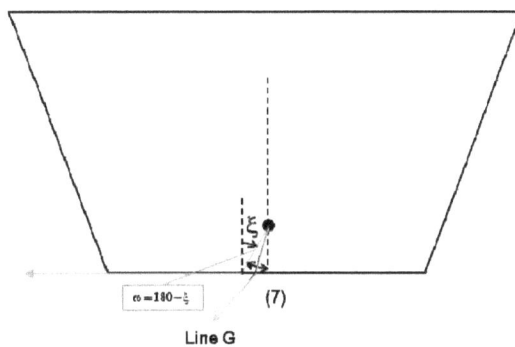

Figure A7

Phys. Chem. Glasses: Eur. J. Glass Sci. Technol. B, February 2014, 55 (1), 1–12

Velocity of sound in and elastic properties of alkali metal borate glasses

Masao Kodama* & Seiji Kojima

Graduate School of Pure and Applied Sciences, University of Tsukuba, Tsukuba, Ibaraki 305-8573, Japan

Manuscript received 3 October 2011
Revised version received 14 May 2013
Accepted 13 January 2014

For each of the five binary alkali metal borate glass systems, the mass density and the velocities of longitudinal and transverse ultrasonic waves have been measured at 298 K as a function of composition over their respective bulk glass formation ranges. On the basis of the mass density and the velocity of sound, the following properties have been studied, and are found to have a composition-dependence which is either M- or W-shaped: (a) the elastic constants (longitudinal modulus, shear modulus, and bulk modulus); (b) the number density of alkali metal, boron, and oxygen atoms; (c) the volume of glass containing one mole of boron or oxygen atoms; (d) the pressure derivative of the molar volume of glass; (e) Poisson's ratio. Properties (a), (b), (c), and (d) are analysed in terms of borate structural units to conclude that the M- or the W-shape is caused by the formation of an elastically soft borate structural unit with one nonbridging oxygen atom near the middle of each binary system. Property (e) is analysed in terms of glass network connectivity; the W-shaped behaviour can be explained adequately if the presence of pentaborate groups near the middle of the glass forming range of each binary system is assumed.

1. Introduction

Mass density, velocity of sound, and hence elastic constants, are essential properties for characterising a glass. We have reported the mass density and the velocities of longitudinal and transverse ultrasonic waves for the following binary glass systems: lithium borates (LiB),[1] sodium borates (NaB),[2] potassium borates (KB),[3] rubidium borates (RbB),[4] and caesium borates (CsB),[5] where their respective abbreviated notations are shown in parentheses. The composition ranges studied are the respective bulk glass formation ranges.

We show that the composition dependence of many different physical quantities in each binary alkali metal borate glass system exhibits an M-shape (a succession of a convex curve, a concave curve, and another convex curve; typically, a maximum, a minimum, and another maximum in succession), or inversely a W-shape (a succession of a concave curve, a convex curve, and another concave curve; typically, a minimum, a maximum, and another minimum in succession). The M-shaped physical quantities in this study are the mass density, the velocity of sound, the elastic constants (longitudinal modulus, shear modulus, and bulk modulus), and the number density of alkali metal atoms, boron atoms, or oxygen atoms. The W-shaped physical quantities in this study are the molar volume, the pressure derivative of the molar volume, and Poisson's ratio. This paper considers

all of the binary systems as a whole, and elucidates the cause of the M- or the W-shape in terms of (a) basic borate structural units, and (b) glass network connectivity. The so-called borate anomaly,[6] in the form of a single minimum in the thermal expansivity, has been analysed previously[7,8] in terms of vibrational anharmonicity, and so the present paper does not consider the anharmonic vibrational properties. The present paper follows a review paper by Wright[9] with respect to (a) the notation of (super)structural units, (b) the chemical equilibrium between borate structural units, and (c) the average number density of the constituent atom as a physical quantity relating to glass structure.

2. Number density, molar volume, and pressure derivative of molar volume

In order to gain insights into the structure and properties of alkali metal borate glasses, we derive expressions for the number density, the molar volume, and the pressure derivative of the molar volume. Alkali metal borate glasses can be represented by the chemical formula

$$x\mathrm{M_2O}.(1-x)\mathrm{B_2O_3} = (1-x)(R\mathrm{M_2O}.\mathrm{B_2O_3})$$
$$= 2(1-x)(R\mathrm{MO_{1/2}}.\mathrm{BO_{3/2}}) \quad (1)$$

where M stands for Li, Na, K, Rb, or Cs, and $R = x/(1-x)$. Then the glass composition can be denoted interchangeably by $x\mathrm{M_2O}.(1-x)\mathrm{B_2O_3}$, $R\mathrm{M_2O}.\mathrm{B_2O_3}$, or $R\mathrm{MO_{1/2}}.\mathrm{BO_{3/2}}$.

The number density, n, of a constituent atom is

* Corresponding author. Email masaokodama@arion.ocn.ne.jp
Original version presented at VII Int. Conf. on Borate Glasses, Crystals and Melts, Dalhousie University, Halifax, Nova Scotia, Canada, 21–25 August 2011

defined by

$$n = \frac{\rho N_A}{M} \tag{2}$$

where ρ is the mass density, M is the mass of one mole of the unit which contains one constituent atom, and N_A is Avogadro's number. Let $M[xM_2O.(1-x)B_2O_3]$ be the molar mass of the entity $xM_2O.(1-x)B_2O_3$. The number density, n_M, of an alkali metal atom, M, can be calculated by the following equation:

$$n_M = \frac{2\rho x N_A}{M[xM_2O \cdot (1-x)B_2O_3]} \tag{3}$$

The quantity n_M represents the number of alkali metal atoms contained in unit volume of glass; this quantity will be discussed in section 5.5.1.

The number density, n_B, of boron atoms can be calculated by the following equation:

$$n_B = \frac{2\rho(1-x)N_A}{M[xM_2O \cdot (1-x)B_2O_3]} \tag{4}$$

The quantity n_B represents the number of boron atoms contained in unit volume of glass; this quantity is of importance in analysing glass properties[9] and will be discussed in section 5.5.2.

The number density, n_O, of oxygen atoms can be calculated by the following equation:

$$n_O = \frac{\rho(3-2x)N_A}{M[xM_2O \cdot (1-x)B_2O_3]} \tag{5}$$

The quantity n_O represents the number of oxygen atoms contained in unit volume of glass; this quantity is also of importance in analysing glass properties[9] and will be discussed in section 5.5.3.

By taking the inverse of the number density, we obtain the expression for the molar volume of a constituent atom. Let $M(RMO_{1/2}.BO_{3/2})$ and $M(RMO_{1/2}.BO_{3/2})/\rho = V_m(RMO_{1/2}.BO_{3/2})$ be the molar mass and the molar volume of the entity $RMO_{1/2}.BO_{3/2}$, respectively. Expressing n_B in terms of $V_m(RMO_{1/2}.BO_{3/2})$, we have

$$n_B = \frac{2\rho(1-x)N_A}{M[xM_2O \cdot (1-x)B_2O_3]} \tag{6}$$

$$= \frac{\rho N_A}{M(RMO_{1/2} \cdot BO_{3/2})}$$

$$= \frac{N_A}{V_m(RMO_{1/2} \cdot BO_{3/2})}$$

so that

$$\frac{N_A}{n_B} = \frac{M[xM_2O \cdot (1-x)B_2O_3]}{2\rho(1-x)} \tag{7}$$

$$= \frac{M(RMO_{1/2} \cdot BO_{3/2})}{\rho}$$

$$= V_m(RMO_{1/2} \cdot BO_{3/2})$$

In the above equation, the quantity, $V_m(RMO_{1/2}.BO_{3/2})$, represents the volume of glass containing one mole

of boron atoms; this quantity will be discussed in section 5.6.1.

In the same way, the quantity, N_A/n_O, can be expressed as

$$\frac{N_A}{n_O} = \frac{M[xM_2O \cdot (1-x)B_2O_3]}{\rho(3-2x)} \tag{8}$$

$$= \frac{2}{R+3}V_m[RMO_{1/2}-BO_{3/2}]$$

$$= V_m\left(\frac{2R}{R+3}M + \frac{2}{R+3}B + O\right)$$

$$\equiv V_m(\text{oxygen})$$

In the above equation, the quantity, $V_m(\text{oxygen})$, represents the volume of glass containing one mole of oxygen atoms; this quantity will be discussed in section 5.6.2.

We consider the derivative of the molar volume with respect to pressure. The compressibility, κ, is defined as

$$\kappa = -\frac{1}{V}\frac{dV}{dP} \tag{9}$$

where V and P are volume and pressure, respectively. Substituting $V_m(RMO_{1/2}.BO_{3/2})$ for V leads to the derivative of $V_m(RMO_{1/2}.BO_{3/2})$ with respect to pressure

$$-\frac{dV_m(RMO_{1/2} \cdot BO_{3/2})}{dP} = \kappa V_m(RMO_{1/2} \cdot BO_{3/2}) \tag{10}$$

where the quantities κ and $V_m(RMO_{1/2}.BO_{3/2})$ on the right-hand side of this equation can be determined experimentally. The quantity on the left-hand side of this equation represents the degree to which the molar volume, $V_m(RMO_{1/2}.BO_{3/2})$, (i.e. the volume of glass which contains one mole of boron atoms) is compressed elastically by applying pressure. The quantity $\kappa V_m(RMO_{1/2}.BO_{3/2})$ will be discussed in section 5.7.1.

Substituting $V_m(\text{oxygen})$ for V leads to the derivative of $V_m(\text{oxygen})$ with respect to pressure

$$-\frac{d \ (\text{oxygen})}{dP} = \ (\text{oxygen}) \tag{11}$$

where the quantities κ and $V_m(\text{oxygen})$ on the right-hand side of this equation can be determined experimentally. The quantity on the left-hand side of this equation represents the degree to which the molar volume, $V_m(\text{oxygen})$, (i.e. the volume of glass which contains one mole of oxygen atoms) is compressed elastically by applying pressure. The quantity $\kappa V_m(\text{oxygen})$ will be discussed in section 5.7.2.

3. Experiment

3.1. Glass preparation
Alkali metal borate glasses, $xM_2O.(1-x)B_2O_3$, from five binary systems with M=Li, Na, K, Rb, or Cs, were prepared in their respective bulk glass formation ranges (LiB, $0 \leq x \leq 0.28$; NaB, $0 \leq x \leq 0.36$; KB, $0 \leq x \leq 0.34$;

RbB, $0 \leq x \leq 0.37$; CsB, $0 \leq x \leq 0.39$).[1–5] In order to obtain sufficient transmission of ultrasonic waves, it is necessary to prepare glasses with high homogeneity and without strains or bubbles. With the aim of preparing glasses with high homogeneity, the starting materials were initially made to react in an aqueous solution. The following analytical reagent grade chemicals were used for producing the alkali metal oxides of Li_2O, Na_2O, K_2O, Rb_2O, and Cs_2O: $LiOH.H_2O$, and 50 wt% aqueous solution of NaOH, KOH, RbOH, and CsOH. Each of these chemicals was first diluted, for example to a concentration of 0.5 mol/dm^3, by adding distilled water, and this solution was used for the starting material of glass preparation. After the concentration of the diluted solution had been determined with a potentiometric titration method, the required volume was taken from a burette. As the starting material for B_2O_3, analytical reagent grade boric acid was used. Amounts of the starting materials calculated to give, for example 30 g, in the melts were dissolved by adding distilled water in a beaker made of polytetrafluoroethylene. Then the mixed solution was transferred to a dry box, and after the complete evaporation of the water a chemically synthesized powder was obtained.

The powder was fused in a 20 cm^3 platinum crucible at temperatures from 900 to 1300°C for about 4 h by heating in an SiC resistance electric furnace. The melt was then poured into a cylindrical graphite mould, 15 mm in diameter and 30 mm deep, which had been preheated at 300°C in an electric muffle furnace. Subsequently, the glass cast in the mould was cooled at a rate of 1 K min^{-1} to room temperature. The residual melt was poured onto an aluminum plate, and later used for the determination of chemical composition and for the determination of glass transition temperature by differential thermal analysis (DTA) or by differential scanning calorimeter

(DSC). The chemical compositions of all the prepared glasses were analysed with respect to the contents of M_2O and B_2O_3 with a potentiometric titration method described in a previous paper.[10] Each cast glass was annealed at the glass transition temperature determined by DTA or DSC.[5]

3.2. Mass density

The mass density of each annealed glass was measured at 298 K by a hydrostatic weighing method. In brief, a silicon single crystal was used as a mass density standard[11] for calibrating the mass density of the immersion liquid (carbon tetrachloride or cyclohexane), so that the mass density of each glass could be determined within an accuracy of 1.0×10^{-3} g/cm^3.

3.3. Velocity of sound

By measuring the travel times, τ_l and τ_t, for one round trip of longitudinal and transverse ultrasonic waves in a specimen whose length is denoted by l, we obtained the velocities, c_l and c_t, of longitudinal and transverse ultrasonic waves, according to $c_l = 2l/\tau_l$ and $c_t = 2l/\tau_t$, respectively. In order to measure the ultrasonic velocity, each annealed glass was ground and polished to give a pair of end faces that were flat and parallel. Inspection with a strain viewer showed that all specimens were transparent and free from strains or bubbles. The ultrasonic travel time was measured at a frequency of 10 MHz and at a temperature of 298 K by means of the pulse-echo overlap method.[12] X-cut and Y-cut quartz transducers resonating at a fundamental frequency of 10 MHz were used in order to generate and detect the longitudinal ultrasonic wave and the transverse ultrasonic wave, respectively. The transducer was bonded to the specimen on one of the two parallel faces with phenyl benzoate, and this single transducer was used for both the generation and the detection of ultrasound. Two adjacent echoes were overlapped according to the McSkimin criterion[13] in order that the ultrasonic travel time could be measured within an error of 0.02%. The velocities, c_l and c_t, were determined at 298 K from τ_l and τ_t measured in this way and from the specimen length measured with a micrometer reading to 1 μm. It should be noted that the error is within the point symbol for all experimental plots of the present paper.

4. Results

4.1. Mass density

Figure 1 shows the mass density, ρ, of alkali metal borate glasses against composition, x. The mass density values of LiB, NaB and KB differ only slightly at a given x, and the NaB and KB curves cross at $x=0.10$, as noted by Shartsis et al.[14] The mass densities of NaB, KB, RbB and CsB show an M-shaped curve which

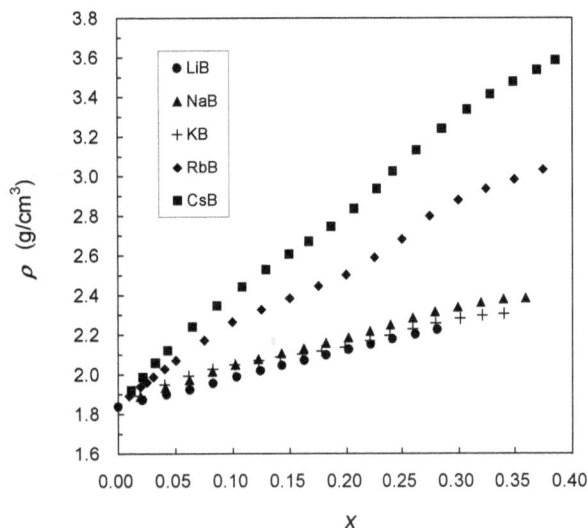

Figure 1. Mass density, ρ, of alkali metal borate glasses against composition, x

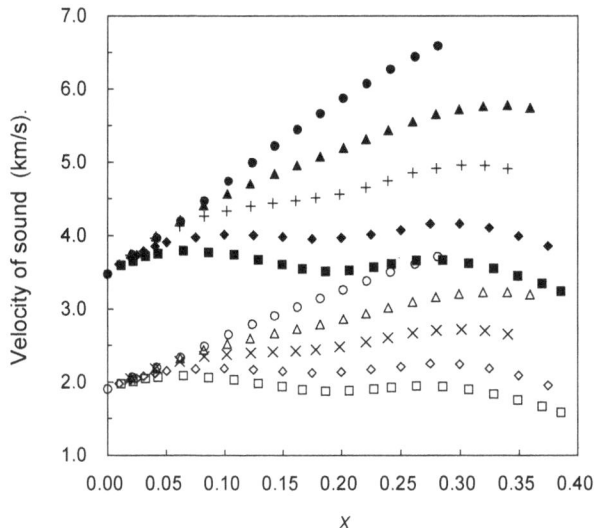

Figure 2. Velocity of sound in alkali metal borate glasses against composition, x
LiB, c_l (●), c_t (O); NaB, c_l (▲), c_t (△); KB, c_l (+), c_t (×); RbB, c_l (◆), c_t (◇); CsB, c_l (■), c_t (□)

is curved downwards with a slight concavity in the composition range $0 \cdot 10 \leq x \leq 0 \cdot 30$. For LiB, however, the mass density increases approximately linearly with x, and no concavity or convexity can be seen.

4.2. Velocity of sound

Figure 2 shows the velocities of both longitudinal and transverse ultrasonic waves in alkali metal borate glasses against composition, x. The upper five curves represent the longitudinal velocities and the lower five curves represent the transverse velocities. In each branch of the longitudinal and the transverse velocities, the velocities in the composition range $0 \leq x < 0 \cdot 05$ increase with x in the same way, irrespective of the alkali metal ion, whilst the velocities above $x = 0 \cdot 05$ depend both upon the composition, x, and upon the kind, M, of alkali metal ion. The longitudinal and transverse velocities both show the following characteristics: for LiB the velocity increases monotonically with increasing x; for NaB and KB the velocity shows an M-shaped curve with a single maximum at around $x = 0 \cdot 30$ to $0 \cdot 35$; for RbB and CsB the velocity shows an M-shaped curve with a maximum at $x \approx 0 \cdot 07$ to $0 \cdot 10$, a minimum at $x \approx 0 \cdot 18$ to $0 \cdot 19$, and another maximum at $x \approx 0 \cdot 28$ to $0 \cdot 29$.

5. Discussion

5.1. Borate (super)structural units
The composition ranges studied in the present paper are the bulk glass formation ranges (see section 3.1 above). We describe borate (super)structural units using the notation given by Wright.[9] The five different basic borate structural units[9] present in alkali metal borate crystals and glasses are $BØ_3$, $BØ_4^-$, $BOØ_2^-$,

$BO_2Ø^{2-}$, and BO_3^{3-}, where the symbol Ø denotes bridging oxygen atoms that are shared between adjacent (super)structural units and the symbol, O, denotes both bridging oxygen atoms that are situated completely within a superstructural unit, and negatively charged nonbridging oxygen atoms that form part of a (super)structural unit. When an alkali metal oxide and boron oxide combine to form an alkali metal borate glass, two types of chemical reactions can be considered:[9]

$$1/2 M_2O + BØ_3 \rightarrow M^+ BØ_4^- \tag{12}$$

$$1/2 M_2O + BØ_3 \rightarrow M^+ BOØ_2^- \tag{13}$$

In the above equations, $M^+ BØ_4^-$ and $M^+ BOØ_2^-$ do not exist alone, but coexist by disproportionation reactions between the structural units:[9]

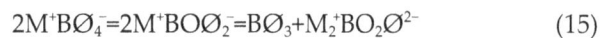

$$M^+ BØ_4^- = M^+ BOØ_2^- \tag{14}$$

$$2 M^+ BØ_4^- = 2 M^+ BOØ_2^- = BØ_3 + M_2^+ BO_2Ø^{2-} \tag{15}$$

and

$$M_4^+ B_2O_5^{4-} = 2 M_2^+ BO_2Ø^{2-} = M^+ BOØ_2^- + M_3^+ BO_3^{3-} \tag{16}$$

In Equations (14), (15), and (16), the equality denotes a chemical equilibrium between the adjacent chemical species. The volumes of the structural units $BØ_3$, $M^+ BØ_4^-$, $M^+ BOØ_2^-$, and $M_2^+ BO_2Ø^{2-}$, which are present in the composition range $0 \leq x \leq 0 \cdot 66$, have been determined by Lower et al.[15] In the composition range $0 \leq x \leq 0 \cdot 50$, the three structural units, $BØ_3$, $M^+ BØ_4^-$, and $M^+ BOØ_2^-$, are assumed to be present in their analysis. The fact that the volume of the $M^+ BØ_4^-$ unit is smaller than that of $M^+ BOØ_2^-$ unit is referred to throughout the discussion of the present paper.[15] The chemical equilibrium expressed by Equation (14), whether the reaction proceeds to the right-hand side or to the left-hand side, depends both on the composition and on the kind of alkali metal ion. The equilibrium proceeds to the right-hand side for a large cation like caesium, while it proceeds to the left-hand side for a small cation like lithium (see Lower et al[15] for a comprehensive table of ionic radii). As a result, the structure and properties of alkali metal borate glasses depend not only upon the composition, x, but also upon the kind of alkali metal ion, M^+. The structural units which are present in alkali metal borate *crystals* in the composition range $0 \leq x \leq 0 \cdot 50$ are confined to the three species $BØ_3$, $BØ_4^-$, and $BOØ_2^-$ (cf. Table 1 of Ref. 9). We may then consider the elasticity of the present alkali metal borate glasses in terms of the three structural units $BØ_3$, $M^+ BØ_4^-$, and $M^+ BOØ_2^-$. Thus, the effects of two structural units, $M_2^+ BO_2Ø^{2-}$ and $M_3^+ BO_3^{3-}$, are not taken into account in the present discussion, since these two structural units are present only in the composition range $x > 0 \cdot 50$.

The structural units further combine to form various superstructural units. Typical superstructural units (using Wright's[9] notation) are boroxol ring, $B_3O_3\emptyset_3$; pentaborate group, $B_5O_6\emptyset_4^-$; triborate group, $B_3O_3\emptyset_4^-$; di-pentaborate group, $B_5O_6\emptyset_5^{2-}$; diborate group, $B_4O_5\emptyset_4^{2-}$; cyclic metaborate, $B_3O_6^{3-}$; metaborate chain, $(BO\emptyset_2^-)_\infty$. It should be noted that the $B\emptyset_4^-$ unit contained in the pentaborate group does not contribute to the connectivity with the surrounding borate network; this fact will be used subsequently when discussing Poisson's ratio.

5.2. Brief review of spectroscopic studies

The change in the respective mole fractions (x_3, x_4 and x_2) of $B\emptyset_3$, $M^+B\emptyset_4^-$, and $M^+BO\emptyset_2^-$ against composition has previously been studied by NMR techniques. An early NMR study by Bray & O'Keefe[16] proposed a model of structural change common to all alkali metal borate glasses; that is, on addition of M_2O to B_2O_3, the structural unit $B\emptyset_3$ is converted only into the structural unit $M^+B\emptyset_4^-$ up to $x\approx0.30$ and, with further addition of M_2O, the structural unit $M^+B\emptyset_4^-$ is converted into the structural unit, $M^+BO\emptyset_2^-$. Thus, the findings by Bray & O'Keefe mean that the relation $x_4=x/(1-x)$ holds in the composition range $0\leq x\leq0.30$.

On the other hand, an NMR study by Zhong & Bray[17] of the five binary alkali metal borate glass systems with compositions $x=0.17$, 0.25, 0.33, 0.40 and 0.45 showed experimental evidence that (a) $x_4<x/(1-x)$ always for the composition range above $x/(1-x)=0.20$; (b) x_4 is a function not only of $x/(1-x)$ but also of the size of alkali metal ions; (c) the difference, $x/(1-x)-x_4$, becomes larger as the size of alkali metal ion increases. Since the difference, $x/(1-x)-x_4$, is equal to the mole fraction, x_2, of $M^+BO\emptyset_2^-$, the findings of Zhong & Bray[17] indicate that nonbridging oxygens due to the formation of $M^+BO\emptyset_2^-$ are already formed above the composition $x=0.17$, and the amount of nonbridging oxygens increases as the size of the alkali metal ion increases. Subsequent NMR studies of alkali metal borate glasses have been reviewed by Ratai et al,[18] but the discrepancy between the study by Bray & O'Keefe[16] and the study by Zhong & Bray[17] as to whether or not nonbridging oxygens are formed in the composition range $0<x<0.30$ is not yet solved completely. Moreover, we cannot ascertain from the study by Zhong & Bray[17] whether the structural unit $B\emptyset_3$ is converted only into the structural unit $M^+B\emptyset_4^-$, or into both structural units $M^+B\emptyset_4^-$ and $M^+BO\emptyset_2^-$ when a slight amount of M_2O is added to B_2O_3 at a composition below $x=0.17$.

A Raman spectroscopic study on caesium borate glasses by Kamitsos et al[19,20] has shown, first, that nonbridging oxygens begin to form at a composition less than $x=0.10$ and, second, that the chain-type metaborate group, $(BO\emptyset_2^-)_\infty$, forms for compositions of $x=0.10$, 0.14 and 0.17, while the ring-type metaborate group, $B_3O_6^{3-}$, forms for compositions $x\geq0.30$. Since these metaborate groups are formed from $Cs^+BO\emptyset_2^-$ units, we see that $Cs^+BO\emptyset_2^-$ is already present above the composition $x=0.10$. Subsequently, Meera & Ramakrishna[21] ascertained the presence of nonbridging oxygens in caesium borate glasses; the chain-type metaborate group in the range $0.14\leq x<0.20$ and the ring-type metaborate group in the range $30\leq x\leq50$.

A neutron diffraction study of caesium borate glasses by Shaw et al[22] has shown that nonbridging oxygens are formed in the bulk glass formation range ($x=0.10$, 0.17, 0.29; $x/(1-x)=0.11$, 0.20, 0.40; $x_4=0.06\pm0.04$, 0.14 ± 0.06, 0.15 ± 0.08, respectively).

5.3. Longitudinal modulus, shear modulus and bulk modulus

With the mass density, ρ, and both sound velocities, c_l and c_t, we can calculate elastic constants, such as the longitudinal modulus, L, shear modulus, G, and bulk modulus, K, from the relations:[23]

$$L=\rho c_l^2 \tag{17}$$

$$G=\rho c_t^2 \tag{18}$$

$$K=L-(4/3)G \tag{19}$$

Figures 3, 4 and 5 show plots of L, G, and K against composition, x, respectively. For LiB, L increases monotonically with increasing x, while NaB shows a slight M-shaped curve, and KB, RbB and CsB show distinct M-shaped curves (Figure 3). The same characteristics can be found for the plots of G and K against x. Compressibility, κ, is defined as the inverse of bulk modulus:

$$\kappa = \frac{1}{K} = \frac{1}{L-(4/3)G} \tag{20}$$

The compressibility will be discussed in the section 5.7. in relation to the pressure derivative of molar volumes.

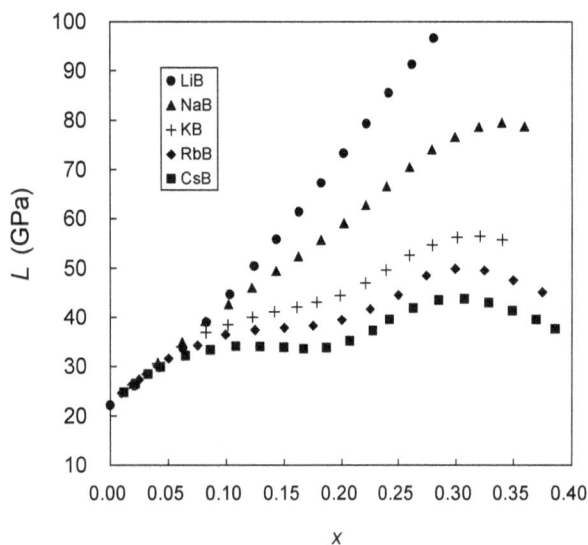

Figure 3. Longitudinal modulus, L, of alkali metal borate glasses against composition, x

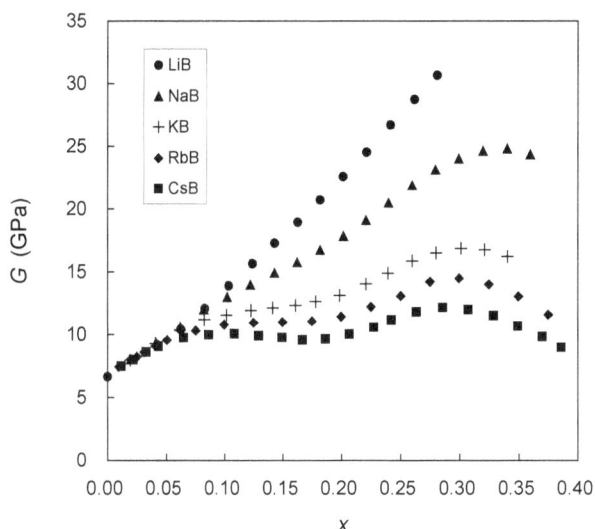

Figure 4. Shear or shear modulus, G, of alkali metal borate glasses against composition, x

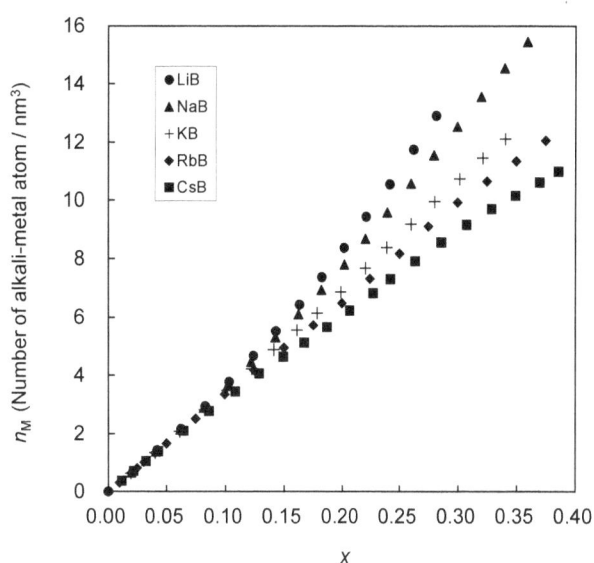

Figure 6. Number density of alkali metal atom, n$_M$, against composition, x

Each plot of L, G, or K shown in Figure 3, 4 or 5 has a close similarity with the plot of x_4 shown by Zhong & Bray.[17] Especially, the behaviour of the elastic constant, L, G or K, for lithium borate glasses coincides with the behaviour of x_4 determined from the NMR study by Jellison *et al*,[24] in which the relation $x_4 = x/(1-x)$ holds strictly over the composition range $0 \leq x \leq 0.29$.

We then assume that the elastic constants, L, G, and K, are increased by the presence of $M^+B\varnothing_4^-$ while they are decreased by the presence of $M^+B O\varnothing_2^-$. This assumption is based upon the fact that (a) the elastic constants, particularly the shear or shear modulus in which no change in volume is accompanied during shear deformation, represent the rigidity of glass; (b) due to the difference in the number of bridging oxygens, the presence of $M^+B\varnothing_4^-$ (which has four bridging oxygens) increases the rigidity of

glass compared with $B\varnothing_3$ (which has three bridging oxygens), while the presence of $M^+B O\varnothing_2^-$ which has two bridging oxygens decreases the rigidity of glass compared with $B\varnothing_3$.

Upon initial addition of M_2O to B_2O_3 in the composition range of $0 \leq x \leq 0.10$, each plot of L, G and K against composition increases monotonically in the same way irrespective of the difference in alkali metal ions, which means that all species of alkali metal ion fill the voids of the initial B_2O_3 glass network in the same way. Above this composition, the elastic constants depend both on the species of alkali metal ion, M^+, and on the composition, x. The behaviour of L, G or K against composition is characterised in terms of the three structural units in the next section.

5.4. Interpretation of M- or W-shaped curves in terms of basic borate structural units

The M-shaped curve is formed from a convex curve, a concave curve, and another convex curve in succession (Figures 1 to 8); the typical M-shaped curve exhibits a maximum, a minimum and another maximum in succession, as can be seen in the velocity of sound in rubidium or caesium borate glasses (Figure 2). The W-shaped curve has an inverted shape of the M-shaped curve and is formed from a concave curve, a convex curve, and another concave curve in succession (Figures 9 to 13).

We consider that the M- or W-shaped behaviour is caused by a change in the kind and amount of (a) basic borate structural units, and (b) superstructural borate groups. Elastic constants, such as the shear modulus, are governed by short-range borate structural units. Among the three elastic constants, longitudinal modulus, shear modulus, and bulk modulus, neither expansion nor compression is involved in shear deformation

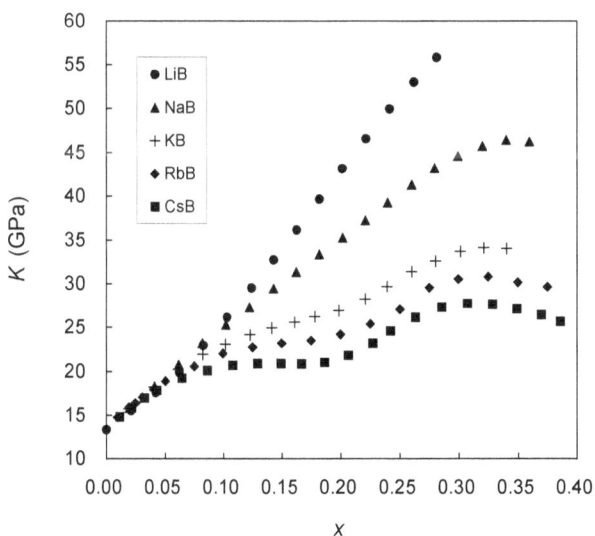

Figure 5. Bulk modulus, K, of alkali metal borate glasses against composition, x

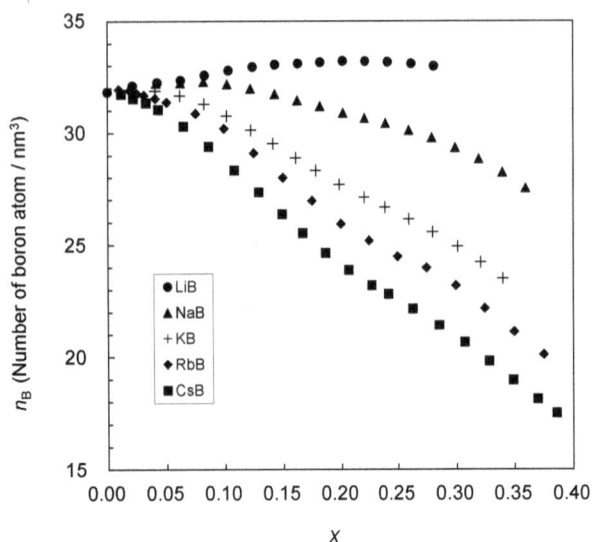

Figure 7. Number density of boron atom, n_B, against composition, x

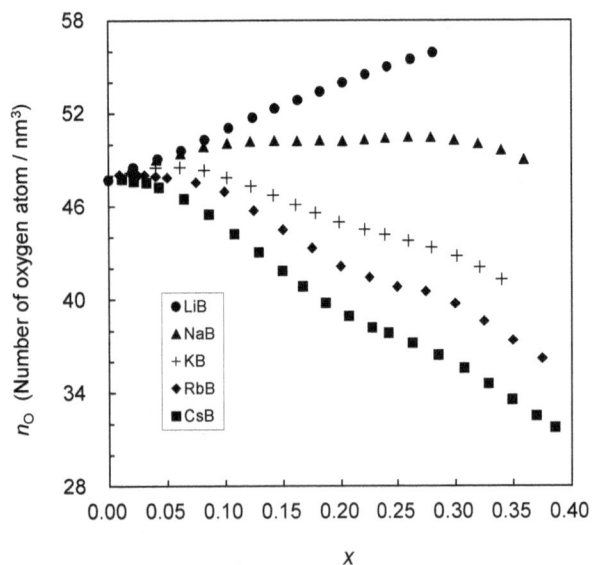

Figure 9. Volume of glass containing one mole of boron atoms against composition, x

so that we can choose the shear modulus as one of the most representative elastic constants in order to characterise the short range glass structure. Poisson's ratio represents glass network connectivity and hence it is governed by middle range superstructural borate groups. The W-shaped behaviour of Poisson's ratio will be discussed in relation to the network connectivity in section 5.8.

We now discuss the M- shaped behaviour of the shear modulus in terms of the basic borate structural units. In the plot of shear modulus against composition shown in Figure 4, the slope of shear modulus increases when $M^+B\varnothing_4^-$ is formed, while the slope decreases when $M^+B O\varnothing_2^-$ is formed. Thus, there is a close relationship between the shear modulus and the mole fraction of $M^+B\varnothing_4^-$ or $M^+B O\varnothing_2^-$, denoted by x_4 or by x_2, respectively.

In lithium borate glasses, the shear modulus increases monotonically with increasing x over the bulk glass formation range. This means that the formation of $Li^+B\varnothing_4^-$, expressed by Equation (12), dominates over the formation of $Li^+B O\varnothing_2^-$, expressed by Equation (13). Owing to the chemical equilibrium between $Li^+B\varnothing_4^-$ and $Li^+B O\varnothing_2^-$ expressed by Equation (14), both $Li^+B\varnothing_4^-$ and $Li^+B O\varnothing_2^-$ may be formed in principle. If $Li^+B O\varnothing_2^-$ is formed, then the shear modulus should decrease. However, any decrease in shear modulus ascribable to the formation of $Li^+B O\varnothing_2^-$ is not observed. The chemical equilibrium expressed by Equation (14) proceeds essentially to the left-hand side, and hence no nonbridging oxygens are formed in this bulk glass formation range.

On the other hand, the shear modulus of caesium borate glasses, for example, exhibits a typical M-

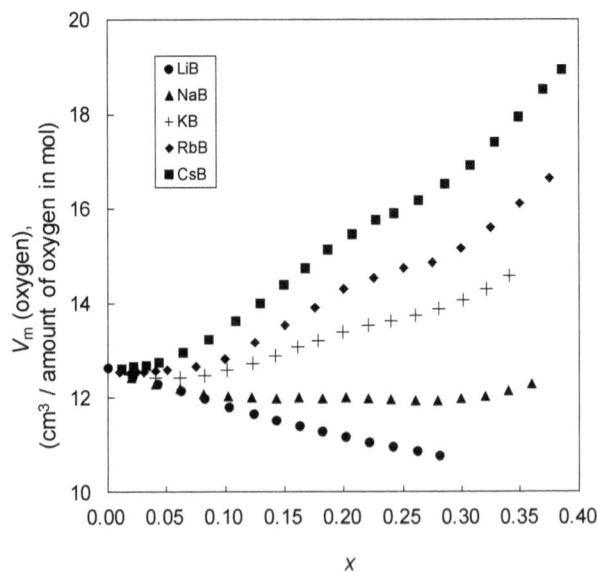

Figure 8. Number density of oxygen atom, n_O, against composition, x

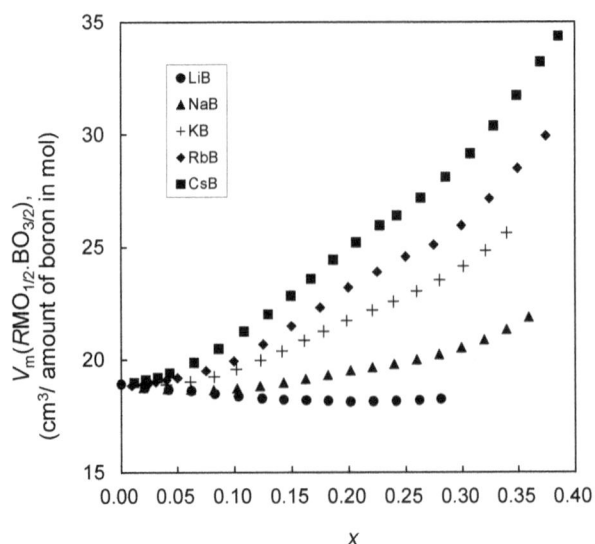

Figure 10. Volume of glass containing one mole of oxygen atoms against composition, x

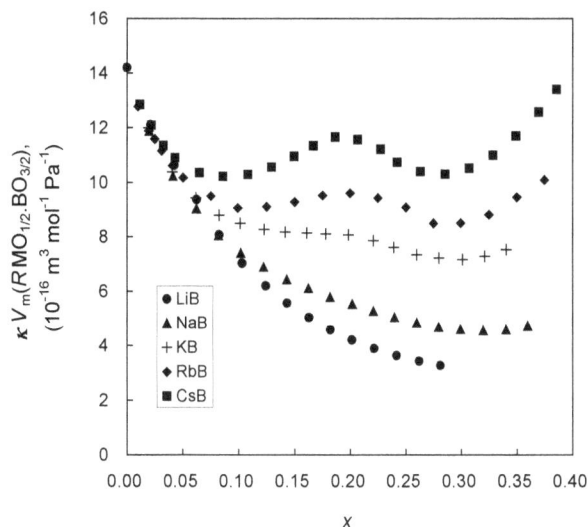

Figure 11. $\kappa V_m(RMO_{1/2}.BO_{3/2})$ *against composition,* x

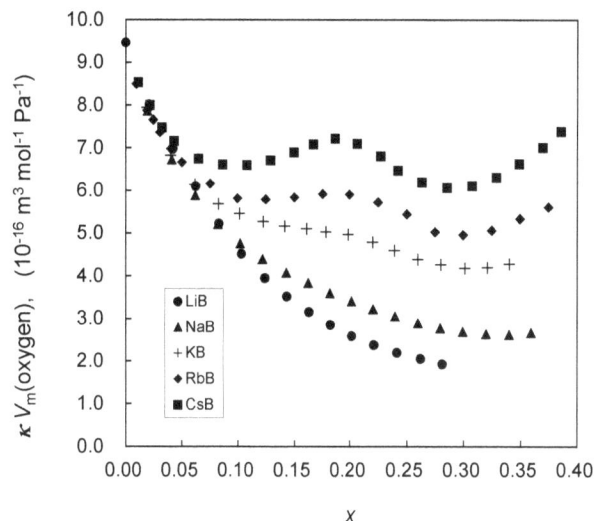

Figure 12. $\kappa V_m(oxygen)$ *against composition,* x

shaped behaviour. From the difference in the slope of the shear modulus, the properties of caesium borate glasses can be divided, approximately, into four characteristic composition ranges: (a) the first composition range of $x=0$ to 0.10, which is defined as the range from pure B_2O_3 to the first maximum (or the middle of the first convex curve) at $x\approx0.10$, (b) the second composition range of $x=0.10$ to 0.20, which is defined as the range from the first maximum (or the middle of the first convex curve) to the minimum (or the middle of the concave curve) at $x\approx0.20$, (c) the third composition range of $x=0.20$ to 0.30, which is defined as the range from the minimum (or the middle of concave curve) to the second maximum (or the middle of the second convex curve) at $x\approx0.30$, and (d) the fourth composition range, $x\geq0.30$, which is defined as the range beyond the second maximum (or the middle of the second convex curve) at $x\approx0.30$. In this way, the properties of alkali metal borate glasses in their respective bulk glass formation ranges can be divided into four composition ranges. In the case of lithium borate glasses, the composition ranges can be defined in the same way from the plot of Poisson's ratio (Figure 13) by making use of a minimum at $x\approx0.10$, a maximum at $x\approx0.20$, and another minimum at $x\approx0.30$. Although the composition range of lithium borate glasses is limited to the bulk glass formation range, $x\leq0.28$, the three composition ranges determined from the Poisson ratio coincide with those determined for the other alkali metal borate glasses. As a result, all of the alkali metal borate glasses studied in the present paper can be divided into these four characteristic composition ranges.

The M-shaped behaviour of the shear modulus can be characterised by considering the composition range, the kind of alkali metal borate glasses, and the equation denoting the type of conversion:
(a) the first composition range, $0\leq x\leq0.10$:
The slope of the shear modulus is approximately the same for M=Li, Na, K, Rb, and Cs. With the addition of M_2O (M=Li, Na, K, Rb, and Cs) to B_2O_3, $B\varnothing_3$ is converted only into $M^+B\varnothing_4^-$ according to Equation (12). The chemical equilibrium denoted by Equation (14) proceeds essentially to the left-hand side and no $M^+BO\varnothing_2^-$ is formed in this composition range.
(b) the second composition range, $0.10\leq x\leq0.20$:
The slope of the shear modulus decreases as the alkali metal is changed from LiB to CsB. As a result, the mole fraction of $M^+BO\varnothing_2^-$, formed according to Equation (13), increases in the order NaB to CsB, while the mole fraction of $M^+B\varnothing_4^-$, which is formed according to Equation (12), decreases in the order NaB to CsB. The resulting chemical equilibrium defined by Equation (14) tends to proceed to the right-hand side as the alkali metal is changed from NaB to CsB.
(c) the third composition range, $0.20\leq x\leq0.30$:
The slope of the shear modulus decreases as the alkali metal is changed from LiB to CsB, but the slope in the third composition range is greater than that in the second composition range. This fact indicates that, in Equation (14), the chemical equilibrium in the third composition range shifts slightly to the left-hand side compared with that in the second composition range. As a result, x_2 takes a maximum value at the boundary composition, $x\approx0.20$.
(d) the fourth composition range, $x\geq0.30$:
Each slope of the shear modulus becomes negative in this composition range. As a result, no more $M^+B\varnothing_4^-$ is formed in this composition range. The second maximum at a composition of $x\approx0.30$ is caused by the conversion of the residual $B\varnothing_3$ into only $M^+BO\varnothing_2^-$ according to Equation (13). A Raman spectroscopic study by Kamitsos *et al*[19,20] has shown that $M^+BO\varnothing_2^-$ in the second and third composition ranges forms a chain-type metaborate group, and that in the fourth composition range it forms a ring-type metaborate group.
Denoting $M^+B\varnothing_4^-$, $B\varnothing_3$, and $M^+BO\varnothing_2^-$ by B(4),

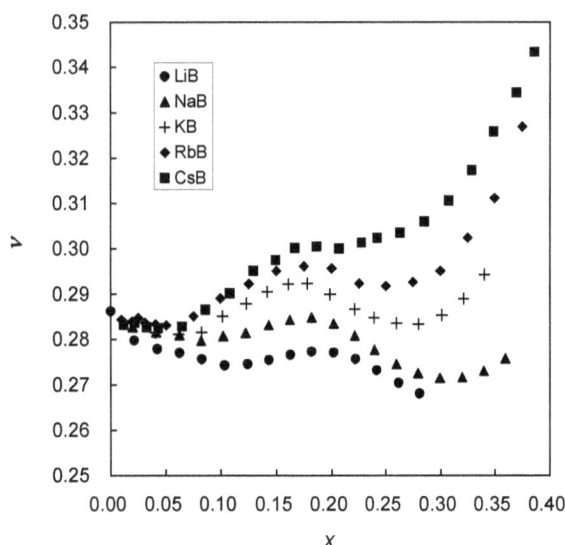

Figure 13. Poisson's ratio, v, of alkali metal borate glasses against composition, x

B(3), and B(2), respectively, the characteristics of the structural changes, e.g. in caesium borate glasses, can be expressed in terms of the abbreviated borate structural units as follows:

(a) the first composition range, $0 \leq x \leq 0 \cdot 10$: B(3) changes only into B(4) without formation of B(2).

(b) the second composition range, $0 \cdot 10 \leq x \leq 0 \cdot 20$: B(3) changes into both B(4) and B(2), but the effect of B(2) is predominant over that of B(4).

(c) the third composition range, $0 \cdot 20 \leq x \leq 0 \cdot 30$: B(3) changes into both B(4) and B(2), but the effect of B(4) is predominant over that of B(2).

(d) the fourth composition range, $x \geq 0 \cdot 30$: B(3) changes only into B(2) without further formation of B(4).

5.5. Number density

We discuss (a) the number density of alkali metal atoms, (b) the number density of boron atoms, and (c) the number density of oxygen atoms, on the basis of Equations (3), (4), and (5), respectively. The mass density used in the calculation of these quantities is shown in Figure 1.

5.5.1. Number density of alkali metal atoms

Figure 6 shows the number density, n_M, of alkali metal atoms against composition, x. In the composition range $0 \leq x \leq 0 \cdot 10$, each number density increases monotonically with x in the same way irrespective of the alkali metal ion, indicating that the alkali metal ions initially fill the voids of the B_2O_3 glass network in the same way. Above the composition $x=0 \cdot 10$, a difference depending on the alkali metal ion can be clearly seen. At a given composition $x \geq 0 \cdot 10$, the number density increases as the size of alkali metal ion decreases, in the order Cs to Li. As a result, the alkali metal ions are packed in a more compact way as the alkali metal

ion is changed from caesium to lithium.

Giri *et al*[25] have shown, from an analysis of the packing fraction of the constituent atoms, that lithium and sodium borate glasses have the nature of covalent packing, while potassium, rubidium and caesium borate glasses have the nature of ionic packing. The plot of the number density of alkali metal atoms shown in Figure 6 may be divided into these two groups according to the difference in their respective slopes; steep slopes for LiB and NaB, and gentle slopes for KB, RbB, and CsB. A closer inspection of Figure 6 shows that a slight concavity is present for KB, RbB, and CsB in the range $0 \cdot 10 \leq x \leq 0 \cdot 30$, which corresponds to the second and the third composition ranges defined from the shear modulus. The cause of the M-shape has been explained in section 5.4 in terms of the borate structural units.

5.5.2. Number density of boron atoms

The number density of boron atoms has already been discussed by Wright.[9] In the present paper, the characteristics will be shortly summarised.

Figure 7 shows the number density, n_B, of boron atoms against composition, x. At a given composition, the number density increases as the size of alkali metal ion decreases, in the order Cs to Li, which indicates that the boron atoms are packed in a more compact way as the alkali metal ion is changed from caesium to lithium. In each binary system, the boron atoms are packed most compactly at the following compositions: LiB, $x=0 \cdot 20$; NaB, $x=0 \cdot 08$; KB, $x=0 \cdot 02$; RbB, $x=0 \cdot 01$; CsB, none.

In the extreme case of lithium borate glasses, the number density of boron atoms is always larger than that of B_2O_3 glass throughout the whole composition range studied, indicating that the boron atoms in lithium borate glasses are packed in a more compact way than in B_2O_3 glass. In the other extreme case of caesium borate glasses, the number density of boron atoms decreases with increasing x, indicating that the boron atoms in caesium borate glasses of any composition are packed in a less compact way than in B_2O_3 glass. Figure 7 also shows the M-shaped behaviour for NaB, KB, RbB, and CsB, with a concavity in the range $0 \cdot 10 \leq x \leq 0 \cdot 30$, which corresponds to the second and third composition ranges determined from the shear modulus. The cause of the M-shape has been explained in section 5.4 in terms of borate structural units.

5.5.3. Number density of oxygen atoms

The number density of oxygen atoms has been discussed by Wright[9] in relation to the packing density of oxygen atoms. In the present paper, the characteristics will be shortly summarised.

Figure 8 shows the number density, n_O, of oxygen atoms against composition, x. At a given composition, the number density increases as the size of alkali metal ion decreases, in the order Cs to Li, which indicates that the oxygen atoms are packed in a more compact way as the alkali metal ion is changed from caesium to lithium. In lithium borate glasses, n_O increases steadily with increasing x, indicating that the boron-oxygen network becomes more compact as Li_2O is added to B_2O_3. Figure 8 again shows the M-shaped behaviour, with a concavity in the range $0.10 \leq x \leq 0.30$ for NaB, KB, RbB, and CsB. The cause of the M-shape is the same as explained in section 5.4.

5.6. Molar volumes

5.6.1. Boron molar volume

The volume of glass, $V_m(RMO_{1/2}.BO_{3/2})$, which contains one mole of boron atoms has been defined by Equation (7) as the physical quantity, N_A/n_B. Thus, $V_m(RMO_{1/2}.BO_{3/2})$ is inversely proportional to the number density, n_B, of boron atoms. Thus the behaviour of $V_m(RMO_{1/2}.BO_{3/2})$ is inverse to the behaviour of n_B. The plot (Figure 9) of $V_m(RMO_{1/2}.BO_{3/2})$ versus composition, x, shows a W-shape for NaB, KB, RbB, and CsB; the presence of convexity in the composition range $0.10 \leq x \leq 0.30$ indicates that a small amount of $M^+BO\emptyset_2^-$ is formed in this composition range, in addition to the formation of $M^+B\emptyset_4^-$. This interpretation is based upon the fact that the volume of the $M^+BO\emptyset_2^-$ unit is larger than that of the $M^+B\emptyset_4^-$ unit.[15] For lithium borate glasses, the plot of $V_m(RMO_{1/2}.BO_{3/2})$ exhibits a slight minimum at $x=0.20$, indicating that the most compact boron–oxygen network is formed at this composition. Above $x=0.30$, the slope of $V_m(RMO_{1/2}.BO_{3/2})$ versus x becomes steeper for NaB, KB, RbB, and CsB, indicating that a large amount of $M^+BO\emptyset_2^-$ is now formed. Although the W-shape and the M-shape are inversely related, the cause for $V_m(RMO_{1/2}.BO_{3/2})$ to exhibit the W-shape is the same as that for the shear modulus to exhibit the M-shape, as has been explained in section 5.4.

5.6.2. Oxygen molar volume

The volume of glass, $V_m(oxygen)$, which contains one mole of oxygen atoms has been defined by Equation (8) as the physical quantity, N_A/n_O. Thus, $V_m(oxygen)$ is inversely proportional to the number density, n_O, of oxygen atoms. Figure 10 shows the plot of $V_m(oxygen)$ against composition, x, and has a close similarity with Figure 9. This similarity is caused by the fact that the boron and oxygen atoms do not behave in separate ways, but combine to form various borate structural units. The plot of $V_m(oxygen)$ shows the W-shape for NaB, KB, RbB, and CsB; the presence of convexity in the composition range $0.10 \leq x \leq 0.30$ indicates that a small amount of $M^+BO\emptyset_2^-$ is formed in this composition range, in addition to the formation of $M^+B\emptyset_4^-$. For lithium borate glasses, $V_m(oxygen)$

decreases steadily with increasing x, indicating that the oxygen atoms are packed in a more compact way when Li_2O is added to B_2O_3. Above $x=0.30$, the slope of $V_m(oxygen)$ versus x becomes steeper for NaB, KB, RbB, and CsB, indicating that a large amount of $M^+BO\emptyset_2^-$ is now formed. The cause for $V_m(oxygen)$ to exhibit the W-shape is the same as that for the shear modulus to exhibit the M-shape, as has been explained in section 5.4.

5.7. Pressure derivative of molar volumes

5.7.1. Pressure derivative for boron molar volume

In Equation (10), $-dV_m(RMO_{1/2}.BO_{3/2})/dP$ or $\kappa V_m(RMO_{1/2}.BO_{3/2})$ represents the effect of boron atoms on the elasticity of alkali metal borate glasses. Figure 11 shows the plot of $\kappa V_m(RMO_{1/2}.BO_{3/2})$ against composition, x, and exhibits the characteristics of Figure 9 more clearly. We can gain several insights into the relationship between the borate structural units and the elasticity of alkali metal borate glasses:
(a) The plot for CsB, RbB, or KB shows the W-shape clearly, and the plot for NaB slightly shows this shape, while the plot for LiB decreases monotonically with increasing x and does not show any non-monotonic behaviour.
(b) The plot for LiB indicates that the structure of LiB becomes more rigid with the addition of Li_2O to B_2O_3, since the value of $\kappa V_m(RMO_{1/2}.BO_{3/2})$ decreases monotonically, and hence the glass becomes more incompressible with the addition of Li_2O.
(c) The W-shaped behaviour for CsB, for example, can be interpreted by dividing the whole composition range again into the same four composition ranges.

The first composition range, $0 \leq x \leq 0.10$, is caused by the conversion of $B\emptyset_3$ units into $Cs^+B\emptyset_4^-$ units, since the value of $\kappa V_m(RMO_{1/2}.BO_{3/2})$ decreases and hence the glass becomes more incompressible with the addition of Cs_2O. In the second and third composition ranges, $0.10 \leq x \leq 0.30$, the plot of $\kappa V_m(RMO_{1/2}.BO_{3/2})$ against x curves upwards with a convex form, which indicates that a compressible and hence an elastically soft structure is formed in this composition range. This indicates that the presence of $M^+BO\emptyset_2^-$ units causes the softness in elasticity, since the volume of the $M^+BO\emptyset_2^-$ unit is larger than that of the $M^+B\emptyset_4^-$ unit, and therefore the $M^+BO\emptyset_2^-$ unit is easily deformed by stress compared with the $M^+B\emptyset_4^-$ unit. At the composition $x=0.30$, an incompressible and therefore a compact structure is formed, since the value of $\kappa V_m(RMO_{1/2}.BO_{3/2})$ shows a minimum at this composition. In the fourth composition range, $x \geq 0.30$, $\kappa V_m(RMO_{1/2}.BO_{3/2})$ increases more steeply with x, indicating that the glass becomes more compressible and therefore elastically softer with the addition of Cs_2O. This softness in elasticity is caused by the formation of the $M^+BO\emptyset_2^-$ unit to a large amount.
(d) The relationship between the borate structural

units and the elasticity of RbB, KB and NaB is essentially the same as for CsB.

5.7.2. Pressure derivative for the oxygen molar volume

In Equation (11), $-dV_m(\text{oxygen})/dP$ or $\kappa V_m(\text{oxygen})$ represents the effect of oxygen atoms on the elasticity of alkali metal borate glasses. Figure 12 shows the plot of $\kappa V_m(\text{oxygen})$ against composition, x, and the behaviour is nearly identical to that in Figure 11. This similarity is caused by the fact that the boron and oxygen atoms combine to form various (super) structural units, and therefore both of these atoms act not individually but as a whole. As a result, the conclusions drawn from Figure 11 can be applied also to Figure 12.

5.8. Poisson's ratio

Poisson's ratio, v, is defined as the ratio of transverse contraction strain to longitudinal extension strain in a stretched bar, and it is expressed as[23]

$$v=(L-2G)/2(L-G) \qquad (21)$$

Poisson's ratio takes a value in the range $-1<v<0.5$, where a negative value is permissible in a peculiar material, such as re-entrant foam materials.[26] All of the glasses known so far, however, have positive values of Poisson's ratio in the range $0.1\leq v\leq 0.4$.[27,28] Bridge et al[29] expressed the glass network connectivity in terms of the crosslink density (network connectivity−2=crosslink density); they also showed that the Poisson ratio is inversely proportional to the crosslink density. Subsequently, the network connectivity of a wide variety of glasses has been examined on the basis of the Poisson ratio.[27]

Figure 13 shows Poisson's ratio of alkali metal borate glasses against composition, x. The Poisson ratio shows the W-shape, except for LiB in which case the W-shape is not completely formed. Two causes can be considered for the behaviour of Poisson's ratio against composition:

(a) One cause is the disproportionation reaction between the borate structural units given by Equation (14). The presence of $M^+B\varnothing_4^-$, whose network connectivity is four, increases the connectivity with the surrounding (super)structural units, while the presence of $M^+BO\varnothing_2^-$, whose network connectivity is two, decreases the connectivity with the surrounding (super)structural units.

(b) The other cause is the presence of superstructural units, especially the pentaborate group. The $B\varnothing_4^-$ unit contained in a pentaborate group, $B_5O_6\varnothing_4^-$, whose network connectivity is also four, does not contribute to the connectivity with the surrounding (super)structural units, while the $B\varnothing_4^-$ unit itself in the pentaborate group contributes to the shear of

glass. Thus, the rigidity and the network connectivity are not the same concept. A Raman spectroscopic study[21] on alkali metal borate glasses has revealed the presence of pentaborate groups in the following composition ranges: LiB, $0\leq x\leq 0.36$; NaB, $0\leq x\leq 0.25$; KB, $0\leq x\leq 0.30$; RbB, $0\leq x\leq 0.30$; CsB, $0\leq x\leq 0.30$.

Since Poisson's ratio is inversely proportional to the crosslink density, and hence to the glass network connectivity, we gain the following insights into the network connectivity of alkali metal borate glasses:

(a) For lithium borate glasses, Poisson's ratio decreases in the first composition range, $0\leq x\leq 0.10$, and then curves upwards with a convex shape in the second and the third composition ranges, $0.10\leq x\leq 0.30$. The decrease in Poisson's ratio in the first composition range is caused by the increase in network connectivity due to the formation of $Li^+B\varnothing_4^-$ from $B\varnothing_3$. The behaviour in the second and third composition ranges is not ascribable to the decrease in network connectivity due to the formation of $Li^+BO\varnothing_2^-$, since the foregoing analysis (section 5.4) on the shear modulus, G, of lithium borate glasses denies the presence of $Li^+BO\varnothing_2^-$ in the second and the third composition ranges. Thus, the decrease in network connectivity in the convex range, $0.10\leq x\leq 0.30$, is ascribable to the formation of pentaborate groups.

(b) For sodium borate glasses, the Poisson ratio decreases in the first range, $0\leq x\leq 0.08$, then curves upwards with a convex shape in the second and third composition ranges, $0.08\leq x\leq 0.30$, and finally increases in the fourth composition range, $x\geq 0.30$. The behaviour in the first composition range is caused by the formation of $Na^+B\varnothing_4^-$. The behaviour in the second and the third composition ranges is caused by the major formation of pentaborate groups, accompanied by a minor formation of $Na^+BO\varnothing_2^-$, since a slight amount of $Na^+BO\varnothing_2^-$ is present in the second and third composition ranges, as shown in the analysis of the shear modulus, G (section 5.4). The behaviour in the fourth composition range is caused by the decrease in network connectivity, owing to the formation of $Na^+BO\varnothing_2^-$.

(c) Potassium, rubidium, and caesium borate glasses have similar characteristics in the behaviour of Poisson's ratio. For the typical case of caesium borate glasses, Poisson's ratio decreases in the first composition range, $0\leq x\leq 0.07$, then curves upwards with a convex shape in the second and second composition ranges, $0.07\leq x\leq 0.30$, and finally increases steeply in the fourth composition range, $x\geq 0.30$. A neutron diffraction study on caesium borate glasses by Shaw et al[22] has shown that superstructural units such as triborate and pentaborate groups, in addition to the boroxol ring, play a significant role in the structures of caesium borate glasses. The behaviour in the first composition range is caused by the formation of $Cs^+B\varnothing_4^-$. The behaviour in the second and third composition ranges is caused mainly by the formation of pentaborate groups, accompanied by a moderate formation of $Cs^+BO\emptyset_2^-$, since a moderate amount of $Cs^+BO\emptyset_2^-$ is already present, as shown in the analysis of the shear modulus, G (section 5.4). The behaviour in the fourth composition range is caused by a rapid decrease in network connectivity, owing to an active formation of $Cs^+BO\emptyset_2^-$.

(d) At a given composition, x, the glass network connectivity increases in the order CsB, RbB, KB, NaB and LiB.

(e) Minima are present at compositions from $x=0.07$ for CsB to $x=0.10$ for LiB, indicating that the glasses have high connectivity at these compositions.

(f) Maxima are present at compositions from $x=0.17$ for CsB to $x=0.20$ for LiB, indicating that the glasses have low connectivity at these compositions. One cause for this low connectivity is the formation of pentaborate groups in the convex composition range.

(g) Minima or abrupt changes in slope are present in the neighbourhood of $x=0.30$, indicating that the glasses again have high network connectivity in the neighbourhood of this composition.

(h) For the composition of $x>0.30$, the Poisson ratio again increases with increasing x for NaB, KB, RbB and CsB, indicating that the glass network connectivity decreases, owing to the formation of $M^+BO\emptyset_2^-$.

6. Conclusions

The mass density and velocities of both longitudinal and transverse ultrasonic waves have been measured at 298 K for each of the five binary alkali metal borate glass systems, $xM_2O.(1-x)B_2O_3$, where M=Li, Na, K, Rb, or Cs. The glass specimens were prepared in their respective bulk glass formation ranges: LiB, $0\leq x\leq0.28$; NaB, $0\leq x\leq0.36$; KB, $0\leq x\leq0.34$; RbB, $0\leq x\leq0.37$; CsB, $0\leq x\leq0.39$.

On the basis of the mass density and the velocity of sound, the following glass properties have been studied: (a) elastic constants, such as longitudinal modulus, shear modulus, and bulk modulus; (b) the number densities of alkali metal ions, boron atoms, and oxygen atoms; (c) the volume of glass which contains one mole of boron atoms or oxygen atoms; (d) the pressure derivative of these molar volumes; (e) Poisson's ratio. The composition dependence of the properties (a) and (b) exhibit M-shapes, while the properties (c), (d), and (e) exhibit W-shapes.

The causes for these properties to show the M- or the W-shapes have been analysed in terms of the three borate structural units, BO_3, $M^+BO\emptyset_4^-$, and $M^+BO\emptyset_2^-$, or in terms of glass network connectivity, by dividing the whole composition range into four composition ranges, $0\leq x\leq0.10$, $0.10\leq x\leq0.20$, $0.20\leq x\leq0.30$, and $0.30\leq x<0.50$.

(1) In lithium borate glasses, the shear modulus shows that BO_3 units are converted into only $M^+BO\emptyset_4^-$ units, without the formation of $M^+BO\emptyset_2^-$ units, over the whole composition range ($0\leq x<0.30$) studied in the present paper. Poisson's ratio shows the presence of pentaborate groups in the composition range $0.10\leq x<0.30$.

(2) In NaB, KB, RbB, and CsB, the behaviour in the first composition range, $0\leq x\leq0.10$, is caused by the conversion of BO_3 units into $M^+BO\emptyset_4^-$ units, without formation of $M^+BO\emptyset_2^-$ units. The behaviour in the second and third composition ranges, $0.10\leq x\leq0.30$, indicates the presence of $M^+BO\emptyset_2^-$ units, in addition to $M^+BO\emptyset_4^-$ units. The Poisson ratio indicates that the glass network connectivity decreases in the composition range $0.10\leq x\leq0.30$, owing to the formation of pentaborate groups. Thus, the convexity arising in the composition range $0.10\leq x\leq0.30$ for NaB, KB, RbB, and CsB results from two causes: the formation of $M^+BO\emptyset_2^-$ units and the formation of pentaborate groups.

(3) In the fourth composition range, $0.30\leq x<0.50$, for NaB, KB, RbB, and CsB, $M^+BO\emptyset_4^-$ units transform into $M^+BO\emptyset_2^-$ units, and no more $M^+BO\emptyset_4^-$ units are formed from BO_3 units. This behaviour is ascertained from the decrease in the shear modulus and from the increase in Poisson's ratio.

References

1. Kodama, M., Matsushita, T. & Kojima, S. *Jpn. J. Appl. Phys.*, 1995, **34**, 2570.
2. Kodama, M. *J. Mater. Sci.*, 1991, **26**, 4048.
3. Kodama, M. *J. Non-Cryst. Solids*, 1991, **127**, 65.
4. Kodama, M. *J. Am. Ceram. Soc.*, 1991, **74**, 2603.
5. Kodama, M., Hirashima, T. & Matsushita, T. *Phys. Chem. Glasses*, 1993, **34**, 129.
6. Shelby, J. E. *J. Am. Ceram. Soc.*, 1983, **66**, 225.
7. Kodama, M., Kojima, S., Feller, S. & Affatigato, M. *Phys. Chem. Glasses*, 2005, **46**, 190.
8. Kodama, M., Ono, A., Kojima, S., Feller, S. A. & Affatigato, M. *Phys. Chem. Glasses: Eur. J. Glass Sci. Technol. B*, 2006, **47**, 465.
9. Wright, A. C. *Phys. Chem. Glasses: Eur. J. Glass Sci. Technol. B*, 2010, **51**, 1.
10. Kodama, M., Iizuka, K., Miyashita, M., Nagai, N., Clarida, W., Feller, S. A. & Affatigato, M. *Glass Technol.*, 2003, **44**, 50.
11. Bowman, H. A., Schoonover, R. M. & Carroll, C. L. *Metrologia*, 1974, **10**, 117.
12. Papadakis, E. P. *Physical acoustics.* Vol. 12. 1976. Edited by W. P. Mason & R. N. Thurston. Academic Press, New York. Chap. 5, Pp 277–374.
13. McSkimin, H. J. *J. Acoust. Soc. Am.*, 1961, **33**, 12.
14. Shartsis, L., Capps W. & Spinner, S. *J. Am. Ceram. Soc.*, 1953, **36**, 35.
15. Lower, N. P., McRae, J. L., Feller, H. A., Betzen, A. R., Kapoor, S., Affatigato, M. & Feller, S. A. *J. Non-Cryst. Solids*, 2001, **293-295**, 669.
16. Bray, P. J. & O'Keefe, J. G. *Phys. Chem. Glasses*, 1963, **4**, 37.
17. Zhong, J. & Bray, P. J. *J. Non-Cryst. Solids*, 1989, **111**, 67.
18. Ratai, E. M., Janssen, M., Epping, J. D., Chan, J. C. C. & Eckert, H. *Phys. Chem. Glasses*, 2003, **44**, 45.
19. Kamitsos, E. I., Karakassides, M. A. & Chryssikos, G. D. *Phys. Chem. Glasses*, 1989, **30**, 229.
20. Chryssikos, G. D., Kamitsos, E. I. & Karakassides, M. A. *Phys. Chem. Glasses*, 1990, **31**, 109.
21. Meera, B. N. & Ramakrishna, J. *J. Non-Cryst. Solids*, 1993, **159**, 1.
22. Shaw, J. L., Wright, A. C., Sinclair, R. N., Frueh, J. R., Williams, R. B., Nelson, N. D., Affatigato, M, Feller, S. A. & Scales, C. R. *Phys. Chem. Glasses*, 2003, **44**, 256.
23. Mason, W. P. *Physical Acoustics and the Properties of Solids*, Van Nostrand, Princeton, 1958, pp. 12–16.
24. Jellison, G. E. Jr, Feller, S. A. & Bray, P. J. *Phys. Chem. Glasses*, 1978, **19**, 52
25. Giri, S., Gaebler, C., Helmus, J., Affatigato, M., Feller, S. & Kodama, M. *J. Non-Cryst. Solids*, 2004, **347**, 87.
26. Lakes, R. S. *Adv. Mater.*, 1993, **5**, 293.
27. Rouxel, T. *J. Am. Ceram. Soc.*, 2007, **90**, 3019.
28. Annapurna, K., Tarafder, A. & Phani, K. K. *J. Appl. Phys.*, 2007, **102**, 083542.
29. Bridge, B., Patel, N. D. & Waters, D. N. *Phys. Status Solidi (a)*, 1983, **77**, 655.

BORATE PHOSPHATE 2014

2014 June 30 - July 4

Pardubice Czech Republic
The 8th International Conference
on Borate Glasses, Crystals and Melts
& The International Conference
on Phosphate Glasses

FIRST ANNOUNCEMENT

The American Ceramic Society
www.ceramics.org

Official Conference Organizer:
Czech Glass Society

ČSS

The 8th International Conference on
BORATE GLASSES, CRYSTALS AND MELTS
June 30 - July 2, 2014

The eighth conference follows the previous meetings held at Alfred, New York, USA (1977); Abingdon, UK (1996); Sofia, Bulgaria (1999); Cedar Rapids, USA (2002), Trento, Italy (2005), Himeji, Japan (2008) and Halifax, Canada (2011). The conference will be dedicated to Professor Stanislav Filatov from the Saint Petersburg State University to honor his achievements in Borate Crystallography.

The International Conference on
PHOSPHATE GLASSES
July 2 - 4, 2014

This conference will cover all aspects of phosphate- and phosphate-containing glasses, including basic science, applications and technologies. A common session on July 2 on borophosphate glasses will form a bridge between the two conferences. Dr. Doris Ehrt from the Otto-Schott Institute, University of Jena (Germany) will be honored for her contributions to the study of phosphate optical glasses.

Conference Site – University of Pardubice

Pardubice (German: Pardubitz) is a regional capital of **Pardubice** region, southern part of East Bohemia, Czech Republic, located on both banks of the Elbe river about 100 km to the East of Prague, the Czech capital, and 20 km to the South of Hradec Kralove. The city can be easily reached by train from Prague in 60 min

The area was first mentioned in historical chronicles in connection with a new monastery built by the crusaders in 1295. One of the first rulers, Arnost of **Pardubice**, the first Archbishop of Prague and advisor to the Czech king and Holy Roman Emperor Charles IV., turned the village around the castle into a prosperous town. In the late 15th century, under the rule William of Pernstejn, the Gothic style castle was rebuilt into a magnificent Renaissance chateau, still well defendable with the castle walls and the wide moat remaining unchanged to this day. Both conferences will take place in lecture halls of the Faculty of Chemical Technology of the University of Pardubice

Contacts

ICARIS Ltd., Conference Management
Karlova 48, 110 00 Praha 1, Czech Republic
Phone: +420 226 220 010, Fax: +420 226 220 012

Contact person:
Romana Kočová, Romana@icaris.cz

BORATE PHOSPHATE 2014 June 30 Pardubice July 4 Czech Republic

www.icaris.cz/conf/borate-phosphate-2014

CONFERENCE PARTICIPANTS

Mario Affatigato
Coe College
United States
maffatig@coe.edu

Bruce Aitken
Corning Inc.
United States
aitkenbg@corning.com

Oliver Alderman
University of Warwick
United Kingdom
o.alderman@warwick.ac.uk

Nathan Barnes
Coe College
United States
npbarnes@coe.edu

Anna Barseghyan
Scientific Production Enterprise of Material Science
Armenia
annbars@gmail.com

Richard Brow
Missouri University of Science and Technology
United States
brow@mst.edu

Courtney Calahoo
Dalhousie University
Canada
court@dal.ca

Sébastien Chenu
Dalhousie University
Canada
sebastien.chenu@dal.ca

Imen Chermiti
Centre d'optique, photonique et laser, Université Laval
Canada
imen.chermiti.1@ulaval.ca

Laurent Cormier
IMPMC, Université Pierre & Marie Curie
France
cormier@impmc.upmc.fr

Giuseppe Dalba
Dept. of Physics, University of Trento
Italy
giuseppe.dalba@unitn.it

Andrea De Camargo
Physics Institute, University of Sao Paulo, IFSC/USP
Brazil
andreasc@ifsc.usp.br

Hellmut Eckert
WWU Münster
Germany
eckerth@uni-muenster.de

Doris Ehrt
Otto-Schott-Institut Uni Jena
Germany
doris.ehrt@uni-jena.de

Roland Ehrt
IGK Roland Ehrt
Germany
roland.ehrt@glaskeram.de

Margit Fabian
Research Institute for Solid State Physics and Optics
Hungary
fabian@szfki.hu

Steve Feller
Coe College
United States
sfeller@coe.edu

Guillaume Ferlat
Université Pierre & Marie Curie
France
ferlat@impmc.upmc.fr

Shingo Fuchi
Nagoya University
Japan
fuchi@mercury.numse.nagoya-u.ac.jp

Justine Galbraith
Dalhousie University
Canada
justine.galbraith@dal.ca

Jaime George
Missouri University of Science and Technology
United States
jgcc3@mst.edu

Kathryn Goetschius
Missouri University of Science and Technology
United States
kathryn.goetschius@gmail.com

Alex Hannon
ISIS Facility, Rutherford Appleton Laboratory
United Kingdom
a.c.hannon@rl.ac.uk

Akitoshi Hayashi
Osaka Prefecture University
Japan
hayashi@chem.osakafu-u.ac.jp

Hubert Huppertz
Leopold-Franzens-Universität
Austria
hubert.huppertz@uibk.ac.at

Steven Jung
Mo-Sci Corporation
United States
sjung@mo-sci.com

Efstratios Kamitsos
National Hellenic Research Foundation
Greece
eikam@eie.gr

Victor Khristenko
Coe College
United States
vdkhristenko@coe.edu

Yiannis Kipouros
Dalhousie University
Canada
yn843798@dal.ca

Shunichi Kobayashi
Nagoya University
Japan
kobayasi@mercury.numse.nagoya-u.ac.jp

Masao Kodama
University of Tsukuba
Japan
masaokodama@arion.ocn.ne.jp

Ladislav Koudelka
University of Pardubice
Czech Republic
ladislav.koudelka@upce.cz

Scott Kroeker
University of Manitoba
Canada
kroekers@cc.umanitoba.ca

Nattapol Laorodphan
University of Warwick
United Kingdom
N.Laorodphan@warwick.ac.uk

Yannick Ledemi
Centre d'optique, photonique et laser, Université Laval
Canada
yannick.ledemi@copl.ulaval.ca

Sung Keun Lee
Seoul National University
Korea, Republic of
sungklee@snu.ac.kr

Gerald Lelong
IMPMC, Université Pierre & Marie Curie
France
gerald.lelong@impmc.upmc.fr

Seiichi Mamiya
University of Tsukuba
Japan
mamiya@ims.tsukuba.ac.jp

Jun Matsuoka
University of Shiga Prefecture
Japan
matsuoka.j@mat.usp.ac.jp

John Mauro
Corning, Inc.
United States
MauroJ@corning.com

Younes Messaddeq
Université Laval
Canada
younes.messaddeq@copl.ulaval.ca

Florent Michel
Saint Gobain Recherche
France
flo_michel@hotmail.fr

Tsutomu Minami
Osaka Prefecture University
Japan
minamit@tri.pref.osaka.jp

Doris Möncke
National Hellenic Research Foundation
Greece
dorismoencke@web.de

Tyler Munhollon
Coe College
United States
tlmunhollon@coe.edu

Toshihiro Okajima
Kyushu Synchrotron Light Research Center
Japan
okajima@saga-ls.jp

Koji Oshima
Nagoya University
Japan
ohshima@mercury.numse.nagoya-u.ac.jp

Annie Pradel
Université Montpellier 2
France
annie.pradel@univ-montp2.fr

Sindy Reibstein
Department of Materials Science, University of
Erlangen-Nuremberg
Germany
sindy.reibstein@ww.uni-erlangen.de

Francesco Rocca
Fondazione Bruno Kessler and IFN-CNR
Italy
rocca@science.unitn.it

Mauricio Rodriguez
Facultad de Quimcia
Uruguay
mrodrig@fq.edu.uy

Silvia Santagneli
Chemistry Institute - UNESP
Brazil
santagneli@iq.unesp.br

David Schubert
U.S. Borax Inc.
United States
david.schubert@borax.com

Atsuko Shinomiya
Osaka Prefecture University
Japan
shinomiya@chem.osakafu-u.ac.jp

Elanor Steffee
Rio Tinto Minerals
United States
elanor.steffee@riotinto.com

Akira Takada
Asahi Glass Company
Japan
akira-takada@agc.com

Yoshikazu Takeda
Nagoya University
Japan
takeda@numse.nagoya-u.ac.jp

Norimasa Umesaki
JASRI/SPring-8
Japan
umesaki@spring8.or.jp

Deborah Watson
Coe College
United States
dewatson@coe.edu

Ulrike Werner-Zwanziger
Dalhousie University
Canada
ulli.zwanziger@Dal.ca

Adrian Wright
University of Reading
United Kingdom
a.c.wright@reading.ac.uk

Randall Youngman
Corning Incorporated
United States
youngmanre@corning.com

Josef Zwanziger
Dalhousie University
Canada
jzwanzig@gmail.com

Author Index

Affatigato, M.	24, 58	Palles, D.	93
Alderman, O. L. G.	44	Petrosyan, B. V.	133
Aleksanyan, H. A.	133		
		Revel, B.	67
Barnes, N.	24	Rodríguez Chialanza, M.	133
Barseghyan, A. H.	133	Rösslerová, I.	67
Berger, H. A.	44		
Bista, S.	58	Shaw, J. L.	44
		Shinomiya, A.	31
Castiglioni, E.	133	Sinclair, R. N.	44
Castiglioni, J.	133	Smith, M. E.	58
Černošek, Z.	67	Stansberry, J.	58
Crist, D.	24	Starkenberg, D.	58
		Stone, C. E.	44
Dupree, R.	35		
		Tadanaga, K.	31
Ehrt, D.	83, 122	Takahashi, Y.	133
Feller, S. A.	24, 44, 58	Takeda, Y.	101, 111, 139
Fischer, H. E.	44	Tatsumisago, M.	31, 105
Fornaro, L.	133	Tholen, K.	24
Franke, M.	58	Toroyan, V. P.	115
Fuchi, S.	133, 139	Tricot, G.	67
Furusawa, D.	133	Troendle, E.	24, 58
Hayashi, A.	11	Vedishcheva, N. M.	44
Hayashi, A.	31, 133		
Holland, D.	44, 58	Williams, R. B.	44
Hopkins, K.	58	Wondraczek, L.	93
Hovhannisyan, R. M.	133	Wren, J. E. C.	58
Howes, A. P.	44	Wright, A. C.	44
Iuga, D.	44	Zacharias, N.	93
		Zwanziger, J. W.	20
Kachhadia, P.	76		
Kamitsos, E. I.	83, 93		
Kaparou, M.	93		
Khristenko, V.	24		
Kiczenski, T. J.	44		
Kobayashi, S.	133		
Kobayashi, S.	133, 139		
Kodama, M.	146		
Kojima, S.	146		
Koudelka, L.	67		
Kroeker, S.	58		
Kroeker, S.	76		
Leipply, D.	58		
McCoy, J.	58		
Michaelis, V. K.	58, 76		
Minami, K.	133		
Minami, T.	1		
Möncke, D.	83, 93		
Montagne, L.	67		
Mošner, P.	67		
Mullenbach, T.	58		
O'Donovan-Zavada, A.	58		
Oshima, K.	133, 139		

Subject Index

Page

A

Alkali borate glasses	44, 146
Alkali borophosphate glasses	76
Alkaline earth borate glasses	44
Archaeology	93

B

B_2O_3–Sb_2O_3–Bi_2O_3 glasses	111
BaO–B_2O_3	31
Barium borate glass ceramics	31
Barium borate glasses	31
Barium borates	35
Barium borosilicate glasses	58
Bi_2O_3–B_2O_3 glasses	101, 139
Bismuth borate glasses	101, 115
Borate glasses	1, 20, 44, 83
Borosilicate glasses	83, 93

C

Calcium bismuth borate glasses	115
Calcium borosilicate glasses	58
CaO–Bi_2O_3–B_2O_3 glasses	115
Cation size	44
Clustering	76, 83
Computational study	20
Computer modelling	20
Computer program	24
Computer simulation	24
Crystal structure	122
Crystalline phases	115
Crystallisation	122, 133

D

Density	58, 146
Devitrification	35
Double rotation NMR	35

E

Edge-sharing	20
Elastic properties	146
Electron paramagnetic resonance	83
Europium ion doped glasses	31

F

Four-fold coordination	20

G

Glass electrolytes	105
Glass formation	115
Glass phosphors	101, 111, 139
Glass structure	20
Glass transition temperature	1, 58, 67

H

History	11

I

Ionic conductivity	1, 11, 105

L

Late Bronze Age	93
Lead borate glasses	133
Lead borophosphate glasses	67
Light extraction efficiency	139
Lithium borate glasses	24, 105
Luminescence	111

M

Manganese borate glasses	122
Manganese-containing glasses	83
Mechanical milling	11, 105
Mechanochemical synthesis	31, 105
Melt quenching	1
Molybdenum doped glasses	67
Mycenaean fragments	93

N

Neodymium doped glasses	101, 139
Network structure	1, 76
Neutron diffraction	44
NMR powder patterns	24
Nuclear magnetic resonance	24, 35, 58, 67, 76

P

Phase separation	76, 83, 122
Phosphors	31
Photoluminescence	31, 122
Poisson's ratio	146
Polycrystals	35
Professor Tsutomu Minami	11

R

Radiation pattern	101
Research	11

S

Samarium doped glasses	111
Secondary batteries	1
Short range order	76
Sidewall angle	101, 139
Silver iodoborate glasses	1
Sol-gel method	11
Solid state ionics	11
Sound velocity	146
SpectraFit	24
SrO–Bi_2O_3–B_2O_3 glasses	115
Strontium bismuth borate glasses	115
Structural coordination	44
Structure–property relationships	58
Superstructural units	44, 146
Surface layer	93

T

Ternary systems	115
Thermal expansion	67
Thermoluminescence	133
Trigonal borons	24

V

Vitreous relief fragments	93
Vitreous semiconductors	11

Y

Ytterbium doped glasses	101, 139

Z

Zinc borate glasses	122

www.ingramcontent.com/pod-product-compliance
Lightning Source LLC
Chambersburg PA
CBHW070039240326

41598CB00083B/4457